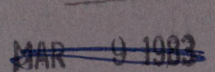

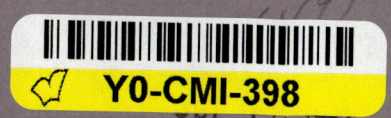

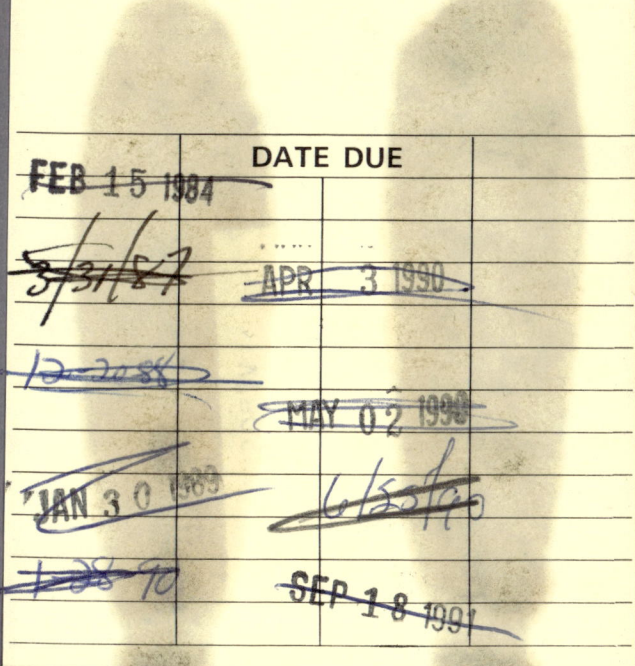

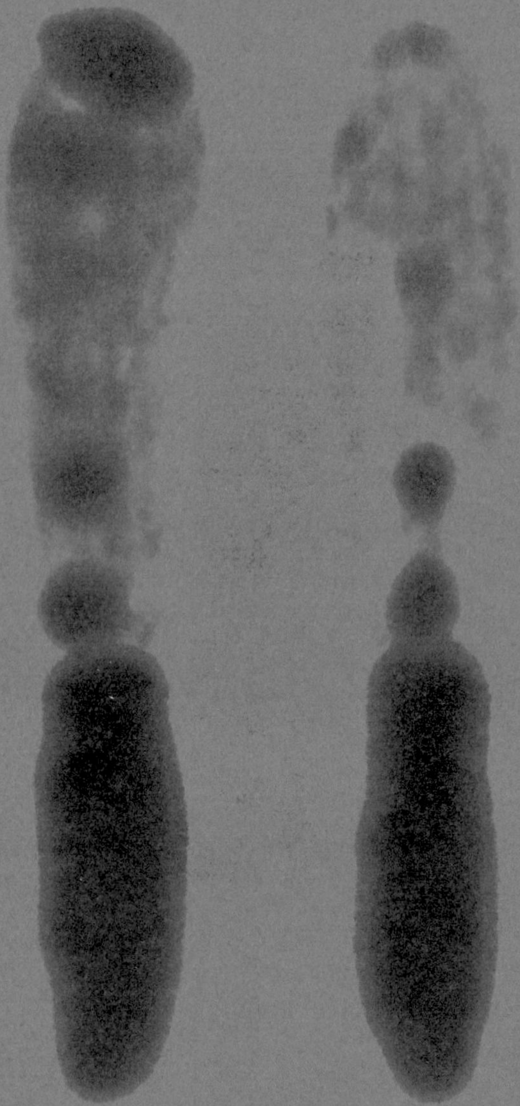

MEASURE THEORY AND INTEGRATION

ELLIS HORWOOD SERIES IN MATHEMATICS AND ITS APPLICATIONS

Series Editor: Professor G. M. BELL, Chelsea College, University of London

The works in this series will survey recent research, and introduce new areas and up-to-date mathematical methods. Undergraduate texts on established topics will stimulate student interest by including present-day applications, and the series can also include selected volumes of lecture notes on important topics which need quick and early publication.

In all three ways it is hoped to render a valuable service to those who learn, teach, develop and use mathematics.

MATHEMATICAL THEORY OF WAVE MOTION
G. R. BALDOCK and T. BRIDGEMAN, University of Liverpool.
MATHEMATICAL MODELS IN SOCIAL MANAGEMENT AND LIFE SCIENCES
D. N. BURGHES and A. D. WOOD, Cranfield Institute of Technology.
MODERN INTRODUCTION TO CLASSICAL MECHANICS AND CONTROL
D. N. BURGHES, Cranfield Institute of Technology and A. DOWNS, Sheffield University.
CONTROL AND OPTIMAL CONTROL
D. N. BURGHES, Cranfield Institute of Technology and A. GRAHAM, The Open University, Milton Keynes.
TEXTBOOK OF DYNAMICS
F. CHORLTON, University of Aston, Birmingham.
VECTOR AND TENSOR METHODS
F. CHORLTON, University of Aston, Birmingham.
TECHNIQUES IN OPERATIONAL RESEARCH
VOLUME 1: QUEUEING SYSTEMS
VOLUME 2: MODELS, SEARCH, RANDOMIZATION
B. CONNOLLY, Chelsea College, University of London
MATHEMATICS FOR THE BIOSCIENCES
G. EASON, C. W. COLES, G. GETTINBY, University of Strathclyde.
HANDBOOK OF HYPERGEOMETRIC INTEGRALS: Theory, Applications, Tables, Computer Programs
H. EXTON, The Polytechnic, Preston.
MULTIPLE HYPERGEOMETRIC FUNCTIONS
H. EXTON, The Polytechnic, Preston
COMPUTATIONAL GEOMETRY FOR DESIGN AND MANUFACTURE
I. D. FAUX and M. J. PRATT, Cranfield Institute of Technology.
APPLIED LINEAR ALGEBRA
R. J. GOULT, Cranfield Institute of Technology.
MATRIX THEORY AND APPLICATIONS FOR ENGINEERS AND MATHEMATICIANS
A. GRAHAM, The Open University, Milton Keynes.
APPLIED FUNCTIONAL ANALYSIS
D. H. GRIFFEL, University of Bristol.
GENERALISED FUNCTIONS: Theory, Applications
R. F. HOSKINS, Cranfield Institute of Technology.
MECHANICS OF CONTINUOUS MEDIA
S. C. HUNTER, University of Sheffield.
GAME THEORY: Mathematical Models of Conflict
A. J. JONES, Royal Holloway College, University of London.
USING COMPUTERS
B. L. MEEK and S. FAIRTHORNE, Queen Elizabeth College, University of London.
SPECTRAL THEORY OF ORDINARY DIFFERENTIAL OPERATORS
E. MULLER-PFEIFFER, Technical High School, Ergurt.
SIMULATION CONCEPTS IN MATHEMATICAL MODELLING
F. OLIVEIRA-PINTO, Chelsea College, University of London.
ENVIRONMENTAL AERODYNAMICS
R. S. SCORER, Imperial College of Science and Technology, University of London.
APPLIED STATISTICAL TECHNIQUES
K. D. C. STOODLEY, T. LEWIS and C. L. S. STAINTON, University of Bradford.
LIQUIDS AND THEIR PROPERTIES: A Molecular and Macroscopic Treatise with Applications
H. N. V. TEMPERLEY, University College of Swansea, University of Wales and D. H. TREVENA, University of Wales, Aberystwyth.
GRAPH THEORY AND APPLICATIONS
H. N. V. TEMPERLEY, University College of Swansea.

MEASURE THEORY AND INTEGRATION

G. de BARRA, Ph.D.
Department of Mathematics
Royal Holloway College, University of London

ELLIS HORWOOD LIMITED
Publishers · Chichester

Halsted Press: a division of
JOHN WILEY & SONS
New York · Brisbane · Chichester · Toronto

First published in 1981 by

ELLIS HORWOOD LIMITED

Market Cross House, Cooper Street, Chichester, West Sussex, PO19 1EB, England

The publisher's colophon is reproduced from James Gillison's drawing of the ancient Market Cross, Chichester.

Distributors:

Australia, New Zealand, South-east Asia:
Jacaranda-Wiley Ltd., Jacaranda Press,
JOHN WILEY & SONS INC.,
G.P.O. Box 859, Brisbane, Queensland 40001, Australia

Canada:
JOHN WILEY & SONS CANADA LIMITED
22 Worcester Road, Rexdale, Ontario, Canada.

Europe, Africa:
JOHN WILEY & SONS LIMITED
Baffins Lane, Chichester, West Sussex, England.

North and South America and the rest of the world:
Halsted Press: a division of
JOHN WILEY & SONS
605 Third Avenue, New York, N.Y. 10016, U.S.A.

British Library Cataloguing in Publication Data
De Barra, G.
Measure theory and integration. – (Ellis Horwood series in mathematics and its applications)
1. Measure theory 2. Integrals, Generalized
I. Title
515.4'2 QA312

Library of Congress Card No. 81-6571 AACR2

ISBN 0-85312-337-3 (Ellis Horwood Ltd., Publishers – Library Edn.)
ISBN 0-85312-363-2 (Ellis Horwood Ltd., Publishers – Student Edn.)
ISBN 0-470-27232-5 (Halsted Press)

Typeset in Great Britain by Activity, Salisbury.
Printed in Great Britain by R. J. Acford, Chichester.

Contents

Preface

This book has a dual purpose, being designed for a University level course on measure and integration, and also for use as a reference by those more interested in the manipulation of sums and integrals than in the proof of the mathematics involved. Because it is a textbook there are few references to the origins of the subject, which lie in analysis, geometry and probability. The only prerequisite is a first course in analysis and what little topology is required has been developed within the text. Apart from the central importance of the material in pure mathematics, there are many uses in different branches of applied mathematics and probability.

In this book I have chosen to approach integration via measure, rather than the other way round, because in teaching the subject I have found that in this way the ideas are easier for the student to grasp and appear more concrete. Indeed, the theory is set out in some detail in Chapters 2 and 3 for the case of the real line in a manner which generalizes easily. Then, in Chapter 5, the results for general measure spaces are obtained, often without any new proof. The essential L^p results are obtained in Chapter 6; this material can be taken immediately after Chapters 2 and 3 if the space involved is assumed to be the real line, and the measure Lebesgue measure.

In keeping with the role of the book as a first text on the subject, the proofs are set out in considerable detail. This may make some of the proofs longer than they might be; but in fact very few of the proofs present any real difficulty. Nevertheless the essentials of the subject are a knowledge of the basic results and an ability to apply them. So at a first reading proofs may, perhaps, be skipped. After reading the statements of the results of the theorems and the numerous worked examples the reader should be able to try the exercises. Over 300 of these are provided and they are an integral part of the book. Fairly detailed solutions are provided at the end of the book, to be looked at as a last resort.

Different combinations of the chapters can be read depending on the student's interests and needs. Chapter 1 is introductory and parts of it can be read in detail according as the definitions, etc., are used later. Then Chapters 2 and 3 provide a basic course in Lebesgue measure and integration. Then Chapter 4 gives essential results on differentiation and functions of bounded variation, all for functions on the real line. Chapters 1, 2, 3, 5, 6 take the reader as far as general measure spaces and the L^p results. Alternatively Chapters 1, 2, 3, 5, 7 introduce the reader to convergence in measure and almost uniform convergence. To get to the Radon-Nikodym results and related material the reader needs Chapters 1, 2, 3, 5, 6, 8. For a course with the emphasis on differentiation and Lebesgue-Stieltjes integrals one reads Chapters 1, 2, 3, 4, 5, 8, 9. Finally, to get to measure and integration on product spaces the appropriate route is Chapters 1, 2, 3, 5, 6, 10. Some sections can be omitted at a first reading for example: Section 2.6 on Hausdorff measures; Section 4.6 on the Lebesgue set; Sections 8.5 and 9.6 on Riesz Representation Theorems and Section 9.2 on Hausdorff measures.

Much of the material in the book has been used in courses on measure theory at Royal Holloway College (University of London). This book has developed out of its predecessor *Introduction to Measure Theory* by the same author (1974), and has now been rewritten in a considerably extended, revised and updated form. There are numerous new proofs and a reorganization of structure. The important new material now added includes Hausdorff measures in Chapters 2 and 9 and the Riesz Representation Theorems in Chapters 8 and 9.

G. de Barra

Notation

Notation is listed in the order in which it appears in the text.

$\square$: end of proof.
iff: if, and only if.
$\exists$: there exists.
$\forall$: given any.
$x \in A$: x is a member of the set A.
$A \subseteq B, (A \supseteq B)$: set A is included in (includes) the set B.
$A \subset B$: set A is a proper subset of the set B.
$[x: P(x)]$. the set of those x with property P.
$\complement A$: the complement of A.
$\emptyset$: the empty set.
$\cup, \cap$: union, intersection (of sets).
$A - B$. the set of elements of A not in B.
$A \Delta B = (A - B) \cup (B - A)$: the symmetric difference of the sets A, B.
Z: integers (positive or negative).
N: positive integers.
Q: rationals.
R: real numbers.
$\mathcal{P}(A)$: the power set of A, i.e. the set of subsets of A.
$A \times B$: the Cartesian product of the sets, A, B.
$[x]$: the equivalence class containing x (Chapter 1), or, in Chapter 2, etc., the closed interval consisting of the real number x.
$[\![X, \rho]\!]$: the metric space consisting of the space X with metric ρ.
$\bar{A}$: the closure of the set A.
G_δ set: one which is a countable intersection of open sets.
F_σ set: one which is a countable union of closed sets.

$\inf A$, $\sup A$: infimum and supremum of the set A.
$\limsup x_n$, $\liminf x_n$: upper and lower limits of the sequence $\{x_n\}$.
$x(\alpha-)$, $x(\alpha+)$: left-hand, right-hand limits of x at α. So $x(\alpha+)$ is the function whose value at α is $\lim x(\alpha_n)$, $\alpha_n \to \alpha$, $\alpha_n > \alpha$. Similarly $f(x+)$, $f(x-)$, etc.
$x_n = o(n^p)$: $n^{-p} x_n \to 0$.
$x_n = O(n^p)$: $\{n^{-p} x_n\}$ is bounded.
χ_A: characteristic function of the set A (= 1 on A, = 0 on $\mathbf{C}A$).
Card A: the cardinality of A.
$\aleph_0$: the cardinality of $\mathbf{N}$.
$C(a)$: (in Theorem 9, Chapter 1) the equivalence class containing a.
$P, P_\xi, P_\xi, P^{(\alpha)}$: Cantor-like sets.
$I_{1,1}, I_{1,1}^{(\alpha)}$, etc.: the 'removed intervals';
$J_{n,k}, J_{n,k}^{(\alpha)}$: the 'residual intervals', for the Cantor-like sets.
$N(x, \epsilon)$: the set $[t: |t-x| < \epsilon]$.
L: the Lebesgue function.
m^*: Lebesgue outer measure.
m: Lebesgue measure.
$A + x = [y + x: y \in A]$.
$l(I)$: the length of the interval I.
σ-algebra (usually $\mathcal{S}$): a class closed under countable unions and complements and containing the whole space.
Intervals: of the form $[a, b)$ unless stated otherwise.
$\mathcal{M}$: the σ-algebra of Lebesgue measurable sets.
a.e.: almost everywhere; i.e. except on a set of zero measure.
$\mathcal{B}$: σ-algebra of Borel sets.
$f^+ = \max(f, 0)$, $f^- = -\min(f, 0)$.
ess sup $f = \inf[\alpha: f \leqslant \alpha \text{ a.e.}]$. ess inf $f = \sup[\alpha: f \geqslant \alpha \text{ a.e.}]$.
$\lim A_i$, $\limsup A_i$, $\liminf A_i$: the limit, upper limit, lower limit of the sequence of sets $\{A_i\}$.
$f^{-1}(A) = [x: f(x) \in A]$.
$T^* = [x - y: x, y \in T]$ (Chapter 2).
$d(A, B) = \inf [|x - y|: x \in A, y \in B]$.
h: a Hausdorff measure function.
$H^*_{p,\delta}$: the 'approximating measure' to Hausdorff measure.
H^*_p: Hausdorff outer measure when $h(t) = t^p$.
ω_f: modulus of continuity of the function f.
$H(A)$: Hausdorff measure corresponding to the Hausdorff measure function h.
$\int f\,\mathrm{d}x$: integral (over the whole line) of f with respect to Lebesgue measure.
$\int_E f\,\mathrm{d}x$: integral of f over the set E.
ϕ, ψ: usually simple functions, taking only a finite number of non-negative values.
$\mathbf{R}\int f\,\mathrm{d}x$: Riemann integral of f.
$|f|$: absolute value of the real (or in Chapter 10, complex) function f.

$\log x$. the natural logarithm of x.

S_D, s_D: upper and lower Riemann sums given by the dissection D.

$L(a, b)$: functions integrable on (a, b).

$f_h(x) = f(x + h)$.

D^+, D_+: upper and lower right derivates.

D^-, D_-: upper and lower ledt derivates.

$P_f[a, b]$, $N_f[a, b]$, $T_f[a, b]$: positive, negative and total variations of f over $[a, b]$.

p (or p_f), n (or n_f), t (or t_f) the corresponding sums for a partition.

$BV[a, b]$: set of functions of bounded (total) variation on $[a, b]$.

$l(\pi)$: length of the polynomial π.

$\delta f(x)$: the 'jump' of f at x.

$f(c, d) = (f(d) - f(c))/(d - c)$, where f is a function of a single variable.

$f'(x) = \mathrm{d}f/\mathrm{d}x$.

F: Conventionally, the indefinite integral of f.

$\mathcal{R}$: ring of sets (closed under unions and differences).

$\mathcal{S}(\mathcal{R})$: σ-ring generated by $\mathcal{R}$.

$\mathcal{H}(\mathcal{R})$: hereditary σ-ring generated by $\mathcal{R}$.

σ-finite measure: one for which the space is a countable union of sets of finite measure.

μ^*: any outer measure, *or* the outer measure defined by μ.

$\mathcal{S}^*$: class of μ^*-measurable sets.

$\bar{\mu}$: measure obtained by restricting μ^* to $\mathcal{S}^*$, also the completion of the measure μ.

$\bar{\mathcal{S}}$: σ-ring obtained on completing measure μ on $\mathcal{S}$.

$[\![X, \mathcal{S}]\!]$: measurable space.

$[\![X, \mathcal{S}, \mu]\!]$: measure space.

$f = \lim f_n$: pointwise limit, $f(x) = \lim f_n(x)$, each x.

$\int f \, \mathrm{d}\mu$: integral of f with respect to μ.

$\int_E f \, \mathrm{d}\mu$: integral of f over the set E.

$L(X, \mu)$: class of functions integrable with respect to μ.

$L^p(X, \mu)$ or $L^p(\mu)$: the class of measurable functions with $\int |f|^p \, \mathrm{d}\mu < \infty$ functions equal a.e. being identified.

$\|f\|_p = (\int |f|^p \, \mathrm{d}\mu)^{1/p}$, the L^p-norm of f.

$\psi \circ f$: composite function, $(\psi \circ f)(x) = \psi(f(x))$.

$f_n \to f$ a.u.: $f_n \to f$ almost uniformly (uniform convergence with an exceptional set of arbitrarily small measure).

$\nu \perp \mu$: measures ν, μ mutually singular, i.e. $\nu(A) = \mu(\mathbf{C}A) = 0$ for some measurable A.

$\nu = \nu^+ - \nu^-$: Jordan decomposition of the signed measure ν.

$|\nu|$: total variation of the signed measure ν.

$\nu \ll \mu$: ν is absolutely continuous with respect to μ, i.e. $\mu(E) = 0 \Rightarrow \nu(E) = 0$.

$d\nu/d\mu$: Radon-Nikodym derivative of ν with respect to μ.

$[\mu]$: (Chapter 8) indicates that an identity holds except on a set of zero μ-measure, (or zero $|\mu|$-measure for a signed measure μ).

$\|x\|$: norm of the vector x.

$\|f\|_\infty = \sup\{|f(x)|\}$ = supremum norm of f.

$\bar{\mu}_g$: Lebesgue-Stieltjes measure, with g a monotone increasing left-continuous function and $\bar{\mu}_g([a, b)) = g(b) - g(a)$.

$\bar{\mathcal{S}}_g$: the $\bar{\mu}_g$-measurable sets.

$\int f \, d\bar{\mu}_g$ or $\int f \, dg$: integral of f with respect to the Lebesgue-Stieltjes measure derived from g. Also $\int f \, dg$ when $g \in BV[a, b]$ (Definition 4, Chapter 9).

μf^{-1}: the measure such that $\mu f^{-1}(E) = \mu(f^{-1}(E))$.

$C(I)$: the set of functions continuous on the interval I with supremum norm $\|f\|_\infty$.

$\mathcal{E}$: elementary sets, i.e. union of a finite number of disjoint measurable rectangles.

$\mathcal{M}_0$, $\mathcal{M}$. monotone classes (Chapter 10).

$[\![X \times Y, \mathcal{S} \times \mathcal{T}]\!]$ product of measurable spaces.

E_x: x-section of $E = [y: (x, y) \in E]$.

E^y: y-section of $E = [x: (x, y) \in E]$.

Ω: class of sets, depending on context (Chapter 10).

$\mu \times \nu$: the product measure. (So $(\mu \times \nu)(A \times B) = \mu(A)\,\nu(B)$).

CHAPTER 1

Preliminaries

In the chapter we collect for reference the various mathematical tools needed in later chapters. As the reader is presumed to be familiar with the content of a first course on real analysis, we are concerned not with setting up the theory from stated axioms but with giving definitions and stating results and theorems about sets, sequences and functions which serve to fix the notation and to make it clear in what form elementary results will be used. Proofs are provided for the less familiar results. In section 1.7 we describe in some detail the special sets of Cantor. These sets and the functions associated with them will be referred to frequently in later chapters.

The standard abbreviations: iff 'if and only if', $\exists$ 'there exists', $\forall$ 'given any', $\Rightarrow$ 'implies', will be used as required. The end of a proof is indicated by the symbol $\square$.

1.1 SET THEORY

Whenever we use set theoretic operations we assume that there is a universal set, X say, which contains all the sets being dealt with, and which should be clear from the context. The empty set is denoted by $\emptyset$; $x \in A$ means that the element x belongs to the set A. By $A \subseteq B$ we mean that $x \in A \Rightarrow x \in B$; $A \subset B$ is strict inclusion, that is, $A \subseteq B$ and there exists x with $x \in B$ and x not in A ($x \notin A$). We denote by $[x:P(x)]$ the set of points or elements x of X with the property P. The **complement** $\complement A$ of A is the set of points x of X not belonging to A; $\complement A$ obviously depends on the sets X implied by the content – in fact X is usually the set of real numbers, except in Chapter 10. We will denote the union of two sets by $A \cup B$ or of a collection of sets by $\bigcup_{\alpha \in I} A_\alpha$, where I denotes some index set, or by $\cup[A_\alpha : P(\alpha)]$ – the union of all sets A_α such that α has the property P. Similarly for intersections $A \cap B$, etc. Then unions and intersections are linked by the De Morgan laws: $\complement(\cup A_\alpha) = \cap \complement A_\alpha$, $\complement(\cap A_\alpha) = \cup \complement A_\alpha$. The **difference** $A - B = A \cap \complement B$; $A \,\Delta\, B = (A - B) \cup (B - A)$ is the **symmetric difference** of A and B, some properties of which are listed in Example 1. The **Cartesian product** $X \times Y$ is the set of ordered pairs $[(x,y): x \in X, y \in Y]$. We will denote the real numbers by R, the integers by N, the set of all integers by Z = [0, ± 1, ± 2, . . .], the set of rationals by Q, and n-dimensional Euclidean space by R^n, so that R^n is the set of n-tuples $(x_1, \ldots, x_n)$ considered as a vector space with the usual inner product. Notations for intervals are $[a,b] = [x: a \leqslant x \leqslant b]$, $[a,b) = [x: a \leqslant x < b]$ etc. $\mathcal{P}(A)$ denotes the set of all subsets of the set A.

Example 1: Show that the following set relations hold:

(i) $E \,\Delta\, F = F \,\Delta\, E$,

(ii) $(E \,\Delta\, F) \,\Delta\, G = E \,\Delta\, (F \,\Delta\, G)$,

(iii) $(E \,\Delta\, F) \,\Delta\, (G \,\Delta\, H) = (E \,\Delta\, G) \,\Delta\, (F \,\Delta\, H)$,

(iv) $E \,\Delta\, F = \emptyset$ if, and only if, $E = F$,

(v) For any sets E, F, G we have $E \,\Delta\, F \subseteq (E \,\Delta\, G) \cup (G \,\Delta\, F)$,

(vi) $\bigcup_{i=1}^{n} E_i \,\Delta\, \bigcup_{i=1}^{n} F_i = \bigcup_{i=1}^{n} (E_i \,\Delta\, F_i)$.

Solution: (i) is obvious from the symmetry of the definition.

To obtain (ii) use the identity $\mathbf{C}(E \,\Delta\, F) = (\mathbf{C}E \cap \mathbf{C}F) \cup (E \cap F)$, to get $((E \,\Delta\, F) \,\Delta\, G) = (E \cap \mathbf{C}F \cap \mathbf{C}G) \cup (\mathbf{C}E \cap F \cap \mathbf{C}G) \cup (\mathbf{C}E \cap \mathbf{C}F \cap G) \cup (E \cap F \cap G)$. By symmetry this must equal the right hand-side.

(iii) $(E \,\Delta\, F) \,\Delta\, (G \,\Delta\, H) = ((F \,\Delta\, E) \,\Delta\, G) \,\Delta\, H$ by (i) and (ii)

$= (F \,\Delta\, (E \,\Delta\, G)) \,\Delta\, H$ by (ii)

$= ((E \,\Delta\, G) \Delta F) \,\Delta\, H$ by (i)

$= (E \,\Delta\, G) \,\Delta\, (F \,\Delta\, H)$ by (ii).

(iv) is obvious.

(v) We have $E - F \subseteq (E - G) \cup (G - F)$ and $F - E \subseteq (F - G) \cup (G - E)$, so taking the union gives the result.

(vi) This follows from the more obvious inclusion

$$\bigcup_{i=1}^{n} E_i - \bigcup_{i=1}^{n} F_i = \bigcup_{i=1}^{n} (E_i - F_i).$$

Example 2: Let $E_1 \supseteq E_2 \supseteq \ldots \supseteq E_n \ldots$. Show that $\bigcup_{i=1}^{\infty} (E_1 - E_i) = E_1 - \bigcap_{i=1}^{\infty} E_i$.

Solution: This is just an application of De Morgan's laws with E_1 as the whole space.

Principle of Finite Induction. Let $P(n)$ be the proposition that the positive integer n has the property P. If $P(1)$ holds and the truth of $P(n)$ implies that of $P(n+1)$, then $P(n)$ holds for all $n \in$ N. It is to this property of positive integers that we are appealing in our frequent 'proofs by induction' or in inductive constructions.

Definition 1: An **equivalence relation** R on a set E is a subset of $E \times E$ with the following properties:

(i) $(x,x) \in R$ for each $x \in E$,
(ii) $(x,y) \in R \Rightarrow (y,x) \in R$,
(iii) if $(x,y) \in R$ and $(y,z) \in R$ then $(x,z) \in R$.

We write $x \sim y$ if $(x,y) \in R$. Then R partitions E into disjoint equivalence classes such that x and y are in the same class if, and only if, $x \sim y$. For, writing $[x] = [z: z \sim x]$, we have $x \in [x]$ by (i), so $\cup[[x] : x \in E] = E$. Also by (ii) and (iii), for any $x, y \in E$, either $[x] = [y]$ or $[x] \cap [y] = \emptyset$, so the sets $[x]$ are the required equivalence classes.

In Chapter 2 we will need the *Axiom of Choice* which states that if $[E_\alpha : \alpha \in A]$ is a non-empty collection of non-empty disjoint subsets of a set X, then there exists a set $V \subseteq X$ containing just one element from each set E_α.

1.2 TOPOLOGICAL IDEAS

A quite broad class of spaces in which we can consider the ideas of convergence and open sets is provided by metric spaces.

Definition 2: A non-negative function ρ on the ordered pairs $[(x,y) = x \in X, y \in X]$ is a **metric** on X if it satisfies

(i) $\rho(x,y) = 0$ if, and only if, $x = y$,
(ii) $\rho(x,y) = \rho(y,x)$,
(iii) $\rho(x,z) \leqslant \rho(x,y) + \rho(y,z)$, for any $x, y, z \in X$.

The function ρ then defines a distance between points of X, and the pair $[\![X,\rho]\!]$ forms a metric space. If we relax the condition that the distance between distinct points be strictly positive so that (i) reads: $\rho(x,y) = 0$ if $x = y$, we obtain a **pseudo-metric**. We will be especially concerned with the space R and, briefly in Chapter 10, with R^n. But the idea of convergence in a metric space is implied in many of the definitions of Chapters 6 and 7.

A set A in a metric space $[\![X,\rho]\!]$ is **open** if given $x \in A$ there exists $\epsilon > 0$ such that $[y: \rho(y,x) < \epsilon] \subseteq A$. That is: A contains an 'ϵ-neighbourhood' of x, denoted $N(x,\epsilon)$. So X and $\emptyset$ are open and it also follows that any union of open sets is open and that the intersection of two open sets is again open. The class of open sets of X forms a **topology** on X. We now define various other ideas which can be derived from that of the metric on X. The properties that follow immediately are assumed known. In the case of the real line we list various properties which will be needed, in Theorem 1 and in the later sections.

A set A is **closed** if $\mathsf{C}A$ is open. The closure $\bar{A}$ of a set A is the intersection of all the closed sets containing A, and is closed. A point x is a limit point of A if given $\epsilon > 0$, there exists $y \in A, y \neq x$, with $\rho(y,x) < \epsilon$. We say that A is a dense set, or A is dense in X if $\bar{A} = X$. The set A is **nowhere dense** if $\bar{A}$ contains no non-empty open set, so that a nowhere dense set in R is one whose closure contains no open interval. A is said to be a **perfect** set if the set $[x{:}x$ a limit point of A] is A itself. If the set A may be written as $A = \bigcap_{i=1}^{\infty} G_i$, where the sets G_i are open, A is said to be a G_δ-set; if $A = \bigcup_{i=1}^{\infty} F_i$ where the sets F_i are closed, then A is an F_σ-set. Clearly the complement of a G_δ-set is an F_σ-set.

In Chapter 10 we will refer to the **relative topology** on a subset A of R^n, namely the class of sets G of the form $H \cap A$ where H is an open set in R^n. This class of sets forms a topology on A.

We will assume the notion of **supremum** (or least upper bound) of a set A of real numbers, denoted usually by $\sup[x: x \in A]$ or by $\sup[x: P(x)]$, where P is the property satisfied by x. In the cases where we use this notation the set in question will be non-empty. For a finite set we will write $\max_{1 \leq i \leq n} x_i$ for the supremum of the relevant set. Similarly we will write $\inf[x: x \in A]$ for the **infimum** (or greatest lower bound) or $\min_{1 \leq i \leq n} x_i$ in the finite case.

We will need the following important property of the real numbers.

Theorem 1 (Heine-Borel Theorem): If A is a closed bounded set in R and $A \subseteq \bigcup_{\alpha \in I} G_\alpha$, where the sets G_α are open and I is some index set, then there exists a finite subcollection of the sets, say $[G_i, i = 1, \ldots, n]$ whose union contains A.

Exercises

1. (i) Let ρ be a pseudo-metric of a space E and write $x \sim y$ if $\rho(x,y) = 0$. Show that this defines an equivalence relation on E and describe the equivalence classes to which it gives rise.
   (ii) In the notation of p. 17, let ρ^* be defined by $\rho^*([x], [y]) = \rho(x,y)$; show that ρ^* is a metric on the set of equivalence classes.
2. Show that A is nowhere dense iff $\mathbf{C}\bar{A}$ is dense.
3. Show by examples that G_δ- and F_σ-sets may be open, closed or neither open nor closed.
4. Show that in a metric space each point is a closed G_δ-set.
5. Show by examples that Theorem 1 can break down if either of the conditions A closed or A bounded is omitted.

1.3 SEQUENCES AND LIMITS

A numerical sequence $\{x_n\}$ is a function from N to R. We define the **upper limit** of $\{x_n\}$ to be $\limsup_{n \to \infty} x_n = \inf[\sup_{m \geq n} x_m : n \in \mathsf{N}]$. If there is no ambiguity possible we will write this as $\limsup x_n$. Similarly: $\liminf x_n = \sup[\inf_{m \geq n} x_m : n \in \mathsf{N}]$ is the **lower limit** of $\{x_n\}$. If $\limsup x_n = \liminf x_n$, we write their common value as $\lim x_n$, the **limit** of $\{x_n\}$. From the definitions we get easily $\limsup x_n = -\liminf(-x_n)$.

If we consider a function from R to R we get the analogous definition: the **upper limit** of x_α at α_0 is $\limsup_{\alpha\to\alpha_0} = \inf[\sup_{0<|\alpha-\alpha_0|<h} x_\alpha : h>0]$. Similarly we define $\liminf_{\alpha\to\alpha_0} x_\alpha$; their common value, if it exists, is $\lim_{\alpha\to\alpha_0} x_\alpha$. In the more usual functional notation we may write $\limsup_{\alpha\to\alpha_0} x(\alpha)$ and $\liminf_{\alpha\to\alpha_0} x(\alpha)$, where $x(\alpha)$ is a real-valued function. A property of upper and lower limits is given by the following example.

Example 3: Prove that if $\lim_{\alpha\to\alpha_0} y_\alpha$ exists, then $\limsup_{\alpha\to\alpha_0}(x_\alpha + y_\alpha) = \limsup_{\alpha\to\alpha_0} x_\alpha + \lim_{\alpha\to\alpha_0} y_\alpha$, $\liminf_{\alpha\to\alpha_0}(x_\alpha + y_\alpha) = \liminf_{\alpha\to\alpha_0} x_\alpha + \lim_{\alpha\to\alpha_0} y_\alpha$, where all the limits are supposed finite.

Solution: We prove the first equation. Write

$$l_1 = \limsup_{\alpha\to\alpha_0}(x_\alpha + y_\alpha),\ l_2 = \limsup_{\alpha\to\alpha_0} x_\alpha,\ l_3 = \lim_{\alpha\to\alpha_0} y_\alpha.$$

Given $\epsilon > 0$, there exists $\delta > 0$ such that $x_\alpha < l_2 + \epsilon$ and $y_\alpha < l_3 + \epsilon$ when $0 < |\alpha - \alpha_0| < \delta$. So $x_\alpha + y_\alpha < l_2 + l_3 + 2\epsilon$ in this range, and as ϵ is arbitrary $l_1 \leqslant l_2 + l_3$. Conversely: there exists $\delta' > 0$ such that $x_\alpha + y_\alpha < l_1 + \epsilon$ and $y_\alpha > l_3 - \epsilon$ when $0 < |\alpha - \alpha_0| < \delta'$, so in this range $x_\alpha = (x_\alpha + y_\alpha) - y_\alpha < l_1 - l_3 + 2\epsilon$ and so $l_2 \leqslant l_1 - l_3$ giving the result.

A similar result holds for sequences $\{x_n\}, \{y_n\}$.

We will be particularly concerned with 'one-sided' upper and lower limits, and express these in functional notation:

$$\limsup_{t\to\alpha-} x(t) = \inf[\sup_{0<u<h} x(\alpha - u): h>0],$$

$$\liminf_{t\to\alpha-} x(t) = \sup[\inf_{0<u<h} x(\alpha - u): h>0].$$

If these quantities are equal, we say that $\lim_{t\to\alpha-} x(t)$ exists and we write this limit as $x(\alpha-)$. Note that $x(\alpha-)$ need not be defined, although $\limsup_{t\to\alpha-} x(t)$ and $\liminf_{t\to\alpha-} x(t)$ always are. Replacing $\alpha - u$ by $\alpha + u$ in these definitions we get $\limsup_{t\to\alpha+} x(t)$, $\liminf_{t\to\alpha+} x(t)$ and, if it exists, $x(\alpha+)$.

The sequence $\{x_n\}$ is **monotone increasing**, and we write $x_n\uparrow$, if for each $n \in$ N, $x_{n+1} \geqslant x_n$ for each n; so if a sequence is both monotone increasing and monotone decreasing it is constant. We will assume the result that if $\{x_n\}$ is a monotone sequence and is bounded, then it has a limit, and we write $x_n \uparrow x$ or

$x_n \downarrow x$ as appropriate; if $\{x_n\}$ is monotone but not bounded, then $x_n \to \infty$ or $x_n \to -\infty$ as the case may be.

A sequence $\{x_n\}$ is a **Cauchy** sequence if for any positive ϵ there exists N such that $|x_n - x_m| < \epsilon$ for $n,m > N$. We will assume the result that a sequence converges if, and only if, it is a Cauchy sequence. We define a Cauchy sequence of elements of a metric space similarly, requiring that $\rho(x_n, x_m) < \epsilon$ for $n,m > N$. Then the space is a **complete** metric space if every Cauchy sequence converges, so that the result assumed above is that R forms a complete metric space with $\rho(x,y) = |x - y|$.

We use the o- and O-notations; so that if $\{x_n\}$ is a sequence of real numbers, then $x_n = o(n^p)$ means that $n^{-p} x_n \to 0$ as $n \to \infty$, and $x_n = o(1)$ means that $\lim x_n = 0$. Similarly, $x_n = O(n^p)$ means that $\{n^{-p} x_n\}$ is bounded; $x_n = O(1)$ means that $\{x_n\}$ is bounded. For functions from R to R: $f(x) = o(g(x))$ as $x \to \alpha$ means that given $\epsilon > 0$, there exists $\delta > 0$ such that $|f(x)| < \epsilon |g(x)|$ for $0 < |x - \alpha| < \delta$; $f(x) = o(g(x))$ as $x \to \infty$ means that given $\epsilon > 0$, there exists $K > 0$ such that $|f(x)| < \epsilon |g(x)|$ for $x > K$. Similarly $f(x) = O(g(x))$ means that there exists $M > 0$ such that $|f(x)| \leqslant M |g(x)|$ as $x \to \alpha$ or $x \to \infty$ as the case may be.

We also need some properties of **double sequences** $\{x_{n,m}\}$ which are functions on N × N, and we recall that $\lim_{n,m\to\infty} x_{n,m} = x$ means that given $\epsilon > 0$ there exists N such that if $n,m > N$, $|x_{n,m} - x| < \epsilon$. We write $\lim_{n,m\to\infty} x_{n,m} = \infty$ if given $M > 0$ there exists N such that $x_{n,m} > M$ for all $n,m > N$.

If one index, m or n, is kept fixed, $\{x_{n,m}\}$ is an ordinary 'single' sequence and we have the usual notation of iterated limits $\lim_{n\to\infty} \lim_{m\to\infty} x_{n,m}$ and $\lim_{m\to\infty} \lim_{n\to\infty} x_{n,m}$.

Theorem 2: If $\{x_{n,m}\}$ is increasing with respect to n and to m, then $\lim_{n,m\to\infty} x_{n,m}$ exists and we have

$$\lim_{n,m\to\infty} x_{n,m} = \lim_{m\to\infty} \lim_{n\to\infty} x_{n,m} = \lim_{n\to\infty} \lim_{m\to\infty} x_{n,m} \tag{1.1}$$

and if any one is infinite, all three are.

Proof: Write $y_m = \lim_{n\to\infty} x_{n,m}$; this is clearly defined for each m. Since $x_{n,m} \leqslant x_{n,m+1}$ for each n, we have $y_m \leqslant y_{m+1}$. So $l_1 = \lim y_m$ exists (it may $= \infty$). Write $l = \sup[x_{n,m} : n, m \in \mathrm{N}]$, where we may have $l = \infty$. Then it is obvious that $\lim_{n,m\to\infty} x_{n,m} = l$. In the case $l < \infty$ we wish to show that $l = l_1$. Since $x_{n,m} \leqslant l$ for each n and m, it is obvious that $l_1 \leqslant l$. Also, given $\epsilon > 0$ let N be such that $|x_{n,m} - l| < \epsilon$ for $n,m \geqslant N$. As $y_N \leqslant l_1$ we have $x_{N,N} \leqslant l_1$ and hence $l \leqslant l_1 + \epsilon$, and as ϵ is arbitrary, $l \leqslant l_1$, so $l = l_1$. Similarly we may show that the third limit in (1.1) exists and equals l. The case $l = \infty$ is considered similarly. □

In the case of a series $\Sigma\, a_i$ we may consider the sequence $\{s_n\}$ of its partial sums, $s_n = \sum_{i=1}^{n} a_i$, and in the case of a double series $\Sigma\, a_{ij}$, the double sequence $s_{n,m} = \sum_{j=1}^{m} \sum_{i=1}^{n} a_{i,j}$. So to each property of sequences there corresponds one of series. For instance, to Theorem 2 corresponds the following result.

Theorem 3: If $a_{i,j} \geqslant 0$ for all $i,j \in \mathsf{N}$, then $\Sigma\, a_{i,j}$ exists and

$$\sum_{i,j=1}^{\infty} a_{i,j} = \sum_{j=1}^{\infty} \sum_{i=1}^{\infty} a_{i,j} = \sum_{i=1}^{\infty} \sum_{j=1}^{\infty} a_{i,j},$$

all three sums being infinite if any one is.

We will also need the following elementary properties of series.

Theorem 4: If $\Sigma\, a_i$ is **absolutely convergent**, that is, $\sum_{i=1}^{\infty} |a_i| < \infty$, then the series $\Sigma\, a_i$ is convergent to a finite sum.

Theorem 5: If $a_i \downarrow 0$, then $\Sigma(-1)^n\, a_n$ is a convergent series, with sum s, say, and for each n, $|s - \sum_{i=1}^{n} (-1)^i\, a_i| \leqslant a_{n+1}$.

Exercise

6. Let ϕ be a monotone function defined on $[a,b]$. Show that $\phi(a+)$ and $\phi(b-)$ exist.

1.4 FUNCTIONS AND MAPPINGS

Functions considered will be real-valued (or, briefly, in Chapter 10, complex-valued) functions on some space X. In many cases the space X will be R. If the function f is defined on X and takes its values in Y we will frequently use the notation $f\colon X \to Y$. If $g\colon X \to Y$ and $f\colon Y \to Z$, then the composite function $f o g\colon X \to Z$ is defined by $(f \circ g)(x) = f(g(x))$.

The **domain** of f is the set $[x\colon f(x)$ is defined$]$. The **range** of f is the set $[y\colon y = f(x)$ for some $x]$. If $f\colon X \to Y$ and $A \subseteq Y$ we write $f^{-1}(A) = [x\colon x \in X, f(x) \in A]$, and if $B \subseteq X$ we write $f(B) = [y\colon y = f(x), x \in B]$. The function f is a **one-to-one mapping** of X onto Y if the domain of f is X, the range of f is Y, and $f(x_1) = f(x_2)$ only if $x_1 = x_2$. In this case f^{-1} is a well-defined function on Y. The **identity** mapping is denoted by 1 so $1\,x = x$. If a function is a one-to-one mapping, then on the domain of f, $f^{-1} \circ f = 1$ and on the range of f, $f \circ f^{-1} = 1$.

The function f extends the function g or is an extension of g if the domain of f contains that of g, and on the domain of g, $f = g$. Frequently, in applications, we will use this definition for set functions, that is, functions whose domains are classes of sets and which take values in R.

If $[f_\alpha: \alpha \in A]$ is a set of functions mapping $X \to \mathsf{R}$, we will write $\sup f_\alpha$ for the function f defined on X by $f(x) = \sup [f_\alpha(x): \alpha \in A]$. The notations $\sup_{1 \leq i \leq n} f_i$ and $\max(f,g)$ have the obvious meanings, and we denote similarly infima and minima.

Definition 3: Let f_n, $n = 1, 2, \ldots$, and f be real-valued functions on the space X. Then $f_n \to f$ **uniformly** if given $\epsilon > 0$, there exists $n_0 \in \mathsf{N}$ such that

$$\sup [|f_n(x) - f(x)|: x \in X] < \epsilon \text{ for } n > n_0.$$

Elementary results concerning continuity and differentiation will be used as required, as will the definitions and more familiar properties of standard functions. A standard result on continuous functions which we will assume known is the following.

Theorem 6: Let $\{f_n\}$ be a sequence of continuous functions, $f_n: X \to \mathsf{R}$ and let $f_n \to f$ uniformly; then f is a continuous function.

Statements about sets can be turned into ones about functions using the following notation.

Definition 4: Let the set A be contained in the space X; then the **characteristic function** of A, written χ_A, is the function on X defined by: $\chi_A(x) = 1$ for $x \in A$, $\chi_A(x) = 0$ for $x \in CA$.

A **step function** on R is one of the form $\sum_{i=1}^{n} a_i \chi_{I_i}$, where $I_i, i = 1, \ldots, n$ denote disjoint intervals. An example of such a function which will be used is sgn x which is defined as: sgn $x = 1$ for $x > 0$, sgn $0 = 0$, sgn $x = -1$ for $x < 0$.

1.5 CARDINAL NUMBERS AND CARDINALITY

Two sets A and B are said to be equipotent, and we write $A \sim B$, if there exists a one-to-one mapping with domain A and range B. A standard result of set theory (cf. [11], p. 99) is that with every set A we may associate a well-defined object, Card A, such that $A \sim B$ if, and only if, Card A = Card B. We say that Card $A >$ Card B if for some $A' \subset A$ we have $B \sim A'$ but there is no set $B' \subset B$ such that $A \sim B'$. We assume the result: for any set A, Card $A <$ Card $\mathcal{P}(A)$. If Card $A = a$, we write Card $\mathcal{P}(A) = 2^a$.

If A is finite, we have Card $A = n$, the number of elements in A. If $A \sim \mathsf{N}$ we write Card $A = \aleph_0$ and describe A as an infinite countable set. If $A \sim \mathsf{R}$ we write Card $A = c$. It is easy to show that if I is any interval, open, closed, or half-open, and if I contains more than one point, then Card $I = c$. Another result is that if $[A_i: i \in \mathsf{N}]$ is a collection of countable sets, then Card $\bigcup_{i=1}^{\infty} A_i = \aleph_0$. Also

Card Q $= \aleph_0$. In Chapter 2, Exercise 45, we need the following result: Card $[f: f: \mathsf{R} \to \mathsf{R}] = 2^c$; for an explicit proof see [8], p. 50.

1.6 FURTHER PROPERTIES OF OPEN SETS

Theorem 7 (Lindelöf's Theorem): If $\mathcal{G} = [I_\alpha: \alpha \in A]$ is a collection of open intervals, then there exists a subcollection, say $[I_i: i = 1, 2, \ldots]$, at most countable in number, such that $\bigcup_{i=1}^{\infty} I_i = \bigcup_{\alpha \in A} I_\alpha$.

Proof: Each $x \in I_\alpha$ is contained in an open interval J_i with rational end-points, such that $J_i \subseteq I_\alpha$, for each α; since the rationals are countable the collection $[J_i]$ is at most countable. Also, it is clear that $\bigcup_{\alpha \in A} I_\alpha = \bigcup_{i=1}^{\infty} J_i$. For each i choose an interval I_i of $\mathcal{G}$ such that $I_i \supseteq J_i$. Then $\bigcup_{\alpha \in A} I_\alpha = \bigcup_{i=1}^{\infty} J_i \subseteq \bigcup_{i=1}^{\infty} I_i$, so we get the identity and the result. □

If the subcollection obtained is finite, we make the obvious changes of notation.

Theorem 8 (Lindelöf's Theorem in R^n): If $\mathcal{G} = [G_\alpha: \alpha \in A]$ is a collection of open sets in R^n, then there exists a subcollection of these, say $[G_i: i = 1, 2, \ldots]$, at most countable in number, such that $\bigcup_{i=1}^{\infty} G_i = \bigcup_{\alpha \in A} G_\alpha$.

Proof: Since $[t: |t - x| < r] \subset G_\alpha$ for some $r > 0$, there exists an open 'cube' T_α with the sides of length $r/\sqrt{n}$ such that $x \in T_\alpha \subset G_\alpha$, and a 'rectangle' J_i with rational coordinates for its vertices and containing x may be chosen within T_α. The proof then proceeds as in R. (We have written $|x - y|$ for the usual distance between points x, y of R^n.) □

Theorem 9: Each non-empty open set G in R is the union of disjoint open intervals, at most countable in number.

Proof: Following Definition 1, p. 16, write $a \sim b$ if the closed interval $[a,b]$, or $[b,a]$ if $b < a$, lies in G. This is an equivalence relation, in particular $a \sim a$ since $[a]$ is itself a closed interval. G is therefore the union of disjoint equivalence classes. Let $C(a)$ be the equivalence class containing a. Then $C(a)$ is clearly an interval. Also $C(a)$ is open, for if $k \in C(a)$, then $(k - \epsilon, k + \epsilon) \subseteq G$ for sufficiently small ϵ. But then $(k - \epsilon, k + \epsilon) \subseteq C(a)$; so G is the union of disjoint intervals. These are at most countable in number by Theorem 7. □

1.7 CANTOR-LIKE SETS

We now describe the Cantor-like sets. These, and the functions defined on them, are particularly useful for the construction of counter-examples. A special case – the Cantor ternary set or Cantor set – is sufficient for many purposes and will be described separately.

The construction is inductive. From [0,1] remove an open interval $I_{1,1}$ with centre at 1/2 and of length < 1. This leaves two 'residual intervals', $J_{1,1}, J_{1,2}$, each of length $< 1/2$. Suppose that the nth step has been completed, leaving closed intervals $J_{n,1}, \ldots, J_{n,2^n}$, each of length $< 1/2^n$. We carry out the $(n+1)$st step by removing from each $J_{n,k}$ an open interval $I_{n+1,k}$ with the same centre as $J_{n,k}$ and of length $< 1/2^n$. Let $\tilde{P}_n = \bigcup_{k=1}^{2^n} J_{n,k}$ and let $\tilde{P} = \bigcap_{n=1}^{\infty} \tilde{P}_n$. Any set $\tilde{P}$ formed in this way is a **Cantor-like set**.

In particular $\tilde{P}$ contains the end-points of each $J_{n,k}$. Since $[0,1] - \tilde{P} = \bigcup_{n=1}^{\infty} \bigcup_{k=1}^{2^{n-1}} I_{n,k}$, an open set, $\tilde{P}$ is closed. Since $\tilde{P}$ contains no interval, indeed each $\tilde{P}_n$ contains no interval of length $\geqslant 2^{-n}$, it follows that $\tilde{P}$ is nowhere dense. The set $\tilde{P}$ is perfect since if $x \in \tilde{P}$, then for each $n, x \in J_{n,k_n}$ for some k_n. So if for any positive ϵ we choose n such that $1/2^n < \epsilon$, then the end-points of J_{n,k_n} lie in $(x - \epsilon, x + \epsilon)$. But these end-points belong to $\tilde{P}$, so x is a limit point of $\tilde{P}$.

A particular case which will be useful is when $l(J_{1,1}) = l(J_{1,2}) = \xi < 1/2$, $l(J_{2,1}) = \ldots = l(J_{2,4}) = \xi^2$, etc., where $l(I)$ denotes the length of the interval I. So at each stage residual interval is divided in the same proportions as the original interval [0,1]. Denote the resulting Cantor-like set, in this case, by P_ξ. Slightly more generally, let $l(J_{1,1}) = l(J_{1,2}) = \xi_1 (< 1/2)$, $l(J_{2,1}) = \ldots = l(J_{2,4}) = \xi_1 \xi_2$ etc., so that at each stage the residual intervals are equal but the proportions are allowed to change from stage to stage. Denote the resulting set in this case by $P_{\boldsymbol{\xi}}$, where $\boldsymbol{\xi} = \{\xi_1, \xi_2, \ldots\}$. Note that $\xi_{n+1} < \xi_n/2$ for each n. Use is made of P_ξ and $P_{\boldsymbol{\xi}}$ in the next chapter.

We may vary the construction by choosing the removed open intervals 'off-centre', with centres a fixed combination γ, $1 - \gamma$ of the end-points of the $J_{n,k}$, where $0 < \gamma < 1$. For a general construction of such perfect sets, including those given, see [7], Chapter 1.

The Cantor Set *P*

From the interval [0,1] first remove (1/3, 2/3), then (1/9, 2/9) and (7/9, 8/9), etc., removing at each stage the open intervals constituting the 'middle thirds' of the closed intervals left at the previous stage. This gives a special case of the previous constructions, with the residual closed intervals at the nth stage, $J_{n,1}, \ldots, J_{n,2^n}$ each of length $1/3^n$, the open intervals $I_{n,r}$ also being of length $1/3^n$. If as before, we write $P_n = \bigcup_{k=1}^{2^n} J_{n,k}$ then $P = \bigcap_{n=1}^{\infty} P_n$ is the **Cantor ternary set** or more briefly the **Cantor set**. That P is uncountable follows from Example 4 below or can be seen directly as follows. It consists of those points x which can be given an expansion to the base 3 in the form $x = 0.x_1x_2 \ldots$ with $x_n = 0$ or 2. Suppose that P is countable and let $x^{(1)}, x^{(2)}, \ldots,$ be an enumeration of P. If $x_n^{(n)} = 0$, let $x_n = 2$; if $x_n^{(n)} = 2$, let $x_n = 0$. Then $x = 0.x_1x_2 \ldots$ differs from each $x^{(n)}$, but $x \in P$. So no enumeration exists.

Example 4: Every non-empty perfect set $E \subseteq \mathsf{R}$ is uncountable.

Solution: Suppose false, so E may be enumerated as a sequence $\{x_n\}$. Form the sequence $\{y_n\}$ in E inductively as follows. Let $y_1 = x_1, y_2 = x_2$, and choose $\epsilon_1, 0 < \epsilon_1 < |x_1 - x_2|$. Since E is perfect we may choose $y_3 \in E, y_3 \in N(x_2,\epsilon_1)$, in the notation of p. 17, $y_3 \neq x_3$, and with a neighbourhood $N(y_3,\epsilon_2) \subset N(x_2,\epsilon_1)$. We may suppose that $0 < \epsilon_2 < \epsilon_1/2$ and that x_1, x_2, x_3 are not in $N(y_3,\epsilon_2)$. Now choose $y_4 \in E, y_4 \in N(y_3,\epsilon_2)$, with a neighbourhood $N(y_4,\epsilon_3) \subset N(y_3,\epsilon_2)$, such that $0 < \epsilon_3 < \epsilon_2/2$ and that $x_1, \ldots, x_4 \notin N(y_4,\epsilon_3)$, etc., by induction. Then $\{y_n\}$ is a Cauchy sequence in E with limit y_0 and $y_0 \in E$ as any perfect set is closed. But for each $n, N(y_n, \epsilon_{n-1})$ contains y_0 but does not contain x_n. So $y_0 \neq x_n$ for any n, and so no such enumeration of E exists.

Clearly the result and its proof apply in any complete metric space.

The Lebesgue function

For each n, let L_n be the monotone increasing function on $[0,1]$ which is linear and increasing by $1/2^n$ on each $J_{n,k}$ and constant on the removed intervals $I_{n,k}$, where the notations $I_{n,k}, J_{n,k}$ refer to the construction of the Cantor set P. So $L_n(0) = 0, L_n(1) = 1$. It is easy to see, from a diagram, that for $n > m$, $|L_n(x) - L_m(x)| < 1/2^m$. Write $L(x) = \lim_{n\to\infty} L_n(x)$. Then the **Lebesgue function** $L(x)$ is well defined for $x \in [0,1]$ since $\{L_n(x)\}$ is a Cauchy sequence for each x. Also, letting $n \to \infty$, $|L_m(x) - L(x)| \leqslant 1/2^m$, so the convergence is uniform and so L is continuous. Clearly L is a monotone increasing function, $L(0) = 0$, $L(1) = 1$, and L is constant on each removed 'middle third' $I_{n,k}$.

The points of P are given by

$$x = \frac{2}{3}\epsilon_1 + \frac{2}{3^2}\epsilon_2 + \ldots + \frac{2}{3^k}\epsilon_k + \ldots,$$

where each ϵ_k equals 0 or 1, and expansions of this form of the points of P are unique. Then, with the same notation, L on P is given by $L(x) = \sum_{k=1}^{\infty} \frac{\epsilon_k}{2^k}$, since by continuity, if $x = \sum_{k=1}^{\infty} \frac{2}{3^k}\epsilon_k$, then

$$L(x) = \lim L\left(\sum_{k=1}^{n} \frac{2}{3^k}\epsilon_k\right) = \lim \sum_{k=1}^{n} \frac{\epsilon_k}{2^n} = \sum_{k=1}^{\infty} \frac{\epsilon_k}{2^k}. \tag{1.2}$$

We may similarly construct the Lebesgue function corresponding to the Cantor-like set P_ξ. As before the function, again denoted by L is continuous and monotone increasing. The expression for $L(x)$ corresponding to (1.2) is now more complicated.

Example 5: Consider the special case of Cantor-like sets such that for some fixed α, $0 < \alpha \leqslant 1$, we have in an obvious notation, $l(I^{(\alpha)}_{n,k}) = \alpha/3^n$, for $k = 1, \ldots, 2^{n-1}$ and for each n. Denote the Cantor-like set obtained in this way by $P^{(\alpha)}$, the Cantor set P being obtained for $\alpha = 1$. Show that there is a continuous increasing function F on [0,1] such that $F(P^{(\alpha)}) = P$.

Solution: Let F_n be the monotone increasing piecewise-linear function, F_n: $[0,1] \to [0,1]$, mapping the end-points of the $J^{(\alpha)}_{n,k}$ onto those of $J_{n,k}$ for $k = 1, \ldots, 2^n$. Then for $n > m$, F_n and F_m differ only on $J^{(\alpha)}_{m,k}$, $k = 1, \ldots, 2^m$; in fact $|F_n - F_m| \leqslant l(J_{m,k}) = 1/3^m$. So for each x, $\{F_m(x)\}$ converges and $\lim_{m\to\infty} F_m(x)$ defines a function F on [0,1]. Since $|F_m(x) - F(x)| = \lim_{n\to\infty} |F_n(x) - F_m(x)| \leqslant 1/3^m$, $\{F_m\}$ converges uniformly to F. So F is continuous and is clearly monotone increasing. We have $F([0,1]) = [0,1]$ and $F(I^{(\alpha)}_{n,k}) = I_{n,k}$ for each n and k, so $F(P^{(\alpha)}) = P$. We need only show that F is one-to-one. Suppose $(x,y) \subset [0,1]$. If either x or y lies in a removed interval it is easy to see that $F(y) > F(x)$. So suppose x and y lie in $P^{(\alpha)}$; than as $P^{(\alpha)}$ is nowhere dense, there is an interval $I^{(\alpha)}_{n,k} \subseteq (x,y)$ for some n,k. But then $F(y) - F(x) \geqslant 1/3^n$, so F is strictly increasing.

Exercises

7. Find the length of the intervals $J^{(\alpha)}_{n,k}$ of Example 5.
8. Using the notation of the construction of the Lebesgue function, p. 25, show that the estimate of $|L_m - L|$ may be improved to $|L_m - L| \leqslant 1/3.2^m$.

CHAPTER 2

Measure on the Real Line

We consider a class of sets (measurable sets) on the real line and the functions (measurable functions) arising from them. It is for this large class of functions that we will construct a theory of integration in the next chapter. On this class of sets, which includes the intervals, we show how to define Lebesgue measure which is a generalization of the idea of length, is suitably additive and is invariant under translations of the set. Apart from integration theory the methods are of independent interest as tools for studying sets on the real line. Indeed for sets which are 'scanty' we further refine the idea of measure in the last section §2.6 and construct Hausdorff measures particularly appropriate for the Cantor-like sets constructed in the last chapter. Sections 2.5 and 2.6 will not be used in the integration theory of the next chapter.

2.1 LEBESGUE OUTER MEASURE

All the sets considered in this chapter are contained in R, the real line, unless stated otherwise. We will be concerned particularly with intervals I of the form $I = [a,b)$, where a and b are finite, and unless otherwise specified, intervals may be supposed to be of this type. When $a = b$, I is the empty set $\emptyset$. We will write $l(I)$ for the length of I, namely $b - a$.

Definition 1: The **Lebesgue outer measure**, or more briefly the **outer measure**, of a set is given by $m^*(A) = \inf \Sigma l(I_n)$, where the infimum is taken over all finite or countable collections of intervals $[I_n]$ such that $A \subseteq \cup I_n$.

For notational convenience we need only deal with countable coverings of A; the finite case is included since we may take $I_n = \emptyset$ except for a finite number of integers n.

Theorem 1: (i) $m^*(A) \geqslant 0$,
(ii) $m^*(\emptyset) = 0$,
(iii) $m^*(A) \leqslant m^*(B)$ if $A \subseteq B$,
(iv) $m^*([x]) = 0$ for any $x \in$ R.

Proof: (i), (ii) and (iii) are obvious. Since $x \in I_n = [x, x + (1/n))$ for each n, and $l(I_n) = 1/n$, (iv) follows. □

Example 1: Show that for any set A, $m^*(A) = m^*(A + x)$ where $A + x = [y + x : y \in A]$, that is: outer measure is translation invariant.

Solution: For each $\epsilon > 0$ there exists a collection $[I_n]$ such that $A \subseteq \cup I_n$ and $m^*(A) \geqslant \Sigma\, l(I_n) - \epsilon$. But clearly $A + x \subseteq \cup(I_n + x)$. So, for each ϵ, $m^*(A + x) \leqslant \Sigma\, l(I_n + x) = \Sigma\, l(I_n) \leqslant m^*(A) + \epsilon$. So $m^*(A + x) \leqslant m^*(A)$. But $A = (A + x) - x$ so we have $m^*(A) \leqslant m^*(A + x)$.

Theorem 2: The outer measure of an interval equals its length.

Proof: Case 1. Suppose that I is a closed interval, $[a,b]$, say. Then, for each $\epsilon > 0$, we have from Theorem 1 and Definition 1 that

$$m^*([a,b]) \leqslant m^*([a,b + \epsilon)) \leqslant b - a + \epsilon,$$

so

$$m^*(I) \leqslant b - a. \tag{2.1}$$

To obtain the opposite inequality: for each $\epsilon > 0$, I may be covered by a collection of intervals $[I_n]$ such that $m^*(I) \geqslant \sum_{n=1}^{\infty} l(I_n) - \epsilon$, where $I_n = [a_n, b_n)$ say.

For each n, let $I'_n = (a_n - \epsilon/2^n, b_n)$, then $\bigcup_{n=1}^{\infty} I'_n \supseteq I$, so by the Heine-Borel Theorem, p. 18, a finite subcollection of the I'_n, say $J_1, \ldots, J_N$ where $J_k = (c_k, d_k)$, covers I. Then, as we may suppose that no J_k is contained in any other, we have, supposing that $c_1 < c_2 < \ldots < c_N$,

$$d_N - c_1 = \sum_{k=1}^{N} (d_k - c_k) - \sum_{k=1}^{N-1} (d_k - c_{k+1}) < \sum_{k=1}^{N} l(J_k).$$

So we have

$$m^*(I) \geqslant \sum_{n=1}^{\infty} l(I_n) - \epsilon \geqslant \sum_{n=1}^{\infty} l(I'_n) - 2\epsilon \geqslant \sum_{k=1}^{N} l(J_k) - 2\epsilon$$

$$\geqslant d_N - c_1 - 2\epsilon > b - a - 2\epsilon > l(I) - 2\epsilon. \tag{2.2}$$

Then (2.1) and (2.2) give the result.

Case 2. Suppose that $I = (a,b]$, and $a > -\infty$. If $a = b$, Theorem 1(ii) gives the result. If $a < b$, suppose that $0 < \epsilon < b - a$ and write $I' = [a + \epsilon, b]$. Then

$$m^*(I) \geqslant m^*(I') = l(I) - \epsilon. \tag{2.3}$$

But $I \subseteq I'' = [a, b + \epsilon)$, so

$$m^*(I) \leqslant l(I'') = l(I) + \epsilon. \tag{2.4}$$

Since (2.3) and (2.4) are true for all small ϵ, $m^*(I) = l(I)$. We can consider similarly the cases $I = (a,b)$ and $I = [a,b)$.

Case 3. Suppose that I is an infinite interval. Four types of interval occur.

Suppose that $I = (-\infty, a]$, the other cases being similar. For any $M > 0$, there exists k such that the finite interval I_M, where $I_M = [k, k+M)$, is contained in I. So $m^*(I) > M$ and hence $m^*(I) = \infty = l(I)$. □

The next theorem asserts that m^* has the property of countable subadditivity.

Theorem 3: For any sequence of sets $\{E_i\}$, $m^*\left(\bigcup_{i=1}^{\infty} E_i\right) \leqslant \sum_{i=1}^{\infty} m^*(E_i)$.

Proof: For each i, and for any $\epsilon > 0$, there exists a sequence of intervals $\{I_{i,j}, j = 1, 2, \ldots\}$ such that $E_i \subseteq \bigcup_{j=1}^{\infty} I_{i,j}$ and $m^*(E_i) \geqslant \sum_{j=1}^{\infty} l(I_{i,j}) - \epsilon/2^i$. Then $\bigcup_{i=1}^{\infty} E_i \subseteq \bigcup_{i=1}^{\infty}\bigcup_{j=1}^{\infty} I_{i,j}$, that is: the sets $[I_{i,j}]$ form a countable class covering $\bigcup_{i=1}^{\infty} E_i$. So

$$m^*\left(\bigcup_{i=1}^{\infty} E_i\right) \leqslant \sum_{i,j=1}^{\infty} l(I_{i,j}) \leqslant \sum_{i=1}^{\infty} m^*(E_i) + \epsilon.$$

But ϵ is arbitrary and the result follows. □

Example 2: Show that, for any set A and any $\epsilon > 0$, there is an open set O containing A and such that $m^*(O) \leqslant m^*(A) + \epsilon$.

Solution: Choose a sequence of intervals I_n such that $A \subseteq \bigcup_{n=1}^{\infty} I_n$ and $\sum_{n=1}^{\infty} l(I_n) - \epsilon/2 \leqslant m^*(A)$. If $I_n = [a_n, b_n)$, let $I'_n = (a_n - \epsilon/2^{n+1}, b_n)$ so that $A \subseteq \bigcup_{n=1}^{\infty} I'_n$.

Hence if $O = \bigcup_{n=1}^{\infty} I'_n$, O is an open set and

$$m^*(O) \leqslant \sum_{n=1}^{\infty} l(I'_n) = \sum_{n=1}^{\infty} l(I_n) + \epsilon/2 \leqslant m^*(A) + \epsilon.$$

Example 3: Suppose that in the definition of outer measure, $m^*(E) = \inf \Sigma\, l(I_n)$ for sets $E \subseteq \mathrm{R}$, we stipulate (i) I_n open, (ii) $I_n = [a_n, b_n)$, (iii) $I_n = (a_n, b_n]$, (iv) I_n closed, or (v) mixtures are allowed, for different n, of the various types of interval. Show that the same m^* is obtained.

Solution: In case (ii) we obtain the m^* of Definition 1, p. 27. Write the corresponding m^* as m^*_o in case (i), m^*_{oc} in case (iii), m^*_c in case (iv), m^*_m in case (v). We show that each equals m^*_m. Consider m^*_o, the proof in the other cases being similar. From the definition, $m^*_m(E) \leqslant m^*_o(E)$. To prove the converse: for each $\epsilon > 0$ and each interval I_n let I'_n be an open interval containing I_n with $l(I'_n) = (1+\epsilon)\, l(I_n)$. Suppose that the sequence $\{I_n\}$ is such that $E \subseteq \bigcup_{n=1}^{\infty} I_n$ and $m^*_m(E) \leqslant \sum_{n=1}^{\infty} l(I_n) - \epsilon$. Then $m^*_m(E) + \epsilon \geqslant (1+\epsilon)^{-1} \sum_{n=1}^{\infty} l(I'_n)$. But $E \subseteq \bigcup_{n=1}^{\infty} I'_n$, a union of open intervals, so $m_o^*(E) \leqslant (1+\epsilon)\, m_m^*(E) + \epsilon(1+\epsilon)$, for any $\epsilon > 0$, so $m_o^*(E) \leqslant m_m^*(E)$, as required.

Exercises

1. Show that if $m^*(A) = 0$, then $m^*(A \cup B) = m^*(B)$ for any set B.
2. Show that every countable set has measure zero.
3. Let $[I_n]$ be a finite set of intervals covering the rationals in $[0,1]$. Show that $\Sigma\, l(I_n) \geqslant 1$.
4. Show that the intervals I_n of Example 3 may be restricted so as to have endpoints in some set dense in R, for example the rationals, and again in each case the function m^* obtained is unaltered.

2.2 MEASURABLE SETS

Definition 2: The set E is **Lebesgue measurable** or, more briefly, **measurable** if for each set A we have

$$m^*(A) = m^*(A \cap E) + m^*(A \cap \complement E). \tag{2.5}$$

As m^* is subadditive, to prove E is measurable we need only show, for each A, that

$$m^*(A) \geqslant m^*(A \cap E) + m^*(A \cap \complement E). \tag{2.6}$$

Example 4: Show that if $m^*(E) = 0$ then E is measurable.

Solution: By Theorem 1(iii), p. 27, (2.6) is satisfied for each A.

Definition 3: A class of subsets of an arbitrary space X is said to be a **σ-algebra** (sigma algebra) or, by some authors, a σ-field, if X belongs to the class and the class is closed under the formation of countable unions and of complements.

Definition 4: If in Definition 3 we consider only finite unions we obtain an **algebra** (or a field).

We will denote by $\mathcal{M}$ the class of Lebesgue measurable sets.

Theorem 4: The class $\mathcal{M}$ is a σ-algebra.

Proof: From Definition 2, $\mathsf{R} \in \mathcal{M}$ and the symmetry in Definition 2 between E and $\complement E$ implies that if $E \in \mathcal{M}$ then $\complement E \in \mathcal{M}$. So it remains to be shown that if $\{E_j\}$ is a sequence of sets in $\mathcal{M}$ then $E = \bigcup_{j=1}^{\infty} E_j \in \mathcal{M}$.

Let A be an arbitrary set. By (2.5) (with E replaced by E_1) we have

$$m^*(A) = m^*(A \cap E_1) + m^*(A \cap \complement E_1),$$

and by (2.5) again (with E replaced by E_2 and A by $A \cap \complement E_1$) we have

$$m^*(A) = m^*(A \cap E_1) + m^*(A \cap E_2 \cap \complement E_1) + m^*(A \cap \complement E_1 \cap \complement E_2).$$

Continuing in this way we obtain, for $n \geqslant 2$,

$$m^*(A) = m^*(A \cap E_1) + \sum_{i=2}^{n} m^*(A \cap E_i \cap \bigcap_{j<i} \mathsf{C}E_j) + m^*(A \cap \bigcap_{j=1}^{n} \mathsf{C}E_j)$$

$$= m^*(A \cap E_1) + \sum_{i=2}^{n} m^*(A \cap E_i \cap \mathsf{C}\bigcup_{j<i} E_j) + m^*(A \cap \mathsf{C}\bigcup_{j=1}^{n} E_j)$$

$$\geqslant m^*(A \cap E_1) + \sum_{i=2}^{n} m^*(A \cap E_i \cap \mathsf{C}\bigcup_{j<i} E_j) + m^*(A \cap \mathsf{C}\bigcup_{j=1}^{\infty} E_j),$$

by Theorem 1(iii).

Therefore

$$m^*(A) \geqslant m^*(A \cap E_1) + \sum_{i=2}^{\infty} m^*(A \cap E_i \cap \mathsf{C}\bigcup_{j<i} E_j) + m^*(A \cap \mathsf{C}\bigcup_{j=1}^{\infty} E_j)$$

$$\geqslant m^*(A \cap \bigcup_{j=1}^{\infty} E_j) + m^*(A \cap \mathsf{C}\bigcup_{j=1}^{\infty} E_j) \geqslant m^*(A), \tag{2.7}$$

using Theorem 3 twice, and using the fact that for any n, $\bigcup_{i=1}^{n} (E_i \cap \mathsf{C}\bigcup_{j<i} E_j) = \bigcup_{i=1}^{n} E_i$. Hence we have equality throughout in (2.7) and we have shown that $\bigcup_{j=1}^{\infty} E_j$ is measurable. □

Example 5: Show that if $F \in \mathcal{M}$ and $m^*(F \,\Delta\, G) = 0$, then G is measurable.

Solution: By Example 4 we have that $F \,\Delta\, G$ is measurable and that its subsets $F - G$ and $G - F$ are measurable. So by Theorem 4, $F \cap G = F - (F - G)$ is measurable. So $G = (F \cap G) \cup (G - F)$ is measurable.

Theorem 5: If $\{E_i\}$ is any sequence of disjoint measurable sets then

$$m^*\left(\bigcup_{i=1}^{\infty} E_i\right) = \sum_{i=1}^{\infty} m^*(E_i), \tag{2.8}$$

that is, m^* is countably additive on disjoint sets of $\mathcal{M}$.

Proof: Take $A = \bigcup_{i=1}^{\infty} E_i$ in (2.7) which we have seen to be an equality, and note that the expression simplifies since the sets E_i are disjoint. □

Note that since the sets E_i in (2.8) may be replaced by $\emptyset$ from a certain stage onward, the same result for finite unions follows as a special case.

If E is a measurable set we will write $m(E)$ in place of $m^*(E)$. Then the set function m is defined on the σ-algebra $\mathcal{M}$ of measurable sets. Theorem 5 states that m is a countably additive set function, and $m(E)$ is called the **Lebesgue measure** of E.

Theorem 6: Every interval is measurable.

Proof: We may suppose the interval to be of the form $[a,\infty)$, as Theorem 4 then gives the result for the other types of interval. For any set A we wish to show that

$$m^*(A) \geqslant m^*(A \cap (-\infty, a)) + m^*(A \cap [a, \infty)). \tag{2.9}$$

Write $A_1 = A \cap (-\infty, a)$ and $A_2 = A \cap [a, \infty)$. Then for any $\epsilon > 0$ there exist intervals I_n such that $A \subseteq \bigcup_{n=1}^{\infty} I_n$ and $m^*(A) \geqslant \sum_{n=1}^{\infty} l(I_n) - \epsilon$. Write $I'_n = I_n \cap (-\infty, a)$ and $I''_n = I_n \cap [a, \infty)$, so that $l(I_n) = l(I'_n) + l(I''_n)$. Then

$$A_1 \subseteq \bigcup_{n=1}^{\infty} I'_n, \quad A_2 \subseteq \bigcup_{n=1}^{\infty} I''_n$$

So
$$m^*(A_1) + m^*(A_2) \leqslant \sum_{n=1}^{\infty} l(I'_n) + \sum_{n=1}^{\infty} l(I''_n)$$
$$\leqslant \sum_{n=1}^{\infty} l(I_n) \leqslant m^*(A) + \epsilon,$$

and (2.9) follows. □

Theorem 7: Let $\mathcal{A}$ be a class of subsets of a space X. Then there exists a smallest σ-algebra $\mathcal{S}$ containing $\mathcal{A}$. We say that $\mathcal{S}$ is the *σ-algebra generated by* $\mathcal{A}$.

Proof: Let $[\mathcal{S}_\alpha]$ be any collection of σ-algebras of subsets of X. Then from Definition 3, p. 30, $\bigcap_\alpha \mathcal{S}_\alpha$ is a σ-algebra. But there exists a σ-algebra containing $\mathcal{A}$, namely the class of all subsets of X. So taking the intersection of the σ-algebras containing $\mathcal{A}$ we get a σ-algebra, necessarily the smallest, containing $\mathcal{A}$. □

This theorem holds with 'σ-algebra' replaced by 'algebra', the class obtained being the generated algebra. The proof is the same.

Definition 5: We denote by $\mathcal{B}$ the σ-algebra generated by the class of intervals of the form $[a,b)$; its members are called the **Borel sets** of R.

Theorem 8: (i) $\mathcal{B} \subseteq \mathcal{M}$, that is: every Borel set is measurable.
(ii) $\mathcal{B}$ is the σ-algebra generated by each of the following classes: the open intervals, the open sets, the G_δ-sets, the F_σ-sets.

Proof: (i) follows immediately from Theorem 4, p. 30, and Theorem 6, p. 32.

(ii) Let $\mathcal{B}_1$ be the σ-algebra generated by the open intervals. Every open interval, since it is the union of a sequence of intervals of the form $[a,b)$, is a Borel set. So $\mathcal{B}_1 \subseteq \mathcal{B}$. But every interval $[a,b)$, is the intersection of a sequence of open intervals and so $\mathcal{B} \subseteq \mathcal{B}_1$. So $\mathcal{B} = \mathcal{B}_1$. Since every open set is the union of a sequence of open intervals the second result follows. Since G_δ-sets and F_σ-sets are formed from open sets using only countable intersections and complements the result in these cases follows similarly. □

It will follow from a later result, Theorem 21, p. 47, that $\mathcal{B} \subsetneq \mathcal{M}$, that is: there exist measurable sets which are not Borel sets.

Example 6: For any set A there exists a measurable set E containing A and such that $m^*(A) = m(E)$.

Solution: In Example 2, p. 29, let $\epsilon = 1/n$ and write 0_n for the corresponding open set. Then the G_δ-set $E = \bigcap_{n=1}^{\infty} 0_n$ has the required properties, since for every n, $m(E) \leqslant m(0_n) \leqslant m^*(A) + 1/n$.

Definition 6: For any sequence of sets $\{E_i\}$

$$\limsup E_i = \bigcap_{n=1}^{\infty} \bigcup_{i \geqslant n} E_i, \liminf E_i = \bigcup_{n=1}^{\infty} \bigcap_{i \geqslant n} E_i.$$

It is easily seen from the definition that $\liminf E_i \subseteq \limsup E_i$. If they are equal, this set is denoted by $\lim E_i$. It also follows from the definition that $\limsup E_i$ is the set of points belonging to infinitely many of the sets E_i and that $\liminf E_i$ is the set of points belonging to all but finitely many of the sets E_i. It is also immediate that if $E_1 \subseteq E_2 \subseteq \ldots$, then $\lim E_i = \bigcup_{i=1}^{\infty} E_i$ and that if $E_1 \supseteq E_2 \supseteq \ldots$, then $\lim E_i = \bigcap_{i=1}^{\infty} E_i$, which is analogous to the result that monotone sequences of numbers have limits.

Theorem 9: Let $\{E_i\}$ be a sequence of measurable sets. Then
(i) if $E_1 \subseteq E_2 \subseteq \ldots$, we have $m(\lim E_i) = \lim m(E_i)$,
(ii) if $E_1 \supseteq E_2 \supseteq \ldots$, and $m(E_i) < \infty$ for each i, then we have $m(\lim E_i) = \lim m(E_i)$.

Proof: (i) Write $F_1 = E_1, F_i = E_i - E_{i-1}$ for $i > 1$. Then $\bigcup_{i=1}^{\infty} E_i = \bigcup_{i=1}^{\infty} F_i$ and the sets F_i are measurable and disjoint.

$$\text{So } m(\lim E_n) = m\left(\bigcup_{i=1}^{\infty} E_i\right) = \sum_{i=1}^{\infty} m(F_i) = \lim \sum_{i=1}^{n} m(F_i)$$

$$= \lim m\left(\bigcup_{i=1}^{n} F_i\right) = \lim m(E_n),$$

as required.

(ii) We have $E_1 - E_1 \subseteq E_1 - E_2 \subseteq E_1 - E_3 \subseteq \ldots$, so by (i)

$$m(\lim (E_1 - E_i)) = \lim (m(E_1 - E_i))$$
$$= m(E_1) - \lim m(E_i). \tag{2.10}$$

But $\lim (E_1 - E_i) = \bigcup_{i=1}^{\infty} (E_1 - E_i) = E_1 - \bigcap_{i=1}^{\infty} E_i = E_1 - \lim E_i$. So taking the measures of both sides the result follows from (2.10) since $m(E_1) < \infty$. □

Example 7: (i) Show that every non-empty open set has positive measure.

(ii) The rationals Q are enumerated as $q_1, q_2, \ldots$, and the set G is defined by

$$G = \bigcup_{n=1}^{\infty} \left(q_n - \frac{1}{n^2}, q_n + \frac{1}{n^2}\right).$$

Prove that, for any closed set F, $m(G \Delta F) > 0$.

Solution: (i) follows immediately from Theorem 9, p. 23, and Theorem 2, p. 28.

(ii) if $m(G - F) > 0$ there is nothing to prove. If $m(G - F) = 0$, then since $G - F$ is open we must, by (i), have $G \subseteq F$. But G contains Q whose closure is R, so $F = $ R and $m(F) = \infty$. But $m(G) < 2 \sum_{n=1}^{\infty} 1/n^2 < \infty$. So $m(F - G) = \infty$, and the result follows.

Example 8: Show that there exist uncountable sets of zero measure.

Solution: We show that the Cantor set P, p. 24, is measurable and $m(P) = 0$. From the construction the sets P_n are measurable for each n, so $P = \bigcap_{n=1}^{\infty} P_n$ is measurable. Also $P^* = [0,1] - P = \bigcup_{n=1}^{\infty} \bigcup_{r=1}^{2^{n-1}} I_{n,r}$, a union of disjoint sets. So

$$m(P^*) = \sum_{n=1}^{\infty} \frac{2^{n-1}}{3^n} = 1$$

giving the result.

Exercises

5. Obtain the interval (c,d) from intervals of the form $[a,\infty)$ using the σ-algebra operations of Definition 3.
6. Show that Theorem 9(ii) will hold if $m(E_i)$ is finite for some i, and that the result will not hold generally in the absence of such a finiteness condition.
7. Show that if $\{E_i\}$ is a sequence of measurable sets, $m\left(\bigcup_{i=1}^{\infty} E_i\right) < \infty$ and $\lim E_i$ exists, then $m(\lim E_i) = \lim m(E_i)$.
8. For $k > 0$ and $A \subseteq$ R let $kA = [x: k^{-1}x \in A]$. Show that (i) $m^*(kA) = km^*(A)$, (ii) A is measurable iff kA is measurable.
9. For $A \subseteq$ R let $-A = [x: -x \in A]$. Show that (i) $m^*(A) = m^*(-A)$, (ii) A is measurable iff $-A$ is measurable.
10. Let $E \subseteq M$ where M is measurable and $m(M) < \infty$. Show that E is measurable iff $m(M) = m^*(E) + m^*(M - E)$.

11. Let $\{E_n\}$ be a sequence of sets such that $E_1 \subseteq E_2 \subseteq \ldots$. Show that $m^*(\lim E_n) = \lim m^*(E_n)$.
12. Show that if E is a measurable set, $0 < m(E) < \infty$ and $0 < \alpha < 1$, then there exists an open interval U such that $m(U \cap E) > \alpha\, l(U)$.
13. The density of a set E at a point x is defined to be

$$\lim_{\delta \to 0} \frac{1}{2\delta}\, m(E \cap I_\delta)$$

where I_δ is the interval $(x - \delta, x + \delta)$, assuming this limit exists. Prove that the set $[x: x \neq 0, \cos 1/x > 1/2]$ has density $1/3$ at $x = 0$.
14. Show that each Cantor-like set P_ξ has measure zero.
15. Show that the Cantor-like set $P^{(\alpha)}$ is measurable with measure $1 - \alpha$.
16. Let G be the set of numbers which can be represented in the form

$$\frac{c_1}{5} + \frac{c_2}{5^2} + \ldots + \frac{c_n}{5^n} + \ldots,$$

where $c_n = 0$ or 4 for each n. Show that $m(G) = 0$.
17. Show that the set of numbers in $[0,1]$ which possess decimal expansions not containing the digit 5 has measure zero.
18. Let k be a positive integer and $\{n_i\}$ a finite sequence of positive integers, all less than k. Show that the set of numbers in $[0,1]$, in whose expansions to base k the sequence $\{n_i\}$ does not occur, has measure zero.
19. Give an example of a set $A \subseteq [0,1]$ such that $m^*(A) > 0$ and $m^*(A \cap I) < l(I)$ for all open intervals $I \subseteq [0,1]$.
20. Show that $[0,1]$ may be written as the union of a countable number of nowhere dense perfect sets and a set of measure zero.
21. Find an upper bound for the number of sets in the σ-algebra generated by n sets.
22. Let $\mathcal{F}$ be a σ-algebra containing an infinite number of distinct sets. Show that $\mathcal{F}$ contains an uncountable number of sets.
23. Let S be a bounded set. Show that every real number is the mid-point of an open interval such that $S \cap I$ and $S \cap \mathbf{C}I$ have outer measure $\frac{1}{2}m^*(S)$.

2.3 REGULARITY

The next results states that the measurable sets are those which can be approximated closely, in terms of m^*, by open or closed sets. A non-negative countably additive set function satisfying the conditions (ii) to (iii)* below is said to be a **regular measure**. For the terminology G_δ, F_σ, used in the theorem, see Chapter 1, p. 18.

Theorem 10: The following statements regarding the set E are equivalent:

(i) E is measurable,

(ii) $\forall\, \epsilon > 0$, $\exists\, O$, an open set, $O \supseteq E$ such that $m^*(O - E) \leqslant \epsilon$,
(iii) $\exists\, G$, a G_δ-set, $G \supseteq E$ such that $m^*(G - E) = 0$,
(ii)* $\forall\, \epsilon > 0$, $\exists\, F$, a closed set, $F \subseteq E$ such that $m^*(E - F) \leqslant \epsilon$,
(iii)* $\exists\, F$, an F_σ-set, $F \subseteq E$ such that $m^*(E - F) = 0$.

Proof: (i) ⇒ (ii): suppose first that $m(E) < \infty$. As in Example 2 there is an open set $O \supseteq E$ such that $m(O) \leqslant m(E) + \epsilon$. So $m(O - E) = m(O) - m(E) \leqslant \epsilon$.

If $m(E) = \infty$, write $\mathsf{R} = \bigcup_{n=1}^{\infty} I_n$, a union of disjoint finite intervals. Then if $E_n = E \cap I_n$, we have $m(E_n) < \infty$ so there is an open set $O_n \supseteq E_n$ such that $m(O_n - E_n) \leqslant \epsilon/2^n$. Write $O = \bigcup_{n=1}^{\infty} O_n$, an open set. Then

$$O - E = \bigcup_{n=1}^{\infty} O_n - \bigcup_{n=1}^{\infty} E_n \subseteq \bigcup_{n=1}^{\infty} (O_n - E_n).$$

So $m(O - E) \leqslant \sum_{n=1}^{\infty} m(O_n - E_n) \leqslant \epsilon$.

(ii) ⇒ (iii): for each n, let O_n be an open set, $O_n \supseteq E$, $m^*(O_n - E) < 1/n$. Then if $G = \bigcap_{n=1}^{\infty} O_n$, G is a G_δ-set, $E \subseteq G$ and $m^*(G - E) \leqslant m^*(O_n - E) < 1/n$ for each n, and the result follows.

(iii) ⇒ (i): $E = G - (G - E)$, the set G is measurable and by Example 4, p. 30, $G - E$ is measurable. So E is measurable.

(i) ⇒ (ii)*: $\mathsf{C}E$ is measurable and so from above there exists an open set O such that $O \geqslant \mathsf{C}E$ and $m(O - \mathsf{C}E) \leqslant \epsilon$. But $O - \mathsf{C}E = E - \mathsf{C}O$, so taking $F = \mathsf{C}O$ gives the result.

(ii)* ⇒ (iii)*: for each n, let F_n be a closed set, $F_n \subseteq E$ and $m^*(E - F_n) < 1/n$. Then if $F = \bigcup_{n=1}^{\infty} F_n$, F is an F_σ-set, $F \subseteq E$, and, for each n, $m^*(E - F) \leqslant m^*(E - F_n) < 1/n$, and the result follows.

(iii)* ⇒ (i): since $E = F \cup (E - F)$, we find E measurable as before. □

Theorem 11: If $m^*(E) < \infty$ then E is measurable if, and only if, $\forall\, \epsilon > 0$, $\exists$ disjoint finite intervals $I_1, \ldots, I_n$ such that $m^*(E \,\Delta \bigcup_{i=1}^{n} I_i) < \epsilon$. We may stipulate that the intervals I_i be open, closed or half-open.

Proof: Suppose that E is measurable. Then by the last theorem $\forall\, \epsilon > 0$, $\exists$ an open set O containing E with $m(O - E) < \epsilon$. As $m(E)$ is finite so is $m(O)$. But by Theorem 9, p. 23, O is the union of disjoint open intervals I_i, $i = 1, 2, \ldots$. So by Theorem 5, p. 31, $\exists\, n$ such that $\sum_{i=n+1}^{\infty} l(I_i) < \epsilon$. Write $U = \bigcup_{i=1}^{n} I_i$. Then $E \,\Delta\, U =$

$(E - U) \cup (U - E) \subseteq (0 - U) \cup (0 - E)$. So $m^*(E \,\Delta\, U) < 2\epsilon$.

If we wish the intervals to be, say, half-open, we first obtain open intervals $I_1, \ldots, I_n$ as above and then for each i choose a half-open interval $J_i \subset I_i$ such that $m(I_i - J_i) < \epsilon/n$. Then the intervals J_i are disjoint and we have by Example 1, p. 16,

$$m(E \,\Delta \bigcup_{i=1}^{n} J_i) \leqslant m(E \,\Delta \bigcup_{i=1}^{n} I_i) + m\left(\bigcup_{i=1}^{n} I_i \,\Delta \bigcup_{i=1}^{n} J_i\right) < 2\epsilon,$$

so the construction goes through for the intervals J_i.

We prove the converse. By Example 2, p. 29, $\forall\, \epsilon > 0$, $\exists\, 0$ open, $0 \supseteq E$ such that

$$m(0) \leqslant m^*(E) + \epsilon. \tag{2.11}$$

If we can show that $m^*(0 - E)$ can be made arbitrarily small, then E is measurable by the last theorem. Write $J = \bigcup_{i=1}^{n} I_i$ and $U = 0 \cap J$. Then by Example 1, p. 16, and subadditivity

$$m^*(0 \,\Delta\, E) \leqslant m^*(0 \,\Delta\, U) + m^*(U \,\Delta\, E). \tag{2.12}$$

Since $U \subseteq J$ we have $U - E \subseteq J - E$ and since $E \subseteq 0$ we have $E - U = E - J$. So $U \,\Delta\, E \subseteq J \,\Delta\, E$ and $m^*(U \,\Delta\, E) < \epsilon$. But $E \subseteq U \cup (U \,\Delta\, E)$, so $m^*(E) < m(U) + \epsilon$. So by (2.11), $m(0 \,\Delta\, U) = m(0 - U) = m(0) - m(U) \leqslant m^*(E) - m(U) + \epsilon < 2\epsilon$. Then, by (2.12), $m^*(0 - E) = m^*(0 \,\Delta\, E) < 3\epsilon$, as required. □

Exercises

24. Show that if $m^*(E) < \infty$ and there exist intervals $I_1, \ldots, I_n$ such that $m^*(E \,\Delta \bigcup_{i=1}^{n} I_i) < \infty$, then each of the intervals I_i is finite.
25. Show that the condition $m^*(E) < \infty$ is necessary in Theorem 11, even if the condition $l(I_i) < \infty$ is removed.
26. Show that in Theorem 11, n will in general depend on ϵ and give an example where $n \to \infty$ as $\epsilon \to 0$.
27. In Theorem 11 take for E the Cantor set P and for a given ϵ find a corresponding n.

2.4 MEASURABLE FUNCTIONS

Sets of infinite measure and functions taking the values ∞ or $-\infty$ occur in a natural way, in for example Theorem 7 of Chapter 3, p. 59. To avoid inconvenient restrictions we use the **extended real-number system**, that is we add ∞ and $-\infty$ to the real number system with the conventions that

$$a + \infty = \infty \quad (a \text{ real, or } a = \infty),$$

$$a \cdot \infty = \infty \quad (a > 0),$$
$$a \cdot \infty = -\infty \ (a < 0), \infty \cdot \infty = \infty,$$
$$0 \cdot \infty = 0$$

and similarly for $-\infty$. We do not define $\infty + (-\infty)$. Of these conventions only the important one $0 \cdot \infty = 0$ is at all special to measure theory, and with them functions or measures with infinite values can be handled in a consistent way.

Definition 7: Let f be an extended real-valued function defined on a measurable set E. Then f is a **Lebesgue-measurable function** or, more briefly, a **measurable function** if, for each $\alpha \in \mathrm{R}$, the set $[x: f(x) > \alpha]$ is measurable.

In practice the domain of definition of f will usually be either R or $\mathrm{R} - F$ where $m(F) = 0$.

Theorem 12: The following statements are equivalent:
(i) f is a measurable function,
(ii) $\forall\, \alpha$, $[x: f(x) \geqslant \alpha]$ is measurable,
(iii) $\forall\, \alpha$, $[x: f(x) < \alpha]$ is measurable,
(iv) $\forall\, \alpha$, $[x: f(x) \leqslant \alpha]$ is measurable.

Proof: Let f be measurable. Then

$$[x: f(x) \geqslant \alpha] = \bigcap_{n=1}^{\infty} [x: f(x) > \alpha - \frac{1}{n}] \text{ is measurable;}$$

so (i) $\Rightarrow$ (ii). Let $[x: f(x) \geqslant \alpha]$ be measurable. Then $[x: f(x) < \alpha] = \mathbf{C}[x: f(x) \geqslant \alpha]$ is measurable and (ii) $\Rightarrow$ (iii). If (iii) holds, then

$$[x: f(x) \leqslant \alpha] = \bigcap_{n=1}^{\infty} [x: f(x) < \alpha + \frac{1}{n}] \text{ is measurable;}$$

so (iii) $\Rightarrow$ (iv). Finally, if $[x: f(x) \leqslant \alpha]$ is measurable, then its complement $[x: f(x) > \alpha]$ is measurable; so (iv) $\Rightarrow$ (i) and the theorem is proved. □

Example 9: Show that if f is measurable, then $[x: f(x) = \alpha]$ is measurable for each extended real number α.

Solution: For finite α, $[x: f(x) = \alpha] = [x: f(x) \geqslant \alpha] \cap [x: f(x) \leqslant \alpha]$ and so is measurable. For $\alpha = \infty$:

$$[x: f(x) = \infty] = \bigcap_{n=1}^{\infty} [x: f(x) > n],$$

a measurable set. Similarly for $\alpha = -\infty$.

Example 10: The constant functions are measurable.

Solution: Depending on the choice of α, the set $[x: f(x) > \alpha]$, where f is constant, is the whole real line or the empty set.

Example 11: The characteristic function χ_A of the set A, is measurable iff A is measurable.

Solution: Depending on α, the set $[x\colon \chi_A(x) > \alpha] = A$, R or $\emptyset$, and the result follows.

Example 12: Continuous functions are measurable.

Solution: If f is continuous, $[x\colon f(x) > \alpha]$ is open and therefore measurable.

Theorem 13: Let c be any real number and let f and g be real-valued measurable functions defined on the same measurable set E. Then $f + c, cf, f + g, f - g$ and fg are also measurable.

Proof: For each α, $[x\colon f(x) + c > \alpha] = [x\colon f(x) > \alpha - c]$, a measurable set. So $f + c$ is measurable. If $c = 0$, cf is measurable as in Example 10 above; otherwise, if $c > 0$, $[x\colon cf(x) > \alpha] = [x\colon f(x) > c^{-1}\alpha]$, a measurable set, and similarly for $c < 0$. So cf is always measurable. To show that $f + g$ is measurable, observe that $x \in A = [x\colon f(x) + g(x) > \alpha]$ only if $f(x) > \alpha - g(x)$, that is, only if there exists a rational r_i such that $f(x) > r_i > \alpha - g(x)$, where $\{r_i, i = 1, 2, \ldots\}$ is an enumeration of Q. But then $g(x) > \alpha - r_i$ and so $x \in [x\colon f(x) > r_i] \cap [x\colon g(x) > \alpha - r_i]$. Hence $A \subseteq B = \bigcup_{i=1}^{\infty} ([x\colon f(x) > r_i] \cap [x\colon g(x) > \alpha - r_i])$, a measurable set. Since A clearly contains B we have $A = B$ and so $f + g$ is measurable. Then $f - g = f + (-g)$ is also measurable.

Finally: $fg = \frac{1}{4}((f + g)^2 - (f - g)^2)$, so it is sufficient to show that f^2 is measurable whenever f is. If $\alpha < 0$, $[x\colon f^2(x) > \alpha] = $ R is measurable. If $\alpha \geqslant 0$, $[x\colon f^2(x) > \alpha] = [x\colon f(x) > \sqrt{\alpha}] \cap [x\colon f(x) < -\sqrt{\alpha}]$, a measurable set. □

Corollary: The results hold for extended real-valued measurable functions except that $f + g$ is not defined whenever $f = \infty$ and $g = -\infty$ or vice versa, and similarly for $f - g$. For $[x\colon f(x) + g(x) > \alpha] = \bigcup_{i=1}^{\infty} ([x\colon f(x) > r_i] \cap [x\colon g(x) > \alpha - r_i]) \cup ([x\colon f(x) = \infty] - [x\colon g(x) = -\infty]) \cup ([x\colon g(x) = \infty] - [x\colon f(x) = -\infty])$, a measurable set. The case of $f - g$ is similar. □

Theorem 14: Let $\{f_n\}$ be a sequence of measurable functions defined on the same measurable set. Then

(i) $\sup_{1 \leqslant i \leqslant n} f_i$ is measurable for each n,
(ii) $\inf_{1 \leqslant i \leqslant n} f_i$ is measurable for each n,
(iii) $\sup f_n$ is measurable,
(iv) $\inf f_n$ is measurable,
(v) $\lim\sup f_n$ is measurable,
(vi) $\lim\inf f_n$ is measurable.

Proof:

(i) Since $[x: \sup_{1 \leqslant i \leqslant n} f_i(x) > \alpha] = \bigcup_{i=1}^{n} [x: f_i(x) > \alpha]$, we have $\sup_{1 \leqslant i \leqslant n} f_i$ measurable.

(ii) $\inf_{1 \leqslant i \leqslant n} f_i = -\sup_{1 \leqslant i \leqslant n} (-f_i)$ and so is measurable.

(iii) $[x: \sup f_n(x) > \alpha] = \bigcup_{n=1}^{\infty} [x: f_n(x) > \alpha]$, so $\sup f_n$ is measurable.

(iv) $\inf f_n = -\sup(-f_n)$ and so is measurable.

(v) $\limsup f_n = \inf(\sup_{i \geqslant n} f_i)$, a measurable function by (iii) and (iv).

(vi) $\liminf f_n = -\limsup(-f_n)$, and so is measurable. □

Definition 8: In line with Definition 5, we say that the function f is **Borel measurable** or a **Borel function** if $\forall\, \alpha$, $[x: f(x) > \alpha]$ is a Borel set.

Theorems 12, 13, 14 and their proofs, apply also to Borel functions when 'measurable function' and 'measurable set' are replaced throughout by 'Borel measurable function' and 'Borel set' respectively. The next theorem cannot be adapted in this way: see Exercise 43, p. 45.

Definition 9: If a property holds except on a set of measure zero, we say that it holds **almost everwhere**, usually abbreviated to a.e.

Theorem 15: Let f be a measurable function and let $f = g$ a.e. Then g is measurable.

Proof: $[x: f(x) > \alpha] \,\Delta\, [x: g(x) > \alpha] \subseteq [x: f(x) \neq g(x)]$ and the result follows immediately from Example 4, p. 30, and Example 5, p. 31. □

Example 13: Let $\{f_i\}$ be a sequence of measurable functions converging a.e. to f; then f is measurable, since $f = \limsup f_i$, a.e.

Example 14: If f is a measurable function, then so are $f^+ = \max(f, 0)$ and $f^- = -\min(f, 0)$.

Solution: Example 10 and Theorem 14, (i) and (ii), give the result.

Example 15: The set of points on which a sequence of measurable functions $\{f_n\}$ converges, is measurable.

Solution: The set in question is $[x: \limsup f_n(x) - \liminf f_n(x) = 0]$ which is measurable by Theorem 14, (v) and (vi), and Example 9, p. 38.

Definition 10: Let f be a measurable function; then $\inf[\alpha: f \leqslant \alpha \text{ a.e.}]$ is called the **essential supremum** of f, denoted by ess sup f.

Example 16: Show that $f \leqslant \operatorname{ess\,sup} f$, a.e.

Solution: If ess sup $f = \infty$, the result is obvious. Suppose ess sup $f = -\infty$. Then

$\forall n \in Z, f \leqslant n$ a.e. from Definition 10. So $f = -\infty$, a.e., as required. So suppose that ess sup f is finite. Write $E_n = [x: f(x) > 1/n + \text{ess sup} f]$ and $E = [x: f(x) >$ ess sup $f]$, so $E = \bigcup_{n=1}^{\infty} E_n$. But from its definition $m(E_n) = 0$, so $m(E) = 0$.

Example 17: Show that for any measurable functions f and g

$$\text{ess sup}\,(f+g) \leqslant \text{ess sup}\, f + \text{ess sup}\, g,$$

and give an example of strict inequality.

Solution: From the last example

$$f + g \leqslant \text{ess sup}\, f + \text{ess sup}\, g, \text{ a.e.,}$$

implying the result. For inequality take $f = \chi_{[-1,0)} - \chi_{[0,1]}$ and $g = -f$. Then the left-hand side is 0 and the right-hand side 2.

Definition 11: Let f be a measurable function; then sup $[\alpha: f \geqslant \alpha$ a.e.$]$ is called the **essential infimum** of f, denoted by ess inf f.

Example 18: Ess sup $f = -$ess inf $(-f)$.

Solution: Ess sup $f = \inf[\alpha: f \leqslant \alpha \text{ a.e.}] = \inf[\alpha: -f \geqslant -\alpha \text{ a.e.}]$
$= -\sup[-\alpha: -f \geqslant -\alpha \text{ a.e.}] = -\text{ess inf}\,(-f)$.

So results analogous to those holding for ess sup f, for example those of Examples 16 and 17, hold also for ess inf f, with the obvious alterations.

Definition 12: If f is a measurable function and ess sup $|f| < \infty$, then f is said to be **essentially bounded.**

Example 19: Let f be a measurable function and B a Borel set; then $f^{-1}(B)$ is a measurable set.

Solution: We have $f^{-1}\left(\bigcup_{i=1}^{\infty} A_i\right) = \bigcup_{i=1}^{\infty} f^{-1}(A_i)$ and $f^{-1}(\complement A) = \complement f^{-1}(A)$, so the class of sets whose inverse images under f are measurable forms a σ-algebra. But this class contains the intervals. So it must contain all Borel sets.

Exercises

28. Let f be defined on $[0,1]$ by $f(0) = 0$, $f(x) = x \sin 1/x$ for $x > 0$. Find the measure of the set $[x: f(x) \geqslant 0]$.
29. Show that monotone functions are measurable.
30. Let $f = g$ a.e. where f is a continuous function. Show that
$$\text{ess sup}\, f = \text{ess sup}\, g = \sup f.$$
31. Show that for any measurable function f, ess sup $f \leqslant \sup f$.
32. Let f and g be measurable functions and $g \geqslant 0$. Show that
$$fg \leqslant (\text{ess sup}\, f)g \text{ a.e.}$$

33. Let f be a measurable function not almost everywhere infinite. Show that there exists a set of positive measure on which f is bounded.
34. Let f be a measurable function on $[a,b]$ and let f be differentiable a.e. Show that there is a function, measurable on $[a,b]$, which equals f' a.e.
35. Let f be a continuous function and g a measurable function. Show that the composite function $f \circ g$ is measurable.
36. Let $x \in [0,1]$ have the expansion to the base l, $x = 0 \cdot x_1 x_2 \ldots x_n \ldots$ for some integer l, the non-terminating expansion being used in cases of ambiguity. Show that $f_n(x) = x_n$ is a measurable function of x for each n.

2.5 BOREL AND LEBESGUE MEASURABILITY

This section considers the relation between the class $\mathcal{B}$ of Borel sets, the class $\mathcal{M}$ of Lebesgue-measurable sets and the class $\mathcal{P}(\mathsf{R})$ of all subsets of R. The section may be omitted without loss of logical continuity, but it provides distinctions between these classes without which the theory, though still valid, would be rather artificial.

From Theorem 8, p. 32, we have that $\mathcal{B} \subseteq \mathcal{M} \subseteq \mathcal{P}(\mathsf{R})$. Using Theorem 16 we show in Theorem 17 that $\mathcal{M} \neq \mathcal{P}(\mathsf{R})$ and in Theorem 18 that $\mathcal{B} \neq \mathcal{M}$. Since the characteristic function χ_A is measurable or Borel measurable if, and only if, A is measurable or is a Borel set, respectively (Example 11, p. 39), we have corresponding relations between the two classes of measurable functions and the class of all real-valued functions.

Theorem 16: Let E be a measurable set. Then for each y the set $E + y = [x + y: x \in E]$ is measurable and the measures are the same.

Proof: By Theorem 10, p. 36, $\forall\, \epsilon > 0$, $\exists$ an open set O, $O \supseteq E$ and $m(O - E) \leqslant \epsilon$. Then the set $O + y$ is open and $O + y \supseteq E + y$. But $(O + y) - (E + y) = (O - E) + y$. By Example 1, p. 28, $m((O - E) + y) \leqslant \epsilon$ and the measurability of $E + y$ follows, using Theorem 10 again. That the measures are equal follows on using Example 1 again. □

Theorem 17: There exists a non-measurable set.

Proof: If $x, y \in [0,1]$, let $x \sim y$ if $y - x \in \mathsf{Q}_1 = \mathsf{Q} \cap [-1, 1]$. Then $\sim$ is easily seen to be an equivalence relation on $[0,1]$ and so by Chapter 1, p. 17, $[0,1] = \cup E_\alpha$, E_α disjoint sets such that x and y are in the same E_α if, and only if, $x \sim y$. Since Q_1 is countable, each E_α is a countable set. Since $[0,1]$ is uncountable there are uncountable many sets E_α. Using the Axiom of Choice, p. 17, we consider a set V in $[0,1]$ containing just one element x_α from each E_α. Let $\{r_i\}$ be an enumeration of Q_1, and for each n write $V_n = V + r_n$. If $y \in V_n \cap V_m$ there exist $x_\alpha, x_\beta \in V$ such that $y = x_\alpha + r_n$ and $y = x_\beta + r_m$. But then $x_\beta - x_\alpha \in \mathsf{Q}_1$, so $x_\beta = x_\alpha$ by definition of V and we have $n = m$. So $V_n \cap V_m = \emptyset$

for $n \neq m$. Also $[0,1] \subseteq \bigcup_{n=1}^{\infty} V_n \subseteq [-1,2]$, since $\forall\, x \in [0,1], x \in E_\alpha$ for some α and then $x = x_\alpha + r_n$ giving $x \in V_n$; the second inclusion is obvious.

If V is measurable, then by the last theorem so is each V_n and $m(V) = m(V_n)$. Then using the measurability of the sets V_n we have

$$1 = m([0,1]) \leqslant \sum_1^{\infty} m(V_n) = m(V) + m(V) + \ldots \leqslant 3.$$

But this sum can only be 0 or ∞. So V is not measurable. □

Theorem 18: Not every measurable set is a Borel set.

Proof: Write each $x \in [0,1]$ in binary form

$$x = \sum_{n=1}^{\infty} \frac{\epsilon_n}{2^n}$$

with $\epsilon_n = 0$ or 1, choosing a non-terminating expansion for each $x > 0$. Define the function f by

$$f(x) = \sum_{n=1}^{\infty} \frac{2\epsilon_n}{3^n}.$$

Then the values of f, which is known as Cantor's function, lie entirely in the Cantor set P, p. 24. Since ϵ_n is a measurable function of x (Exercise 36), f is measurable. Also f is a one-to-one mapping from $[0,1]$ onto its range, since the value $f(x)$ defines the sequence $\{\epsilon_n\}$ in the expansion $\sum_{n=1}^{\infty} \frac{2\epsilon_n}{3^n}$ uniquely, so x is determined uniquely.

If $\mathcal{B}$ and $\mathcal{M}$ were the same, then by Example 19, p. 41, $f^{-1}(B)$ would be measurable for any measurable set B and any measurable function f. Let f be the Cantor function and V a non-measurable set in $[0,1]$. Then $B = f(V)$ lies in P and so has measure zero. So B is measurable. But since f is one-to-one, $f^{-1}(B) = V$ which is non-measurable. We conclude that $\mathcal{B}$ is strictly contained in $\mathcal{M}$. □

We now give two examples showing unexpected implications of measurability.

Example 20: Let T be a measurable set of positive measure and let $T^* = [x - y: x \in T, y \in T]$. Show that T^* contains an interval $(-\alpha,\alpha)$ for some $\alpha > 0$.

Solution: By Theorem 10, p. 36, T contains a closed set C of positive measure. Since $m(C) = \lim m(C \cap [-n,n])$ we may assume that C is a bounded set. By Theorem 10, again, there exists an open set U, $U \supset C$, such that $m(U - C) < m(C)$. Define the distance between two sets A and B to be $d(A,B) = \inf[\,|x - y|: x \in A, y \in B]$. Since $|x - y|$ is a continuous function of x and y, the distance between A and B is positive if A and B are disjoint closed sets one of which is bounded. Let α be the distance between the closed sets C and $\mathbf{C}U$, so that $\alpha > 0$. Let x be any point of $(-\alpha,\alpha)$. We wish to show that $C \cap (C-x) \neq \emptyset$. For then, since $C - x = [y: y + x \in C]$, we have that $\forall\, x \in (-\alpha,\alpha)$, $\exists\, z \in C$ such that $z' = z + x \in C$ and so $x = z' - z \in T^*$.

Since $|x| < \alpha$ we have $C - x \subset U$ from the definition of α. So

$$\begin{aligned} m(C-(C-x)) &\leqslant m(U-(C-x)) \\ &= m(U) - m(C-x) \\ &= m(U) - m(C) \text{ (by Theorem 16)} \\ &< m(C). \end{aligned}$$

Hence $m(C \cap (C-x)) > 0$ and so we must have $C \cap (C-x) \neq \emptyset$, as required.

Example 21: Suppose that f is any extended real-valued function which for every x and y satisfies

$$f(x) + f(y) = f(x+y). \tag{2.13}$$

(i) Show that f is either everywhere finite or everywhere infinite.
(ii) Show that if f is measurable and finite, then $f(x) = xf(1)$ for each x.

Solution: (i) f cannot take both values $\infty, -\infty$ for then (2.13) would be meaningless for some pair x,y. Suppose that $f(x) = \infty$ for some x. Then $f(x+y) = \infty + f(y) = \infty$ for all y, and so $f = \infty$ everywhere. Similarly if $f(x) = -\infty$ for some x.

(ii) By induction (2.13) gives $f(nx) = nf(x)$ for each x and each positive integer n, so $f(x/n) = n^{-1}f(x)$; and hence $f(mx/n) = m\, n^{-1} f(x)$. In particular $f(r) = rf(1)$ for each $r \in \mathrm{Q}$. Since f is finite there exists a measurable set E such that $m(E) > 0$ and $|f| < M$, say, on E (cf. Exercise 34). Let $z \in E^*$, in the notation of the last example, so that $z = x - y$ where $x,y \in E$. Then $|f(z)| = |f(x-y)| = |f(x) - f(y)| \leqslant 2M$. But by the last example E^* contains an interval $(-\alpha,\alpha)$ with $\alpha > 0$. So if $|x| < \alpha/n$ we have $|f(nx)| \leqslant 2M$, and so $|f(x)| \leqslant 2M/n$, for each n. Let x be real and let r be a rational such that $|r - x| < \alpha/n$. Then, since $f(r) = rf(1)$, we have

$$\begin{aligned} |f(x) - xf(1)| &= |f(x) - f(r) + (r-x)f(1)| \\ &= |f(x-r) + (r-x)f(1)| \\ &\leqslant \frac{2M}{n} + \frac{\alpha}{n}\, |f(1)| \end{aligned}$$

for each n. So $f(x) = xf(1)$.

[Note: Equation (2.13) has in fact non-measurable solutions, cf [6], p. 96. The method given above is based on [9].]

Exercises

37. Show that sup $[f_\alpha : \alpha \in A]$ is not necessarily measurable even if each f_α is.
38. Give an example of a function such that $|f|$ is measurable but f is not.

39. Let f be Cantor's function defined in Theorem 18. Show that the range of f does not cover the Cantor set P.
40. Show that a nowhere dense perfect set can contain a non-measurable set.
41. Show that a measurable function of a continuous function is not necessarily measurable.
42. Show that there exist sets of zero measure which are not Borel sets.
43. Show that the result of Theorem 15 does not hold for Borel measurable functions.
44. Find the cardinality of the class of measurable sets.
45. Find the cardinality of the class of measurable functions.

2.6 HAUSDORFF MEASURES ON THE REAL LINE

We have found in Example 8, p. 34, that there exist sets of zero Lebesgue measure which are in some sense large. Using Hausdorff measure, and, especially, Hausdorff dimension we may discriminate between these sets of zero Lebesgue measure. We replace $l(I)$ in Definition 1 by its p-th power, $l(I)^p$, where $p > 0$, or more generally by $h(l(I))$, where the *Hausdorff measure function h* is monotone increasing on $[0, \infty)$, $h(x) > 0$ for $x > 0$ and $h(0+) = h(0) = 0$. For simplicity, we set out the main properties of Hausdorff measures for the special case $h(x) = x^p$, though the results in the main apply to the general case. We return to the measure obtained from more general Hausdorff measure functions in Theorem 28. Further results are given in Chapter 9, but otherwise the section is not essential for the following chapters. The methods generalize to higher dimensions and give there a definition of the dimension of an arbitrary set. For a general account see [7] or the full account in [10] which has a large bibliography.

We continue to suppose that all sets are contained in the real line.

Definition 13: The 'approximating measure' $H^*_{p,\delta}$ of the set A is given by $H^*_{p,\delta}(A) = \inf \Sigma\, l(I_k)^p$ where the infimum is taken over all coverings of A by open intervals, $[I_k]$, with $l(I_k) \leqslant \delta$.

Definition 14: The **Hausdorff outer measure** H^*_p of the set A is given by $H^*_p(A) = \lim_{\delta \to 0} H^*_{p,\delta}(A)$.

This limit must exist (though it may be infinite) since $H^*_{p,\delta}(A)$ can only increase as δ decreases. For a general Hausdorff measure function we have $\inf \Sigma\, h(l(I_k))$ in Definition 13.

Theorem 19:
(i) $H^*_p(A) \geqslant 0$,
(ii) $H^*_p(\emptyset) = 0$,
(iii) $H^*_p(A) \leqslant H^*_p(B)$ if $A \subseteq B$,
(iv) $H^*_p([x]) = 0$ for any $x \in \mathsf{R}$.

Proof: (i), (ii) and (iii) are obvious from the definition. To prove (iv) note that $[x] \subset (x - \delta/2, x + \delta/2)$ for each $\delta > 0$. So $H^*_{p,\delta}([x]) \leqslant \delta^p$. Let $\delta \to 0$ to get $H^*_p([x]) = 0$. □

Example 22: Hausdorff outer measure is invariant under translation, that is $H^*_p(A) = H^*_p(A + x)$.

Solution: Each covering of A by intervals $[I_n]$ of length at most δ corresponds to a cover $[I_n + x]$ of $A + x$ and $\Sigma l(I_n)^p = \Sigma l(I_n + x)^p$. Taking the infimum over such covers $[I_n]$ gives $H_{p,\delta}(A) = H_{p,\delta}(A + x)$ and when $\delta \to 0$ the result follows.

Example 23: $H^*_p(kA) = k^p\, H^*_p(A)$ for any positive k.

Solution: Each covering of A by intervals $[I_n]$ of length at most δ corresponds under the mapping $x \to kx$ to a cover of kA by intervals $[kI_n]$ of length at most $k\delta$ and $\Sigma l(kI_n)^p = k^p\, \Sigma l(I_n)^p$. Taking the infimum over all such covers $[I_n]$ gives $H^*_{p,k\delta}(A) = k^p\, H^*_{p,\delta}(A)$. Letting $\delta \to 0$ gives the result.

Example 24: $H^*_p(A)$ is the same whether we stipulate that the intervals I_k in Definition 13 are open, closed or half-open.

Solution: We show that closed intervals give the same outer measure, the other case being similar. Write $H^{*c}_\delta(A)$ and $H^{*c}(A)$ for the corresponding set functions obtained from closed intervals, p being kept fixed. Write I' for the closure of the open interval I. Then

$$\begin{aligned} H^*_{p,\delta}(A) &= \inf \Sigma l(I_k)^p, l(I_k) \leqslant \delta, A \subseteq \cup I_k \\ &= \inf \Sigma l(I'_k)^p \\ &\geqslant \inf \Sigma l(J_k)^p, l(J_k) \leqslant \delta, A \subseteq \cup J_k, J_k \text{ closed intervals}, \\ &\geqslant H^{*c}_\delta(A). \end{aligned}$$

So $H^*_p(A) \geqslant H^{*c}(A)$. In the opposite direction: every closed interval J of length $\epsilon(\epsilon > 0)$ is contained in an open interval J'' of length $\epsilon(1 + \delta)$. Then

$$\begin{aligned} H^{*c}_\delta(A) &= \inf \Sigma l(J_k)^p, l(J_k) \leqslant \delta, J_k \text{ closed}, A \subseteq \cup J_k \\ &= (1+\delta)^{-p} \inf \Sigma l(J''_k)^p \\ &\geqslant (1+\delta)^{-p} \inf \Sigma l(I_k)^p, l(I_k) \leqslant \delta + \delta^2, I_k \text{ open}, A \subseteq \cup I_k, \\ &= (1+\delta)^{-p}\, H^*_{p,\delta+\delta^2}(A). \end{aligned}$$

Let $\delta \to 0$ to get $H^{*c}(A) \geqslant H^*_p(A)$, as required.

Theorem 20: Let $H^*_p(A) < \infty$ and let $q > p$. Then $H^*_q(A) = 0$.

Proof: Let $\delta > 0$ and let $[I_k]$ be a covering of A by intervals with $l(I_k) \leqslant \delta$ for each k. Then

$$\frac{l(I_k)^q}{l(I_k)^p} = l(I_k)^{q-p} \leqslant \delta^{q-p}.$$

So $H^*_{q,\delta}(A) \leqslant \Sigma\, l(I_k)^q \leqslant \delta^{q-p}\, \Sigma\, l(I_k)^p$. Taking the infimum of the last term over all such coverings we get

$$H^*_{q,\delta}(A) \leqslant \delta^{q-p}\, H^*_{p,\delta}(A) \leqslant \delta^{q-p}\, H^*_p(A).$$

Letting $\delta \to 0$, the result follows. □

Corollary: If $0 < H^*_p(A) < \infty$, then $H^*_q(A) = \infty$ for $q < p$.

The result of this theorem would apply to general Hausdorff measure functions provided we have

$$h_1 \leqslant h_2 \text{ and } \lim_{t \to 0} \frac{h_1(t)}{h_2(t)} = 0,$$

but the functions h cannot be ordered in this way in general. We now show that Hausdorff outer measure has an essential property of outer measures.

Theorem 21: For any sequence of sets $\{E_i\}$ we have $H^*_p\left(\bigcup_{i=1}^{\infty} E_i\right) \leqslant \sum_{i=1}^{\infty} H^*_p(E_i)$.

Proof: For each $\epsilon > 0$ there exists a family of intervals $[I_{i,j}]$, with $l(I_{i,j}) \leqslant \delta$, such that $E_i \subseteq \bigcup_{j=1}^{\infty} I_{i,j}$ for each i, and

$$H^*_{p,\delta}(E_i) \geqslant \sum_{j=1}^{\infty} l(I_{i,j})^p - \frac{\epsilon}{2^i}.$$

Then $\bigcup_{i=1}^{\infty} E_i \subseteq \bigcup_{i=1}^{\infty} \bigcup_{j=1}^{\infty} I_{i,j}$, so

$$\begin{aligned} H^*_{p,\delta}\left(\bigcup_{i=1}^{\infty} E_i\right) &\leqslant \sum_{i,j} l(I_{i,j})^p \\ &\leqslant \sum_i H^*_{p,\delta}(E_i) + \epsilon \\ &\leqslant \sum_i H^*_p(E_i) + \epsilon, \text{ for all } \delta > 0. \end{aligned}$$

So $H^*_p\left(\bigcup_{i=1}^{\infty} E_i\right) \leqslant \sum_i H^*_p(E_i) + \epsilon$ and the result follows. □

We now show that Hausdorff measure includes Lebesgue measure as a special case.

Theorem 22: For any set A, $H^*_1(A) = m^*(A)$.

Proof: From their definitions we have $m^*(A) \leqslant H^*_{1,\delta}(A)$ for all $\delta > 0$, so $m^*(A) \leqslant H^*_1(A)$. So we wish to show that $H^*_1(A) \leqslant m^*(A)$ and clearly we may assume $m^*(A) < \infty$. Then for any given $\epsilon > 0$ we have for some family of intervals $[I_n]$ that $A \subseteq \bigcup_{n=1}^{\infty} I_n$ and $m^*(A) \geqslant \sum_{n=1}^{\infty} l(I_n) - \epsilon$. Now if I is any finite interval and

$\epsilon' > 0$ then

$$l(I) \geqslant H_1^*(I) - \epsilon'. \tag{2.14}$$

For let $\delta > 0$ and choose open intervals $J_i, i = 1, \ldots, m$ with $l(J_i) \leqslant \delta, I \subseteq \bigcup_{i=1}^{m} J_i$ and $\sum_{i=1}^{m} l(J_i) < l(I) + \epsilon'$. Then $H_{1,\delta}^*(I) < l(I) + \epsilon'$ for all $\delta > 0$, and, letting $\delta \to 0$ (2.14) follows. Now put $I = I_n, \epsilon' = \epsilon/2^n$ in (2.14) to get

$$\begin{aligned} m^*(A) &\geqslant \sum_{n=1}^{\infty} H_1^*(I_n) - \sum_{1}^{\infty} \epsilon/2^n - \epsilon \\ &\geqslant \sum_{n=1}^{\infty} H_1^*(I_n) - 2\epsilon \\ &\geqslant H_1^*(A) - 2\epsilon, \end{aligned}$$

by the last theorem, giving the result. □

Corollary 1: Let I be an interval of positive or infinite length. Then $H_p^*(I) = \infty$ for $0 < p < 1, H_1^*(I) = l(I)$ and $H_p^*(I) = 0$ for $p > 1$.

Proof: Suppose that I is a finite interval. Then the result follows from Theorem 20 and its Corollary, and Theorem 21. So suppose that I is an infinite interval and $p > 1$. Since $I \subset \cup I_k$, I_k finite intervals, we have $H_p^*(I) \leqslant \Sigma H_p^*(I_k) = 0$. So the result is true for $p \geqslant 1$ and Theorem 20 ensures that $H_p^*(I) = \infty$ for $0 < p < 1$. □

Corollary 2: $H_p^*(\mathsf{R}) = \infty$ for $0 < p \leqslant 1, H_p^*(\mathsf{R}) = 0$ for $p > 1$.

Corollary 3: Every non-empty open set G has $H_p^*(G) = \infty$ for $p < 1$.

So the result of Example 2, p. 29, will not extend to Hausdorff measures.

Corollary 4: Considering $H_p^*(A)$ as a function of p ($p > 0$), either $H_p^*(A) = 0$ for all $p > 0$, or for some p_0 ($0 < p_0 \leqslant 1$) we have $H_p^*(A) = \infty$ for $0 < p < p_0$ and $H_p^*(A) = 0$ for $p > p_0$.

Proof: By Corollary 2, $H_p^*(A) = 0$ for $p > 1$. Now the result follows immediately from Theorem 20 and its Corollary. □

Examples showing the possibilities are given later. Measurability is defined for H_p^* as for m^* in Definition 2, p. 30.

Theorem 23: The H_p^*-measurable sets form a σ-algebra.

The proof is as for Theorem 4, for Lebesgue outer measure. □

Theorem 24: H_p^* is countably additive on the σ-algebra of H_p^*-measurable sets.

The proof is completely analogous to that of Theorem 5, p. 31. □

Theorem 25: H_p^* is a *metric outer measure*; that is, if A and B are non-empty disjoint sets in R with $d(A,B)>0$ then $H_p^*(A\cup B)=H_p^*(A)+H_p^*(B)$.

Proof: It is sufficient to show that the same identity holds for $H_{p,\delta}^*$ for all small δ. But the identity is immediate for $H_{p,\delta}^*$ provided $\delta<d(A,B)$, for then any covering of $A\cup B$ by intervals of length at most δ decomposes into two general coverings of A and of B. □

In particular, of course, Lebesgue outer measure is a metric outer measure. We now show that the H_p^*-measurable sets include the Borel sets.

Theorem 26: Let I be the interval $(-\infty, a]$. Then I is H_p^*-measurable.

Proof: We need to show that for any set A

$$H_p^*(A)\geqslant H_p^*(A\cap I)+H_p^*(A-I). \tag{2.15}$$

We may suppose that $H_p^*(A)$ is finite; otherwise (2.15) is trivial. Let $A_n = A\cap[a+1/n,\infty)$. Then $A_n\subseteq A_{n+1}$ and $\bigcup_{n=1}^{\infty} A_n=A-I$. Also $\lim_{n\to\infty} H_p^*(A_n)$ exists and is finite. By Theorem 25 we have

$$H_p^*(A)\geqslant H_p^*(A\cap I)+H_p^*(A_n).$$

So if we show that $\lim H_p^*(A_n)=H_p^*(A-I)$, the result follows by (2.15). Write $D_n=A_{n+1}-A_n$. Then

$$\begin{aligned} A-I &= A_{2n}\cup\bigcup_{k=2n}^{\infty} D_k \\ &= A_{2n}\cup\bigcup_{k=n}^{\infty} D_{2k}\cup\bigcup_{k=n}^{\infty} D_{2k+1}\,. \end{aligned}$$

So
$$H_p^*(A-I)\leqslant H_p^*(A_{2n})+\sum_{k=n}^{\infty} H_p^*(D_{2k})+\sum_{k=n}^{\infty} H_p^*(D_{2k+1}) \tag{2.16}$$

and all these outer measures are finite. Suppose that both series in (2.16) converge. Then letting $n\to\infty$ we get $H_p^*(A-I)\leqslant\lim H_p^*(A_{2n})=\lim H_p^*(A_n)$. But since $A_n\subseteq A-I$ we must have equality and (2.15) follows. Suppose that the first series, say, diverges. We have $A_{2n}\supseteq\bigcup_{k=1}^{n-1} D_{2k}$. Also $d(D_{2k},D_{2k+2})>0$. So Theorem 25 gives

$$H_p^*(A_{2n})\geqslant\sum_{k=1}^{n-1} H_p^*(D_{2k})\to\infty$$

contradicting the finiteness of $H_p^*(A)$. Similarly the second series in (2.16) must converge and the result follows. □

Corollary: All Borel sets are H_p^*-measurable.

For a Borel set A we will use the notations $H_p(A)$ and $H_{p,\delta}(A)$.

Definition 15: The **Hausdorff dimension** of E is $\inf[p: H_p^*(E) = 0]$.

By Corollary 4 to Theorem 22 Hausdorff dimension is a well defined number in the interval [0,1] for any set A in the real line. Also, except for the trivial case of sets E with $H_p^*(E) = 0$ for all p, we have that Hausdorff dimension equals $\sup[p: H_p^*(E) = \infty]$.

Theorem 27: The Hausdorff dimension of the Cantor-like set P_ξ is $-\log 2/\log \xi$.

Proof: We refer to the construction of P_ξ given in p. 24. The first residual intervals $J_{1,1}$ and $J_{1,2}$ are translates of multiples of [0,1] and contain subsets, say $P^{(1)}$ and $P^{(2)}$, of P_ξ. Since, to get $P^{(1)}$ we dissect $J_{1,1}$ in the same way that [0,1] was dissected, we have that $P^{(1)}$ (and $P^{(2)}$) is a translate of a multiple (by ξ) of P_ξ. Then, by Theorem 25, $H_p(P_\xi) = H_p(P^{(1)}) + H_p(P^{(2)}) = 2^p H_p(P)$, using Examples 22 and 23. So either $H_p(P_\xi) = 0$ or ∞, or $2\xi^p = 1$ giving $p = -\log 2/\log \xi = p_0$, say. We will show that $0 < H_{p_0}(P_\xi) < \infty$ and it will follow that P_ξ has Hausdorff dimension $-\log 2/\log \xi$.

(i) To show $H_{p_0}(P_\xi) < \infty$. Let $[I_i]$ be a covering of P_ξ by open intervals of length at most δ. The set P_ξ is mapped as above onto the sets $P^{(j)}$ $(j = 1, 2)$ by similarity transformations and the same mappings applies to the intervals $[I_i]$ provide a cover $[J_{i,j}]$ of $P^{(j)}$, for $j = 1, 2$. But $l(I_i)^{p_0} = 2\xi^{p_0}\, l(I_i)^{p_0}$ (by the definition of p_0)

$$= l(J_{i,1})^{p_0} + l(J_{i,2})^{p_0}. \tag{2.17}$$

Taking the infimum over all such coverings $[I_i]$ of P_ξ, we get $H_{p_0,\delta}(P_\xi) \geqslant H_{p_0,\xi\delta}(P_\xi)$, as $l(J_{i,j}) \leqslant \xi\,\delta$. Since $H_{p,\delta}$ does not decrease as δ decreases, and $0 < \xi < 1$, it follows that $H_{p_0,\delta}$ is independent of δ, and since we may take as a cover an open interval of length just greater than 1 and containing P_ξ we have $H_{p_0}(P_\xi) \leqslant 1$.

(ii) To show $H_{p_0}(P_\xi) > 0$. The distance between the sets $P^{(1)}$ and $P^{(2)}$ is at least $1 - 2\xi$. Let $\delta \leqslant 1 - 2\xi$. Then as in Theorem 25, any cover $[I_i]$ of P_ξ by intervals of length at most δ may be decomposed into covers $[I_{i,j}]$, $j = 1, 2$, of the sets $P^{(1)}$, $P^{(2)}$ and

$$\sum_i l(I_i)^{p_0} = \sum_i l(I_{i,1})^{p_0} + \sum_i l(I_{i,2})^{p_0}. \tag{2.18}$$

Suppose the first sum of the right of (2.18) is the lesser. Since $P^{(2)}$ is a translate of $P^{(1)}$ the same translation applied to the intervals $[I_{i,1}]$ gives a cover $[I'_{i,1}]$, say of $P^{(2)}$. Then as for (2.17), but in reverse, we may map the intervals $[I_{i,1}]$ onto intervals $[I'_i]$, say, covering P_ξ and with $\xi\, l(I'_i) = l(I_{i,1})$ for each i. So $\sum_i l(I'_i)^{p_0} = \sum_i l(I_{i,1})^{p_0} + \sum_i l(I'_{i,1})^{p_0}$ as for (2.17)

$$\leqslant \sum_i l(I_i)^{p_0}\text{, by construction.} \tag{2.19}$$

So if any one of the intervals I'_i is of length $\geqslant 1 - 2\xi$ we have $\sum_i l(I_i)^{p_0} \geqslant$

$(1-2\xi)^{p_0}$.

Now since P_ξ is compact we may suppose, by the Heine Borel Theorem, p. 18, that all the coverings considered are finite, so $\min l(I_k) > 0$. Since the intervals $[I_i']$ are multiples (by $1/\xi$) of a subset of the intervals $[I_k]$ we have

$$\min l(I_i') \geqslant \xi^{-1} \min l(I_k). \tag{2.20}$$

If each interval I_i' is of length less than $1-2\xi$ we apply the same process to the cover $[I_1']$ which was applied to the cover $[I_i]$, and we must obtain after a finite number of steps, a cover $[I_i^0]$ with $\max l(I_i^0) \geqslant 1-2\xi$ and $\sum_i l(I_i^0)^{p_0} \leqslant \sum_i l(I_i)^{p_0}$ as in (2.19). So in any case we have $\sum l(I_i)^{p_0} \geqslant (1-2\xi)^{p_0}$. So $H_{p_0}(P_\xi) > 0$, as required. □

Corollary: For each α, $0 \leqslant \alpha \leqslant 1$, there exists a set $Q \subseteq \mathsf{R}$ with Hausdorff dimension α.

Proof: The case $\alpha = 1$ is covered by Corollary 1, p. 48. For $0 < \alpha < 1$ take $\xi = \exp[-(\log 2)/\alpha]$ in the last theorem. For $\alpha = 0$ take $Q = [x]$, for example. □

The example $A = \mathsf{R}$ shows that $H_p(A)$ may equal ∞ for $0 < p \leqslant p_1$, $H_p(A) = 0$ for $p > p_1$ ($p_1 = 1$ in this case). The example $A = P_\xi$ shows that $H_p(A)$ may equal ∞ for $0 < p < p_0$, $0 < H_{p_0}(A) < \infty$ and $H_p(A) = 0$ for $p > p_0$. The following example shows that $H_p(A)$ may be ∞ for $p < p_1$, $H_p(A) = 0$ for $p \geqslant p_1$.

Example 25: For each q, $0 < q \leqslant 1$ there exists a set A in R such that $H_p(A) = \infty$ for $0 < p < q$, $H_p(A) = 0$ for $p \geqslant q$.

Solution: Let $p_n \uparrow q$ where $0 < p_n < q$. Choose A_n such that $H_{p_n}(A_n) = 1$ (a suitable multiple of a set P_ξ, for an appropriate ξ, will suffice). Let $A = \bigcup_{n=1}^{\infty} A_n$.

Since $q > p_n$ we have $H_q(A_n) = 0$ by Theorem 20. So $H_q(A) \leqslant \sum_{q=1}^{\infty} H_q(A_n) = 0$ giving $H_p(A) = 0$ for $p \geqslant q$. But if $p_0 < q$ we have $p_n > p_0$ for some n and then $H_{p_n}(A) \geqslant H_{p_n}(A_n) = 1$. So $H_{p_0}(A) = \infty$.

Example 26: For each q, $0 < q \leqslant 1$ there exists a set A in R such that $H_p(A) = \infty$ for $0 < p \leqslant q$, and $H_p(A) = 0$ for $p > q$.

Solution: The case $q = 1$ has been dealt with above ($A = \mathsf{R}$). For $0 < q < 1$ construct the set P_ξ with $H_p(P_\xi)$ finite and positive. Let $A = \bigcup_{n=0}^{\infty} (P_\xi + n)$, a union of disjoint translates of P_ξ. Then $H_q(A) = \infty$, and for $p > q$ we have $H_p(A) = \sum_{n=0}^{\infty} H_p(P_\xi + n) = 0$ as required.

Finally we show that, with a slight restriction on the measure function h, there will always exist sets whose corresponding Hausdorff measure is positive and finite. First we define the modulus of continuity of a continuous function and apply this definition to the Lebesgue function.

Definition 16: Let $f(x)$ be a function continuous on the closed bounded interval $[a,b]$. Then the function ω_f given by $\omega_f(t) = \sup[|f(x) - f(y)|: |x - y| \leqslant t; x,y \in [a,b]]$ is the **modulus of continuity** of f.

Clearly $\omega_f(t)$ increases with t, and since f is uniformly continuous on the compact set $[a,b]$ we have that $\omega_f(t)$ is bounded and tends to zero as $t \to 0$ through positive values.

Let L be the Lebesgue function corresponding to the Cantor-like set P_ξ of Chapter 1, p. 24. Let b_j be the length of the residual intervals at the jth stage in the construction of P_ξ, so that $b_j = \xi_1 \xi_2 \ldots \xi_j$. Now L increases by 2^{-j} on each such interval. Also any interval in $[0,1]$ of length b_j meets at most two of these residual intervals, and L is constant on the removed intervals. So $\omega_L(b_j) \leqslant 2.2^{-j}$.

Definition 17: The function $f(x)$ is **strictly concave** on (a,b) if for any x,y in (a,b) we have

$$f[tx + (1-t)y] > tf(x) + (1-t)f(y), 0 < t < 1;$$

that is: the graph of f lies strictly above the segment joining any two points on the graph.

Note that x^p $(0 < p < 1)$ is strictly concave on $(0,\infty)$ as its second derivative is negative, so the Hausdorff measure functions so far considered have been, in the main, strictly concave. Many of the results obtained hold for general Hausdorff measure functions; in particular Theorem 19, Example 22, Theorems 21, 23, 24, 25, 26. Also Example 24 holds for a concave measure function h. The first part of the proof is the same and the second half follows from the fact that $h[\eta(1+\delta)] \leqslant (1+\delta)h(\eta)$ for $\eta, \delta > 0$, as can be seen from a diagram.

Theorem 28: Let h be any strictly concave Hausdorff measure function and H the corresponding Hausdorff measure. Then there exists a Borel set A, contained in $[0,1]$, such that $0 < H(A) < \infty$.

Proof: The Hausdorff measure is changed by a factor λ if h is, so we may suppose that $h(1) = 1$. Now choose the sequence $\{b_j\}$ such that $h(b_j) = 2^{-j}$, and construct the Cantor-like set P_ξ with b_j equal to the length of the residual interval at the jth stage. This is possible since $2h(t) > h(2t)$ and so $b_{j+1} < \frac{1}{2}b_j$ for each j. Then $\omega_L(b_j) \leqslant 2^{1-j} \leqslant 4h(b_{j+1})$. So for $t \in [b_{j+1}, b_j]$ we have $\omega_L(t) \leqslant \omega_L(b_j) \leqslant 4h(b_{j+1}) \leqslant 4h(t)$, as h is monotone increasing. Write A for the set P_ξ constructed to correspond in this way to h. Suppose that A is covered by the family of open intervals (u_i, v_i). As A is compact this family may be supposed finite. Then as in Theorem 2, p. 28,

$$\begin{aligned}\Sigma h(v_i - u_i) &\geqslant 4^{-1}\, \Sigma\, \omega_L(v_i - u_i)\\ &\geqslant 4^{-1}\, \Sigma\, [L(v_i) - L(u_i)] \geqslant 4^{-1}\, [L(1) - L(0)] = 4^{-1}.\end{aligned}$$

So $H(A)$ is positive.

Write $H_\delta(A)$ for the approximating measure when the intervals are of length at most δ. Since at the jth stage of the construction A is covered by 2^j intervals of length b_j, we have $H_\delta(A) \leqslant 2^j h(b_j) = 1$, if $\delta > b_j$. (These intervals are closed but the extension of Example 24 shows that the measure H is unaffected.) Letting $\delta \to 0$ we get $H(A) \leqslant 1$, proving the theorem. □

Exercises

46. Show that every countable set has Hausdorff dimension zero.
47. Let $h(t) = \sin t$ $(0 < t < \pi/2)$. Show that, for any Borel set A, $H(A) = m(A)$.
48. Let $\{A_n\}$ be Borel sets and let α_n be the Hausdorff dimension of A_n. Find the Hausdorff dimension of $A = \bigcup_{n=1}^{\infty} A_n$.
49. Show that, for $0 < \delta < 1$, $H^*_{p,\delta}(A)$ is a monotone decreasing function of p.
50. Show that, for $0 < q < 1$, we may construct a set A as in Example 26, but which is compact, so that $H_p(A) = \infty$ $(0 < p \leqslant q)$, $H_p(A) = 0$ $(p > q)$.
51. Find $\omega_f(t)$ for $f(x) = \cos x$, on $[0, 2\pi]$.
52. The class of Hausdorff measure functions may be extended by assuming only that
    (i) $\liminf_{y \to x} h(y) > 0$ and $h(x) > 0$ for $x > 0$,
    (ii) $h(0) = h(0+) = 0$.
    Show that the Hausdorff measure so defined satisfies the theorems given.
53. Show that if we now replace the function h of the last exercise by g where $g(x) = \inf[h(y)\colon x \leqslant y \leqslant 1]$ and $g(x) = h(1)$ for $x > 1$, then g is a monotone increasing function satisfying (i) and (ii) of the last exercise and giving the same Hausdorff measure as h.

CHAPTER 3

Integration of Functions of a Real Variable

In analysis it is often convenient to replace an expression of the form $\int\Sigma f_n\, dx$ by $\Sigma\int f_n\, dx$, or $\int \lim f_n\, dx$ by $\lim \int f_n\, dx$, or $\int \lim_{\alpha\to\alpha_0} f_\alpha\, dx$ by $\lim_{\alpha\to\alpha_0} \int f_\alpha\, dx$. In this chapter we give a definition of an integral which applies to a large class of Lebesgue measurable functions and which allows the interchange of integral and sum or limit in very general circumstances. The results justify the choice of the class of measurable functions in the last chapter. Our results on approximation to measurable sets by intervals or by open sets lead to results on approximation to the integral of a measurable function. In the last section we compare the Lebesgue and Riemann integrals.

3.1 INTEGRATION OF NON-NEGATIVE FUNCTIONS

We consider first the class of non-negative measurable functions, define the integral of such a function and examine the properties of the integral. For the present we will suppose these functions to be defined for all real x.

A non-negative finite-valued function $\varphi(x)$, taking only a finite number of different values, is called a **simple function.** If $a_1, a_2, \ldots, a_n$ are the distinct values taken by φ and $A_i = [x\colon \varphi(x) = a_i]$, then clearly

$$\varphi(x) = \sum_{i=1}^{n} a_i\, \chi_{A_i}(x). \tag{3.1}$$

The sets A_i are measurable if φ is a measurable function, by Example 9, p. 38. The convention $0 \cdot \infty = 0$, introduced in Chapter 2, p. 37, is to be understood in the following definition.

Definition 1: Let φ be a measurable simple function.

Then $\quad \int \varphi\, dx = \sum_{i=1}^{n} a_i m(A_i),$

where $a_i, A_i, i = 1, \ldots, n$ are as in (3.1), is called the **integral** of φ.

Example 1: Let the sets A_i be defined as above. Then $A_i \cap A_j = \emptyset$, $i \neq j$, and $\bigcup_{i=1}^{n} A_i = \mathsf{R}$.

Definition 2: For any non-negative measurable function f, the *integral* of f, $\int f\,dx$, is given by $\int f\,dx = \sup \int \varphi\,dx$, where the supremum is taken over all measurable simple functions φ, $\varphi \leqslant f$.

Definition 3: For any measurable set E, and any non-negative measurable function f, $\int_E f\,dx = \int f \chi_E\,dx$ is the *integral of f over E.*

If the set E in Definition 3 is an interval, such as $[a,b]$, then in place of $\int_E f\,dx$ we write $\int_a^b f\,dx$; if $a > b$ we use the usual convention: $\int_a^b f\,dx = -\int_b^a f\,dx$. When a distinction is necessary, the integral defined above will be referred to as the **Lebesgue integral**; to avoid confusion with the Riemann integral, the latter will be denoted by $\mathbf{R}\int_a^b f\,dx$. The relationship of the integrals will be considered later in Section 3.4.

Example 2: If φ is a measurable simple function, Definition 1 and Definition 2 both give a value for its integral. Show that these values are the same.

Solution: Write $\int^* \varphi\,dx = \sup \int \psi\,dx$, ψ any measurable simple function, $\psi \leqslant \varphi$, that is: write $\int^* \varphi\,dx$ for the value given by Definition 2, and write $\int \varphi\,dx$ for that given by Definition 1. From the definition, $\int \varphi\,dx \leqslant \int^* \varphi\,dx$. Also, if ψ $(\leqslant \varphi)$ is a measurable simple function with distinct values b_j $(j = 1, \ldots, m)$ and $\psi = \sum_{j=1}^{m} b_j \chi_{B_j}$, then $\int \psi\,dx = \sum_{j=1}^{m} b_j m(B_j) = \sum_{j=1}^{m} \sum_{i=1}^{n} b_j m(B_j \cap A_i)$ (cf. Example 1), where $b_j \leqslant a_i$ if $m(B_j \cap A_i) > 0$. So

$$\int \psi\,dx \leqslant \sum_{j=1}^{m} \sum_{i=1}^{n} a_i m(B_j \cap A_i) = \sum_{i=1}^{n} a_i m(A_i) = \int \varphi\,dx.$$

So $\int^* \varphi\,dx \leqslant \int \varphi\,dx$, which completes the proof.

Theorem 1: If φ is a measurable simple function, then in the notation of (3.1)

(i) $\int_E \varphi \, dx = \sum_{i=1}^{n} a_i m(A_i \cap E)$ for any measurable set E,

(ii) $\int_{A \cup B} \varphi \, dx = \int_A \varphi \, dx + \int_B \varphi \, dx$ for any disjoint measurable sets A and B,

(iii) $\int a\varphi \, dx = a \int \varphi \, dx$ if $a > 0$.

Proof: (i) is immediate from Definitions 1 and 3. For (ii):

$$\int_A \varphi \, dx + \int_B \varphi \, dx = \sum_{i=1}^{n} a_i m(A \cap A_i) + \sum_{i=1}^{n} a_i m(B \cap A_i)$$

$$= \sum_{i=1}^{n} a_i m((A \cup B) \cap A_i) = \int_{A \cup B} \varphi \, dx.$$

For (iii): As φ takes the values a_i, $a\varphi$ takes the distinct values aa_i, so that $\int a\varphi \, dx = \sum_{i=1}^{n} aa_i m(A_i) = a \int \varphi \, dx$. □

Example 3: Show that if f is a non-negative measurable function, then $f = 0$ a.e. if, and only if, $\int f \, dx = 0$.

Solution: If $f = 0$ a.e. and φ is a measurable simple function, $\varphi \leqslant f$, then clearly $\int \varphi \, dx = 0$, so Definition 2 gives $\int f \, dx = 0$. Conversely, if $\int f \, dx = 0$ and $E_n = [x: f(x) \geqslant 1/n]$, then $\int f \, dx \geqslant \int n^{-1} \chi_{E_n} \, dx = n^{-1} m(E_n)$. So $m(E_n) = 0$. But $[x: f(x) > 0] = \bigcup_{n=1}^{\infty} E_n$, so $f = 0$ a.e.

Theorem 2: Let f and g be non-negative measurable functions.

(i) If $f \leqslant g$, then $\int f \, dx \leqslant \int g \, dx$.

(ii) If A is a measurable set and $f \leqslant g$ on A, then $\int_A f \, dx \leqslant \int_A g \, dx$.

(iii) If $a \geqslant 0$, then $\int af \, dx = a \int f \, dx$.

(iv) If A and B are measurable sets and $A \supseteq B$, then $\int_A f \, dx \geqslant \int_B f \, dx$.

Proof: (i) and (ii) are immediate from Definitions 2 and 3 (p. 55) respectively. (iii) is obvious if $a = 0$. If $a > 0$, φ is a measurable simple function with $\varphi \leqslant af$ if, and only if, $\varphi = a\psi$, where ψ is simple, $\psi \leqslant f$, and then $\int \varphi \, dx = a \int \psi \, dx$ (Theorem 1(iii), above). So $\int af \, dx = \sup \int \varphi \, dx = a \sup \int \psi \, dx = a \int f \, dx$.

(iv): Note that $\chi_A f \geqslant \chi_B f$ and apply (i). □

The following result will be basic in proving convergence theorems.

Theorem 3 (Fatou's Lemma): Let $\{f_n, n = 1, 2, \ldots\}$ be a sequence of non-negative measurable functions. Then

$$\liminf \int f_n \, dx \geqslant \int \liminf f_n \, dx. \tag{3.2}$$

Proof: Let $f = \liminf f_n$. Then f is a non-negative measurable function. From Definition 2, p. 55, the result follows if, for each measurable simple function φ with $\varphi \leqslant f$, we have

$$\int \varphi \, dx \leqslant \liminf \int f_n \, dx. \tag{3.3}$$

Case 1. $\int \varphi \, dx = \infty$. Then from Definition 1, p. 54, for some measurable set A, we have $m(A) = \infty$ and $\varphi > a > 0$ on A. Write $g_k(x) = \inf_{j \geqslant k} f_j(x)$, and $A_n = [x: g_k(x) > a, \text{ all } k \geqslant n]$, a measurable set. Then $A_n \subseteq A_{n+1}$, each n. But, for each x, $\{g_k(x)\}$ is monotone increasing and $\lim_{k \to \infty} g_k(x) = f(x) \geqslant \varphi(x)$. So $A \subseteq \bigcup_{n=1}^{\infty} A_n$. Hence $\lim m(A_n) = \infty$. But, for each n,

$$\int f_n \, dx \geqslant \int g_n \, dx > a\, m(A_n).$$

So $\liminf \int f_n \, dx = \infty$, and (3.3) holds.

Case 2. $\int \varphi \, dx < \infty$. Write $B = [x: \varphi(x) > 0]$. Then $m(B) < \infty$. Let M be the largest value of φ, and if $0 < \epsilon < 1$, write $B_n = [x: g_k(x) > (1 - \epsilon)\varphi(x), k \geqslant n]$, where g_k is as defined above. Then the sets B_n are measurable, $B_n \subseteq B_{n+1}$ for each n, and $\bigcup_{n=1}^{\infty} B_n \supseteq B$. So $\{B - B_n\}$ is a decreasing sequence of sets, $\bigcap_{n=1}^{\infty} (B - B_n) = \emptyset$. As $m(B) < \infty$, by Theorem 9, p. 33, there exists N such that $m(B - B_n) < \epsilon$ for all $n \geqslant N$. So if $n \geqslant N$,

$$\begin{aligned}
\int g_n \, dx &\geqslant \int_{B_n} g_n \, dx \geqslant (1 - \epsilon) \int_{B_n} \varphi \, dx \\
&= (1 - \epsilon)\left(\int_B \varphi \, dx - \int_{B - B_n} \varphi \, dx\right) \text{ (by Theorem 1)} \\
&\geqslant (1 - \epsilon) \int \varphi \, dx - \int_{B - B_n} \varphi \, dx \\
&\geqslant \int \varphi \, dx - \epsilon \int \varphi \, dx - \epsilon M.
\end{aligned}$$

Since ϵ is arbitrary, $\liminf \int g_n \, dx \geqslant \int \varphi \, dx$, and since $f_n \geqslant g_n$, (3.3) follows. □

Theorem 4 (Lebesgue's Monotone Convergence Theorem): Let $\{f_n, n = 1, 2, \ldots\}$ be a sequence of non-negative measurable functions such that $\{f_n(x)\}$ is monotone increasing for each x. Let $f = \lim f_n$. Then $\int f \, dx = \lim \int f_n \, dx$.

Proof: Fatou's Lemma gives

$$\int f\,dx = \int \liminf f_n\,dx \leqslant \liminf \int f_n\,dx. \tag{3.4}$$

But $f \geqslant f_n$ by hypothesis, so by Theorem 2(i), p. 56, $\int f\,dx \geqslant \int f_n\,dx$, and hence

$$\int f\,dx \geqslant \limsup \int f_n\,dx. \tag{3.5}$$

Relations (3.4) and (3.5) give the result. □

Theorem 5: Let f be a non-negative measurable function. Then there exists a sequence $\{\varphi_n\}$ of measurable simple functions such that, for each x, $\varphi_n(x) \uparrow f(x)$.

Proof: By construction. Write, for each n, $E_{nk} = [x\colon (k-1)/2^n < f(x) \leqslant k/2^n]$, $k = 1, 2, \ldots, n2^n$, and $F_n = [x\colon f(x) > n]$. Put

$$\varphi_n = \sum_{k=1}^{n2^n} \frac{k-1}{2^n}\chi_{E_{nk}} + n\chi_{F_n}.$$

Then the functions φ_n are measurable simple functions. Also, since the dissection of the range of f giving φ_{n+1} is a refinement of that giving φ_n, it is easily seen that $\varphi_{n+1}(x) \geqslant \varphi_n(x)$ for each x. If $f(x)$ is finite, $x \in \mathrm{C}F_n$ for all large n, and then $|f(x) - \varphi_n(x)| \leqslant 2^{-n}$. So $\varphi_n(x) \uparrow f(x)$. If $f(x) = \infty$, then $x \in \bigcap_{n=1}^{\infty} F_n$, so $\varphi_n(x) = n$ for all n, and again $\varphi_n(x) \uparrow f(x)$. □

Corollary: $\lim \int \varphi_n\,dx = \int f\,dx$, where φ_n and f are as in Theorem 5.

This application of Theorem 5, with Theorem 4, gives us a method of evaluating $\int f\,dx$ alternative to that of Definition 2, p. 55.

Theorem 6: Let f and g be non-negative measurable functions. Then

$$\int f\,dx + \int g\,dx = \int (f+g)\,dx. \tag{3.6}$$

Proof: Consider (3.6) for measurable simple functions φ and ψ. Let the values of φ be $a_1, \ldots, a_n$ taken on sets $A_1, \ldots, A_n$, and let the values of ψ be $b_1, \ldots, b_m$ taken on sets $B_1, \ldots, B_m$. Then the simple function $\varphi + \psi$ has the value $a_i + b_j$ on the measurable set $A_i \cap B_j$, so from Theorem 1(i), p. 56, we obtain

$$\int_{A_i \cap B_j} (\varphi + \psi)\,dx = \int_{A_i \cap B_j} \varphi\,dx + \int_{A_i \cap B_j} \psi\,dx. \tag{3.7}$$

But the union of the nm disjoint sets $A_i \cap B_j$ is R, so Theorem 1(ii) applied to both sides of (3.7) gives

$$\int (\varphi + \psi)\,dx = \int \varphi\,dx + \int \psi\,dx. \tag{3.8}$$

Let f and g be any non-negative measurable functions. Let $\{\varphi_n\}$, $\{\psi_n\}$ be sequences of measurable simple functions, $\varphi_n \uparrow f$, $\psi_n \uparrow g$. Then $\varphi_n + \psi_n \uparrow f + g$. But, by (3.8), $\int (\varphi_n + \psi_n)\,dx = \int \varphi_n\,dx + \int \psi_n\,dx$. So, letting n tend to infinity, Theorem 4 gives the result. □

Theorem 7: Let $\{f_n\}$ be a sequence of non-negative measurable functions. Then

$$\int \sum_{n=1}^{\infty} f_n \, dx = \sum_{n=1}^{\infty} \int f_n \, dx.$$

Proof: By induction, (3.6) applies to a sum of n functions. So if $S_n = \sum_{i=1}^{n} f_i$, then $\int S_n \, dx = \sum_{i=1}^{n} \int f_i \, dx$. But $S_n \uparrow f = \sum_{i=1}^{\infty} f_i$, so the result follows from Theorem 4. □

Example 4: Give an example where strict inequality occurs in (3.2) (Fatou's Lemma).

Solution: Let $f_{2n-1} = \chi_{[0,1]}, f_{2n} = \chi_{(1,2)}, n = 1, 2, \ldots$. Then $\liminf f_n(x) = 0$, for all x, but $\int f_n(x) \, dx = 1$, for all n.

Such an example is helpful in remembering the 'direction of the inequality' in (3.2).

Example 5: Show that $\int_1^{\infty} dx/x = \infty$.

Solution: x^{-1} is a continuous function for $x > 0$, and so is measurable. It is positive, so the integral is defined. Also $\int_1^{\infty} x^{-1} \, dx > \int_1^{n} x^{-1} \, dx$. But $x^{-1} > k^{-1}$ on $[k-1, k)$, so $\int_1^{n} x^{-1} \, dx > \sum_{k=2}^{n} \int_1^{n} k^{-1} \chi_{[k-1,k)} \, dx > \sum_{k=2}^{n} k^{-1} \to \infty$ as $n \to \infty$.

Example 6: $f(x)$, $0 \leqslant x \leqslant 1$, is defined by: $f(x) = 0$ for x rational, if x is irrational, $f(x) = n$, where n is the number of zeros immediately after the decimal point, in the representation of x on the decimal scale. Show that f is measurable and find $\int_0^1 f \, dx$.

Solution: For $x \in (0,1]$ let $g(x) = n$ if $10^{-(n+1)} \leqslant x < 10^{-n}, n = 0, 1, \ldots$, and $g(1) = 0$. Then $f \leqslant g, f = g$ a.e., so f is measurable and by Example 3,

$$\int_0^1 f \, dx = \int_0^1 g \, dx.$$

But

$$\int_0^1 g \, dx = \sum_{n=0}^{\infty} n\left(\frac{1}{10^n} - \frac{1}{10^{n+1}}\right)$$

$$= \sum_{n=1}^{\infty} \frac{9n}{10^{n+1}} = \frac{1}{9}.$$

Exercises

1. Prove (3.8) directly from Definition 1 without reference to Theorem 1. (Note that the numbers $a_i + b_j$ are not necessarily distinct.)
2. Let $f, g \geqslant 0$ be measurable, with $f \geqslant g$, $\int g\,dx < \infty$. Show that $\int f\,dx - \int g\,dx = \int (f-g)\,dx$.
3. Let $f_n \geqslant 0$ be measurable, $\lim f_n = f$ and $f_n \leqslant f$ for each n. Show that $\int f\,dx = \lim \int f_n\,dx$.
4. Let $f_n(x) = \min(f(x),n)$ where $f \geqslant 0$ is measurable. Show that $\int f_n\,dx \uparrow \int f\,dx$.
5. Let $f \geqslant 0$, measurable. Construct a sequence ψ_n of measurable simple functions, such that $\psi_n \uparrow f$ and $m[x\colon \psi_n(x) > 0] < \infty$ for each n.
6. Show that Fatou's Lemma and Theorem 4 can be obtained from one another using only the properties of the integral given in Theorems 1 and 2.
7. Fatou's Lemma is sometimes written in the form: if $\{f_n\}$ is a sequence of non-negative measurable functions, and $\lim f_n = f$ a.e., then $\int f\,dx \leqslant \liminf \int f_n\,dx$. Show that this version is equivalent to that given.
8. Let $\{f_n\}$ be a sequence of non-negative finite-valued measurable functions, $f_n \downarrow f$. Show that if $\int f_k\,dx < \infty$ for some k, then $\lim \int f_n\,dx = \int f\,dx$, and that $\int f_k\,dx = \infty$ for all k does not imply $\int f\,dx = \infty$.
9. Let $f(x) = 0$ at each point $x \in P$, the Cantor set in $[0,1]$, $f(x) = p$ in each complementary interval of length 3^{-p}. Show that f is measurable and that $\int_0^1 f\,dx = 3$.
10. Let $\{f_n\}$ be a sequence of non-negative measurable functions such that $\lim f_n = f$ a.e. and $\lim \int f_n\,dx = \int f\,dx < \infty$. Show that for each measurable set E, $\lim \int_E f_n\,dx = \int_E f\,dx$.
11. Show that to every measurable function f there corresponds a Borel-measurable function g such that $f = g$ a.e.
12. The function f is defined on $(0,1)$ by

$$f(x) = \begin{cases} 0, & x \text{ rational} \\ [1/x]^{-1}, & x \text{ irrational,} \end{cases}$$

where $[x]$ = integer part of x. Show that $\int_0^1 f\,dx = \infty$.

3.2 THE GENERAL INTEGRAL

The definition of the integral will now be extended to a wide class of real-valued measurable functions, not necessarily non-negative. The strength of the two main convergence theoreoms (Theorems 10 and 11) shows that the definition is the appropriate one.

Definition 4: If $f(x)$ is any real function,

$$f^+(x) = \max(f(x),0), f^-(x) = \max(-f(x), 0),$$

are said to be the **positive** and **negative parts** of f, respectively.

Theorem 8: (i) $f = f^+ - f^-$; $|f| = f^+ + f^-$; $f^+, f^- \geqslant 0$.
(ii) f is measurable if, and only if, f^+ and f^- are both measurable.

Proof: (i) is clear from Definition 4. (ii) follows from (i) and Example 14, p. 40. □

Definition 5: If f is a measurable function and $\int f^+ \,dx < \infty$, $\int f^- \,dx < \infty$, we say that f is **integrable**, and its integral is given by

$$\int f \,dx = \int f^+ \,dx - \int f^- \,dx.$$

Clearly, a measurable function f is integrable if, and only if, $|f|$ is, and then $\int |f| \,dx = \int f^+ \,dx + \int f^- \,dx$.

Definition 6: If E is a measurable set, f is a measurable function, and $\chi_E f$ is integrable, we say that f is **integrable over** E, and its integral is given by $\int_E f \,dx = \int f \chi_E \,dx$. The notation $f \in L(E)$ is then sometimes used.

Definition 7: If f is a measurable function such that at least one of $\int f^+ \,dx$, $\int f^- \,dx$ is finite, then $\int f \,dx = \int f^+ \,dx - \int f^- \,dx$.

Under Definition 7, integrals are allowed to take infinite values, so this definition is an extension of Definition 2. But f is said to be *integrable* only if the conditions of Definition 5 are satisfied, that is if $|f|$ has a finite integral.

Theorem 9: Let f and g be integrable functions.

(i) af is integrable, and $\int af \,dx = a \int f \,dx$.
(ii) $f + g$ is integrable, and $\int (f + g) \,dx = \int f \,dx + \int g \,dx$.
(iii) If $f = 0$ a.e., then $\int f \,dx = 0$.
(iv) If $f \leqslant g$ a.e., then $\int f \,dx \leqslant \int g \,dx$.
(v) If A and B are disjoint measurable sets, then $\int_A f \,dx + \int_B f \,dx = \int_{A \cup B} f \,dx$.

Proof: (i): Suppose that $a \geqslant 0$. Then $(af)^+ = af^+$, $(af)^- = af^-$. So $\int (af)^+ \,dx < \infty$ and $\int (af)^- \,dx < \infty$. So af is integrable and

$$\int af \,dx = \int af^+ \,dx - \int af^- \,dx = a \int f \,dx.$$

Suppose that $a = -1$. Then $(-f)^+ = f^-$, $(-f)^- = f^+$. So $-f$ is integrable and

$$\int (-f) \,dx = \int f^- \,dx - \int f^+ \,dx = -\int f \,dx.$$

But for $a < 0$, $af = -|a| f$, so

$$\int af \,dx = -\int |a| f \,dx = -|a| \int f \,dx = a \int f \,dx,$$

and (i) follows.

(ii): $(f+g)^+ \leqslant f^+ + g^+$, $(f+g)^- \leqslant f^- + g^-$, so $f+g$ is integrable. Also $(f+g)^+ - (f+g)^- = f+g = f^+ + g^+ - f^- - g^-$, so

$$(f+g)^+ + f^- + g^- = (f+g)^- + f^+ + g^+. \tag{3.9}$$

Apply Theorem 6, p. 58, to both sides of (3.9) and rearrange the terms to get (ii).

(iii): $f^+ = 0$ a.e. and $f^- = 0$ a.e., so by Example 3, p. 56, $\int f^+ \, dx = \int f^- \, dx = 0$.

(iv): $g = f + (g-f)$, so

$$\int g \, dx = \int f \, dx + \int (g-f)^+ \, dx - \int (g-f)^- \, dx.$$

But $(g-f)^- = 0$ a.e., so the result follows by (iii).

(v): $\chi_{A \cup B} = \chi_A + \chi_B$, so Definition 6 and (ii) give the result. □

From Theorem 9, if $f = g$ a.e. and f and g are integrable, then $\int f \, dx = \int g \, dx$. We can extend our results to the case where f is measurable and f is defined except on a set E such that $m(E) = 0$ and $\int_{\mathbf{C}E} |f| \, dx < \infty$. Then we define f arbitrarily on E to get a function g which clearly is necessarily integrable. We have $f = g$ a.e., and so define $\int f \, dx$ to be $\int g \, dx$. This will be relevant in Theorem 11, p. 64 below, for example, where integrals of functions which are defined only a.e. arise naturally.

Example 7: Show that if f and g are measurable, $|f| \leqslant |g|$ a.e., and g is integrable, then f is integrable.

Solution: Redefining f, if necessary, on a set of measure zero, we may suppose that $|f| \leqslant |g|$. Then $f^+ \leqslant |g|$ so that $\int f^+ \, dx \leqslant \int |g| \, dx < \infty$, and similarly for f^-.

Example 8: Show that if f is an integrable function, then $|\int f \, dx| \leqslant \int |f| \, dx$. When does equality occur?

Solution: $|f| - f \geqslant 0$, so $\int |f| \, dx \geqslant \int f \, dx$. Also, $|f| + f \geqslant 0$ so $\int |f| \, dx \geqslant -\int f \, dx$. Hence $\int |f| \, dx \geqslant |\int f \, dx|$. Necessary condition for equality: If $\int f \, dx \geqslant 0$, then $\int |f| \, dx = \int f \, dx$, that is $\int (|f| - f) \, dx = 0$. So by Example 3, p. 56, $|f| = f$ a.e.. If $\int f \, dx < 0$, then $\int |f| \, dx = \int (-f) \, dx$, that is $\int (|f| + f) \, dx = 0$, so $|f| = -f$ a.e. Hence $f \geqslant 0$ a.e. or $f \leqslant 0$ a.e. is a necessary condition.

Clearly this is also a sufficient condition.

Example 9: If f is measurable and g integrable and α, β are real numbers such that $\alpha \leqslant f \leqslant \beta$ a.e., then there exists γ, $\alpha \leqslant \gamma \leqslant \beta$ such that $\int f|g| \, dx = \gamma \int |g| \, dx$.

Solution: $|fg| \leqslant (|\alpha| + |\beta|)|g|$ a.e., so, by Example 7, fg is integrable. Also $\alpha|g| \leqslant f|g| \leqslant \beta|g|$ a.e, so

$$\alpha \int |g| \, dx \leqslant \int f|g| \, dx \leqslant \beta \int |g| \, dx.$$

If $\int |g| \, dx = 0$, then $g = 0$ a.e. and the result is trivial. If $\int |g| \, dx \neq 0$, take $\gamma = (\int f|g| \, dx).(\int |g| \, dx)^{-1}$ and the result follows.

Example 10: Extend Theorem 9, p. 61, to any functions such that the integrals involved are defined in the sense of Definition 7.

Solution: We consider, for example, the extension of (ii):
If $\int (f+g)\,dx$, $\int f\,dx$ and $\int g\,dx$ are defined, then

$$\int(f+g)\,dx = \int f\,dx + \int g\,dx, \tag{3.10}$$

whenever the right hand side is defined. To prove this, suppose for example that $\int f\,dx = \infty = \int g\,dx$. Then $\int f^-\,dx < \infty$ and $\int g^-\,dx < \infty$, so $\int (f+g)^-\,dx < \infty$, and integrating the identity (3.9) gives $\infty = \infty$ in (3.10). The same argument works if, say, $\int f\,dx = \infty$ and $|\int g\,dx| < \infty$.

Example 11: Show that if f is integrable, then f is finite-valued a.e.

Solution: If $|f| = \infty$ on a set E with $m(E) > 0$, then $\int |f|\,dx > n\,m(E)$ for all n, giving a contradiction.

Example 12: If f is measurable, $m(E) < \infty$ and $A \leqslant f \leqslant B$ on E, then $A\,m(E) \leqslant \int_E f\,dx \leqslant B\,m(E)$.

Solution: $A\chi_E, B\chi_E$ are integrable, so Theorem 9(iv) applies.
We now have the main result of this section.

Theorem 10 (Lebesgue's Dominated Convergence Theorem): Let $\{f_n\}$ be a sequence of measurable functions such that $|f_n| \leqslant g$, where g is integrable, and let $\lim f_n = f$ a.e. Then f is integrable and

$$\lim \int f_n\,dx = \int f\,dx. \tag{3.11}$$

Proof: Since, for each n, $|f_n| \leqslant g$, we have $|f| \leqslant g$ a.e., so by Example 7, f_n and f are integrable. Also, $\{g+f_n\}$ is a sequence of non-negative measurable functions, so by Fatou's Lemma

$$\liminf \int (g+f_n)\,dx \geqslant \int \liminf (g+f_n)\,dx.$$

So $\int g\,dx + \liminf \int f_n\,dx \geqslant \int g\,dx + \int f\,dx$. But $\int g\,dx$ is finite so

$$\liminf \int f_n\,dx \geqslant \int f\,dx. \tag{3.12}$$

Again, $\{g-f_n\}$ is also a sequence of non-negative measurable functions, so

$$\liminf \int (g-f_n)\,dx \geqslant \int \liminf (g-f_n)\,dx.$$

So $\int g\,dx - \limsup \int f_n\,dx \geqslant \int g\,dx - \int f\,dx$. So $\limsup \int f_n\,dx \leqslant \int f\,dx \leqslant \liminf \int f_n\,dx$ by (3.12), and (3.11) follows. □

Example 13: With the same hypotheses as Theorem 10, show that

$$\lim \int |f_n - f|\,dx = 0.$$

Solution: $|f_n - f| \leqslant 2g$, for each n, and Theorem 10 applied to $\{f_n - f\}$ gives the result.

The next result is of considerable value in applications.

Theorem 11: Let $\{f_n\}$ be a sequence of integrable functions such that

$$\sum_{n=1}^{\infty} \int |f_n| \,dx < \infty. \tag{3.13}$$

Then the series $\sum_{n=1}^{\infty} f_n(x)$ converges a.e., its sum $f(x)$ is integrable and

$$\int f \,dx = \sum_{n=1}^{\infty} \int f_n \,dx. \tag{3.14}$$

Proof: Let $\varphi(x) = \sum_{n=1}^{\infty} |f_n|$. Then by Theorem 7 and (3.13), $\int \varphi \,dx < \infty$, so φ is finite-valued a.e. by Example 11.

It follows that $\sum_{n=1}^{\infty} f_n(x)$ is absolutely convergent a.e., its sum $f(x)$ is defined a.e., and $|f| \leqslant \varphi$. So f is integrable (see the remarks after Theorem 9, p. 62).

Write $g_n(x) = \sum_{i=1}^{n} f_i(x)$. Then $|g_n(x)| \leqslant \varphi(x)$ and $\lim g_n(x) = f(x)$ a.e., so by Theorem 10, $\lim \int g_n \,dx = \int f \,dx$ and (3.14) follows. □

Example 14: In Theorems 10 and 11 we may suppose that the hypotheses hold only on a measurable set E. Then (3.11) and (3.14), with interals taken over E, follow on replacing throughout f_n, f etc., by $f_n \chi_E, f \chi_E$, etc.

Example 15: Theorem 10 deals with a sequence of functions $\{f_n\}$. State and prove a 'continuous parameter' version of the theorem.

Solution: Theorem: for each $\xi \in [a,b]$, $-\infty \leqslant a < b < \infty$, let f_ξ be a measurable function, $|f_\xi(x)| \leqslant g(x)$ where g is an integrable function, and let $\lim_{\xi\to\xi_0} f_\xi(x) = f(x)$ a.e., where $\xi_0 \in [a,b]$. Then f is integrable and

$$\lim_{\xi\to\xi_0} \int f_\xi \,dx = \int f \,dx. \tag{3.15}$$

Proof: Let $\{\xi_n\}$ be any sequence in $[a,b]$, $\lim \xi_n = \xi_0$. Then the sequence $\{f_{\xi_n}\}$ satisfies the conditions of Theorem 10, and we deduce that f is integrable. Suppose that (3.15) does not hold. Then $\exists\ \delta > 0$ and a sequence $\{\beta_n\}$, with $\lim \beta_n = \xi_0$, such that for all n, $|\int f_{\beta_n} \,dx - \int f \,dx| > \delta$. But, applying Theorem 10 to the sequence $\{f_{\beta_n}\}$, we get a contradiction.

Example 16: (i) If f is integrable, then

$$\int f \,dx = \lim_{a\to\infty} \lim_{b\to-\infty} \int_b^a f \,dx = \lim_{b\to-\infty} \lim_{a\to\infty} \int_b^a f \,dx. \tag{3.16}$$

(ii) If f is integrable on $[a,b]$ and $0<\epsilon<b-a$, then

$$\int_a^b f\,dx = \lim_{\epsilon\to 0}\int_{a+\epsilon}^b f\,dx.$$

Solution: $\int_b^a f\,dx = \int_{-\infty}^a \chi_{[b,\infty)} f\,dx$. But by Example 15,

$$\lim_{b\to-\infty}\int_{-\infty}^a \chi_{[b,\infty)} f\,dx = \int_{-\infty}^a f\,dx.$$

A second application of Example 15 gives the first equation of (3.16) and the second follows in the same way; (ii) is proved similarly.

The following theorem, which will be generalized in Theorem 9, p. 87, allows us to calculate integrals in many cases of importance.

Theorem 12: If f is continuous on the finite interval $[a,b]$, then f is integrable, and $F(x) = \int_a^x f(t)\,dt$ $(a<x<b)$ is a differentiable function such that $F'(x) = f(x)$.

Proof: As f is continuous, it is measurable and $|f|$ is bounded. So f is integrable on $[a,b]$. If $a<x<b$ we have $x+h\in(a,b)$ for all small h, and

$$F(x+h)-F(x) = \int_x^{x+h} f(t)\,dt.$$

But using Example 12 and the continuity of f we have

$$\int_x^{x+h} f(t)\,dt = hf(\xi),\ \xi = x+\theta h,\ 0\leqslant\theta\leqslant 1.$$

So, supposing $h\neq 0$, dividing by h and letting $h\to 0$, we get the result. □

Corollary 1: Integrals of elementary continuous functions over finite intervals can be calculated in the usual way using indefinite integrals.

Corollary 2: From Example 16 it follows that the integral of an integrable continuous function over an infinite interval can be obtained if its indefinite integral is known.

Corollary 3: Techniques involving integration by parts and by substitution can be employed if all the functions involved are continuous and integrable. Infinite intervals can be dealt with in this case as in Example 16.

Corollary 4: In the case of piecewise-continuous functions, if we split the domain appropriately, we can calculate the separate integrals as in Corollary 1.

Using Theorem 12 and its corollaries we can now give specific examples which show some ways in which Lebesgue's Dominated Convergence Theorem (Theorem 10) may be used.

Example 17: Show that if $\alpha > 1$,

$$\int_0^1 \frac{x \sin x}{1+(nx)^\alpha}\,dx = o(n^{-1}) \text{ as } n \to \infty.$$

Solution: We wish to show that

$$\lim_{n\to\infty} \int_0^1 \frac{nx \sin x}{1+(nx)^\alpha}\,dx = 0.$$

Clearly $\lim\limits_{n\to\infty} \dfrac{nx \sin x}{1+(nx)^\alpha} = 0$, so we wish to show that Theorem 10 applies to the sequence

$$f_n(x) = \frac{nx \sin x}{1+(nx)^\alpha}, n = 1, 2, \ldots$$

We consider $h(x) = 1 + (nx)^\alpha - nx^{3/2}$. So $h(0) = 1$, $h(1) = 1 + n^\alpha - n$. For $1 < \alpha \leqslant 3/2$, h has no stationary point in $[0,1]$, for all large n; for $\alpha > 3/2$ it has a stationary point at which its value is easily seen to approach 1 for large n. It follows that for large n, $h(x) > 0$ on $[0,1]$ and so

$$\left|\frac{nx \sin x}{1+(nx)^\alpha}\right| \leqslant \frac{1}{\sqrt{x}},$$

and the result follows.

Example 18: Show that

$$\lim \int_0^\infty \frac{dx}{(1+x/n)^n\, x^{1/n}} = 1.$$

Solution: For $n > 1, x > 0$,

$$(1+x/n)^n = 1 + x + \frac{n(n-1)}{n^2}\frac{x^2}{2} + \ldots > \frac{x^2}{4}.$$

So if we define $g(x) = 4/x^2$ $(x \geqslant 1)$, $g(x) = x^{-1/2}$ $(0 < x < 1)$ we have

$$(1+x/n)^{-n}\, x^{-1/n} < g(x), \quad (n > 1, x > 0).$$

But g is integrable over $(0,\infty)$, so

$$\lim \int_0^\infty (1+x/n)^{-n}\, x^{-1/n}\,dx = \int_0^\infty e^{-x}\,dx = 1.$$

Example 19: Show that

$$\lim_{n\to\infty} \int_a^\infty \frac{n^2 x e^{-n^2x^2}}{1+x^2}\,dx = 0,$$

for $a > 0$, but not for $a = 0$.

Solution: If $a > 0$, substitute $u = nx$ to get

$$\int_a^\infty f_n(x)\,dx = \int_{na}^\infty \frac{u\,e^{-u^2}}{1+u^2/n^2}\,du = \int_0^\infty \chi_{(na,\infty)}\,\frac{u\,e^{-u^2}}{1+u^2/n^2}\,du,$$

and the last integrand is less than $u\,e^{-u^2}$, an integrable function. But, as $a > 0$, $\lim_{n\to\infty} \chi_{(na,\infty)}(1+u^2/n^2)^{-1}\,u\,e^{-u^2} = 0$. So Theorem 10 gives the result.

If $a = 0$, the same substitution gives

$$\int_0^\infty f_n(x)\,dx = \int_0^\infty u\,e^{-u^2}(1+u^2/n^2)^{-1}\,du \to \int_0^\infty u\,e^{-u^2}\,du = 1/2,$$

using Theorem 10.

Example 20: Let f be a non-negative integrable function on $[0,1]$. Then there exists a measurable function $\varphi(x)$ such that φf is integrable on $[0,1]$ and $\varphi(0+) = \infty$.

Solution: It follows easily from Example 15 that $\lim_{a\to 0} \int_0^a f\,dx = 0$. So $\forall\, n$, $\exists\, x_n$ $(0 < x_n < 1)$, such that $\int_0^{x_n} f\,dx < n^{-3}$, and we may suppose that $x_n \downarrow 0$ as $n \to \infty$. Define $\varphi(x) = \sum_{k=2}^{\infty} (k-1)\,\chi_{(x_k, x_{k-1}]}$. So $\varphi(0+) = \infty$. But $\int_{x_k}^{x_{k-1}} \varphi f\,dx = \int_{x_k}^{x_{k-1}} (k-1) f\,dx < (k-1)^{-2}$. So $\int_0^1 \varphi f\,dx \leqslant \sum_{n=1}^{\infty} 1/n^2 < \infty$.

Exercises

13. Show by a counterexample that (3.2) need not hold if, instead of $f_n \geqslant 0$, we are given f_n integrable.
14. Let $\{f_n\}$ be a sequence of integrable functions such that $f_n \uparrow f$. Show that $\int f\,dx = \lim \int f_n\,dx$.
15. Let $\{f_n\}$ be a sequence of integrable functions and let g be an integrable function such that $f_n \geqslant g$ a.e., each n. Then $\liminf \int f_n\,dx \geqslant \int \liminf f_n\,dx$.
16. Show that $\lim \int_0^1 f_n(x)\,dx = 0$, where $f_n(x)$ is

(i) $\dfrac{\log(x+n)}{n} e^{-x} \cos x$, (ii) $\dfrac{nx \log x}{1+n^2x^2}$, (iii) $\dfrac{n\sqrt{x}}{1+n^2x^2}$,

(iv) $\dfrac{nx}{1+n^2x^2}$, (v) $\dfrac{n^{3/2}x}{1+n^2x^2}$,

(vi) $\dfrac{n^p x^r \log x}{1+n^2x^2}$, $r > 0$, $0 < p < \min(2, 1+r)$.

17. Find $\lim \int_0^\infty (1+\frac{x}{n})^{-n} \sin \frac{x}{n} \, dx$.

18. Find $\lim \int_0^1 \frac{1+nx}{(1+x)^n} \, dx$.

19. Show that if $\beta > 0$, then

$$\lim_{n\to\infty} n^{1/\beta} \int_0^1 x^{1/\beta} (1-x)^n \frac{dx}{x} = \beta \int_0^\infty e^{-u^\beta} \, du.$$

20. Show that if $f_n(x) = \left(\frac{n+x}{n+2x}\right)^n$, then $f_n(x) > f_{n+1}(x)$ for $x > 0$ and $n = 1$, 2, . . . Find whether the limit of the integral equals the integral of the limit in the following cases, and evaluate the limits involved.

(i) $\int_0^\infty f_n(x)\, e^{x/2} \, dx$, (ii) $\int_0^\infty f_n(x)\, e^{-x/2} \, dx$.

21. Find $\lim \int_0^n (1+x/n)^n e^{-2x} \, dx$.

22. Show that if $\alpha > 0$, then

$$\lim_{n\to\infty} \int_0^n (1-x/n)^n x^{\alpha-1} \, dx = \int_0^\infty e^{-x} x^{\alpha-1} \, dx.$$

23. Show that

$$\lim_{n\to\infty} \int_\alpha^\infty \sqrt{n}\, e^{-nx^2} \, dx = \int_\alpha^\infty \lim_{n\to\infty} \sqrt{n}\, e^{-nx^2} \, dx,$$

for $\alpha > 0$ but not for $\alpha = 0$.

24. Find the range of values of α for which

$$\int_0^1 \lim_{n\to\infty} n^\alpha (1-x) x^n \, dx = \lim_{n\to\infty} \int_0^1 n^\alpha (1-x) x^n \, dx.$$

Find also the range of values of α for which the conditions of Theorem 10 are satisfied.

3.3 INTEGRATION OF SERIES

In the following examples we wish to write $\int f(x)\, dx$ as a series. We expand $f(x) = \Sigma f_n(x)$ and get $\int f(x)\, dx = \Sigma \int f_n(x)\, dx$, provided we can justify the interchange of Σ and $\int$. If the functions $f_n(x)$ are of constant sign for each n and x we can appeal to Theorem 7, p. 59. If $\Sigma f_n(x)$ is an alternating series, Theorem 10, p. 63, may apply. If $\Sigma \int |f_n(x)|\, dx$ can be shown to converge, Theorem 11, p. 64, applies. In many cases more than one method is available.

Example 21: Show that $\int_0^1 \frac{x^{1/3}}{1-x} \log \frac{1}{x}\, dx = 9 \sum_{n=1}^{\infty} \frac{1}{(3n+1)^2}$.

Solution: $\frac{x^{1/3}}{1-x} \log \frac{1}{x} = x^{1/3} \log \frac{1}{x} \sum_{n=0}^{\infty} x^n$ $(0<x<1)$, and Theorem 7, p. 59, gives $\int_0^1 \frac{x^{1/3}}{1-x} \log \frac{1}{x}\, dx = \sum_{n=0}^{\infty} \int_0^1 x^{n+1/3} \log \frac{1}{x}\, dx = \sum_{n=0}^{\infty} \frac{9}{(3n+4)^2}$.

Example 22: Show that $\int_0^{\infty} \frac{\sin t}{e^t - x}\, dt = \sum_{n=1}^{\infty} \frac{x^{n-1}}{n^2+1}$, $-1 \leqslant x \leqslant 1$.

Solution: The integrand $= \lim_{N\to\infty} \sum_{n=0}^{N} x^n \sin t\, e^{-(n+1)t}$. But for $t>0$,

$$\left| \sum_{n=0}^{N} x^n \sin t\, e^{-(n+1)t} \right| \leqslant t\, e^{-t} \frac{1-x^{N+1}\, e^{-(N+1)t}}{1-x\, e^{-t}} \leqslant \frac{2t}{e^t - x},$$

an integrable function, so Theorem 10 applies to the sequence of partial sums giving

$$\int_0^{\infty} \frac{\sin t}{e^t - x}\, dt = \sum_{n=0}^{\infty} x^n \int_0^{\infty} e^{-(n+1)t} \sin t\, dt = \sum_{n=0}^{\infty} \frac{x^n}{1+(n+1)^2}.$$

Example 23: Show that $\int_0^1 \sin x \log x\, dx = \sum_{n=1}^{\infty} \frac{(-1)^n}{(2n)(2n)!}$.

Solution: $\sin x \log x = \sum_{n=0}^{\infty} \frac{(-1)^n x^{2n+1}}{(2n+1)!} \log x = \sum_{n=0}^{\infty} f_n(x)$, say. But

$$\int_0^1 |f_n(x)|\, dx = (-1)^{n+1} \int_0^1 f_n(x)\, dx = \frac{1}{(2n+2)(2n+2)!},$$

and an application of Theorem 11 gives the result.

Exercises

In the following exercises, an identity displayed without comment is to be proved.

25. Show that if $a>1$, $\int_0^{\infty} \frac{x^{a-1}}{e^x - 1}\, dx = \left(\int_0^{\infty} x^{a-1} e^{-x}\, dx \right) \sum_{n=1}^{\infty} \frac{1}{n^a}$.

26. $\int_0^1 \left(\frac{\log x}{1-x} \right)^2 dx = \pi^2/3$.

27. $\displaystyle\int_0^1 (e^x - 1)(\log x + \frac{1}{x})\,dx = \sum_{n=1}^{\infty} \frac{n^2+n+1}{(n-1)!(n^2+n)^2}$.

28. Prove that if $I(t) = \displaystyle\int_0^{\infty} e^{-t^2x} \frac{\sinh 2tx}{\sinh x}\,dx$, then for $t^2 \neq 1$,

$$I(t) = 4t \sum_{n=0}^{\infty} \frac{1}{(2n+1+t^2)^2 - 4t^2}.$$

Show that $\displaystyle\lim_{t\to 1} \left\{ I(t) - \frac{4t}{(t^2-1)^2} \right\} = 3/4$.

29. Show that if $p > -1$, $\displaystyle\int_0^1 \frac{x^p \log x}{1-x}\,dx = -\sum_{n=1}^{\infty} \frac{1}{(p+n)^2}$.

30. Given that $0 < b < a$, evaluate $\displaystyle\int_0^{\infty} \frac{\sinh bx}{\sinh ax}\,dx$.

31. $\displaystyle\int_0^{\infty} \operatorname{sech} x^2\,dx = \sqrt{\pi} \sum_{n=0}^{\infty} \frac{(-1)^n}{\sqrt{(2n+1)}}$.

32. $\displaystyle\int_0^{\infty} \frac{\cos x}{e^x+1}\,dx = \sum_{n=1}^{\infty} (-1)^{n-1} \frac{n}{n^2+1}$.

33. Show that if $p, q > 0$, then $\displaystyle\int_0^1 \frac{x^{p-1}}{1+x^q}\,dx = \sum_{n=0}^{\infty} (-1)^n \frac{1}{p+nq}$.

34. Show that $\displaystyle\lim_{b\to 1-} \int_0^b \sum_{n=1}^{\infty} \frac{x^{n-1}}{\sqrt{n}}\,dx = \sum_{n=1}^{\infty} \frac{1}{n^{3/2}}$.

35. Show that if $x \geqslant 1$, then $\displaystyle\int_0^{\infty} \frac{\cos^2 t}{1+xe^t}\,dt = \sum_{n=1}^{\infty} \frac{(-1)^{n-1}}{x^n} \frac{n^2+2}{n(n^2+4)}$.

36. Show that $\displaystyle\int_0^{\infty} e^{-x} \cos\sqrt{x}\,dx = \sum_{n=0}^{\infty} (-1)^n \frac{n!}{(2n)!}$.

37. If m is an integer, $m \geqslant 0$, let $J_m(x) = \displaystyle\sum_{n=0}^{\infty} \frac{(-1)^n}{n!(n+m)!} (x/2)^{m+2n}$. (This is the Bessel function of order m.)

(i) Show that if a is a constant, $2 \displaystyle\int_0^{\infty} J_m(2ax)\, x^{m+1} e^{-x^2}\,dx = a^m e^{-a^2}$.

(ii) Show that if $a > 1$, $\displaystyle\int_0^{\infty} J_0(x)\, e^{-ax}\,dx = (1+a^2)^{-1/2}$.

38. Show that $\displaystyle\int_0^1 \frac{(x\log x)^2}{1+x^2}\,dx = 2 \sum_{n=1}^{\infty} \frac{(-1)^{n-1}}{(2n+1)^3}$.

39. Show that if $a > 1$, then $\displaystyle\int_0^{\pi} \sum_{n=1}^{\infty} \frac{n^2 \sin nx}{a^n}\,dx = 2a \frac{(a^2+1)}{(a^2-1)^2}$.

40. If f is integrable over (a, b) and $|r| < 1$,

$$\int_a^b f(x)\left\{\frac{1-r\cos x}{1-2r\cos x+r^2}\right\}\mathrm{d}x=\sum_{n=0}^{\infty} r^n\int_a^b f(x)\cos nx\,\mathrm{d}x.$$

3.4 RIEMANN AND LEBESGUE INTEGRALS

We consider the Riemann integral of a bounded function f over a finite interval $[a,b]$. Let $a=\xi_0<\xi_1<\ldots<\xi_n=b$ be a partition, D, of $[a,b]$. Write

$$S_D=\sum_{i=1}^{n} M_i(\xi_i-\xi_{i-1}),$$

where $M_i=\sup f$ in $[\xi_{i-1},\xi_i]$, $i=1,\ldots,n$. Similarly on replacing M_i by m_i equal to $\inf f$ over the corresponding interval, we obtain $s_D=\sum_{i=1}^{n} m_i(\xi_i-\xi_{i-1})$. Then f is said to be **Riemann integrable** over $[a,b]$ if given $\epsilon>0$, there exists D such that $S_D-s_D<\epsilon$. In this case we have $\inf S_D=\sup s_D$, where the infimum and supremum are taken over all partitions D of $[a,b]$, and we write the common value as $\mathbf{R}\int_a^b f\,\mathrm{d}x$.

Theorem 13: If f is Riemann integrable and bounded over the finite interval $[a,b]$, then f is integrable and $\mathbf{R}\int_a^b f\,\mathrm{d}x=\int_a^b f\,\mathrm{d}x$.

Proof: Let $\{D_n\}$ be a sequence of partitions such that, for each n, $S_{D_n}-s_{D_n}<1/n$. It is easily seen that

$$S_{D_n}=\int_a^b u_n\,\mathrm{d}x \text{ and } s_{D_n}=\int_a^b l_n\,\mathrm{d}x,$$

where u_n and l_n are step functions, $u_n\geqslant f\geqslant l_n$. Indeed we may, for example, define $u_n=M_i$ on (ξ_{i-1},ξ_i), and at a partition point let u_n be the average of the values M_i corresponding to the intervals ending at that point.

Write $U=\inf_n u_n$ and $L=\sup_n l_n$. Now

$$[x\colon U(x)-L(x)>0]=\bigcup_{k=1}^{\infty}[x\colon U(x)-L(x)>1/k].$$

But if $U-L>1/k$, then $u_n-l_n>1/k$ for each n. So if $m[x\colon U(x)-L(x)>1/k]=a$, then $\int(u_n-l_n)\,\mathrm{d}x>a/k$, and so $a/k<1/n$ for each n. So $a=0$. Hence $U-L\leqslant 1/k$ a.e. for each k, so $U=L$ a.e.

But u_n, l_n and hence U, L are measurable. Also $L\leqslant f\leqslant U$, so f is measurable and, being bounded, is integrable. Clearly

$$\int_a^b l_n \, dx \leqslant \int_a^b f \, dx \leqslant \int_a^b u_n \, dx,$$

and letting $n \to \infty$, we get

$$\mathbf{R}\int_a^b f \, dx = \int_a^b f \, dx. \ \square$$

Note: the converse does not hold. Consider for example the function f on $[0,1]$:

$$f(x) = \begin{cases} 0, & x \text{ rational} \\ 1, & x \text{ irrational.} \end{cases}$$

Then f is measurable, indeed $f = 1$ a.e. So $\int_0^1 f \, dx = 1$. But each $S_D = 1$ and each $s_D = 0$, so f is not Riemann integrable.

That the function f of this example is not Riemann integrable can be seen also from the next theorem, since f is discontinuous at each x in $[0,1]$. The theorem shows that the class of Riemann-integrable functions is quite restricted.

Theorem 14: Let f be a bounded function defined on the finite interval $[a, b]$, then f is Riemann integrable over $[a, b]$ if, and only if, it is continuous a.e.

Proof: Suppose that f is Riemann integrable over $[a, b]$. Using the notation of the last theorem, suppose that $U(x) = f(x) = L(x)$, where x is not a partition point of any D_n, the D_n being chosen as before. Then f is continuous at x; for otherwise there would exist $\epsilon > 0$ and a sequence $\{x_k\}$, $\lim x_k = x$, such that for each k, $|f(x_k) - f(x)| > \epsilon$. But then $U(x) \geqslant L(x) + \epsilon$. Now, the set of all partition points of the D_n is countable and so has measure zero, and the set $[x\colon U(x) \neq L(x)]$ has measure zero by the proof of the last theorem. So f is continuous a.e.

Conversely, suppose that f is continuous a.e. Choose a sequence $\{D_n\}$ of partitions of $[a, b]$ such that, for each n, D_{n+1} contains the partition points of D_n and such that the length of the largest interval of D_n tends to zero as $n \to \infty$. Then if u_n, l_n are the corresponding step functions as in the last theorem, we have $u_{n+1} \leqslant u_n$ and $l_{n+1} \geqslant l_n$ for each n. Write $U = \lim u_n$ and $L = \lim l_n$. Now suppose that f is continuous at x. Then, given $\epsilon > 0$, there exists $\delta > 0$ such that $\sup f - \inf f < \epsilon$, where the supremum and infimum are taken over $(x - \delta, x + \delta)$. For all n sufficiently large, an interval of D_n containing x will lie in $(x - \delta, x + \delta)$, and so $u_n(x) - l_n(x) < \epsilon$. But ϵ is arbitrary so $U(x) = L(x)$. So $U = L$ a.e. But then, by Theorem 10, p. 63,

$$\lim \int u_n \, dx = \int U \, dx = \int L \, dx = \lim \int l_n \, dx,$$

and so f is Riemann integrable. $\square$

Definition 8: If, for each a and b, f is bounded and Riemann integrable on $[a, b]$ and

$$\lim_{\substack{a\to-\infty\\ b\to\infty}} \int_a^b f\,dx$$

exists, then f is said to be *Riemann integrable* on $(-\infty, \infty)$, and the integral is written $\mathbf{R}\int_{-\infty}^{\infty} f\,dx$.

Theorem 15: Let f be bounded and let f and $|f|$ be Riemann integrable on $(-\infty, \infty)$. Then f is integrable and

$$\int_{-\infty}^{\infty} f\,dx = \mathbf{R}\int_{-\infty}^{\infty} f\,dx.$$

Proof: From Theorem 13,

$$\int_a^b |f|\,dx = \mathbf{R}\int_a^b |f|\,dx \leqslant \mathbf{R}\int_{-\infty}^{\infty} |f|\,dx.$$

for all a and b. So f is integrable. Theorem 13, applied again, gives $\int_a^b f\,dx = \mathbf{R}\int_a^b f\,dx$ and Example 16, p. 64, gives the result. □

The next result may be used to reduce problems involving integrals of measurable functions to more amenable classes of functions.

Theorem 16: Let f be bounded and measurable on a finite interval $[a, b]$ and let $\epsilon > 0$. Then there exist

(i) a step function h such that $\int_a^b |f-h|\,dx < \epsilon$,

(ii) a continuous function g such that g vanishes outside a finite interval and

$$\int_a^b |f-g|\,dx < \epsilon. \tag{3.17}$$

Proof: (i) As $f = f^+ - f^-$, we may assume throughout that $f \geqslant 0$. Now $\int_a^b f\,dx = \sup \int_a^b \varphi\,dx$, where $\varphi \leqslant f$, φ simple and measurable. So we may assume that f is a simple measurable function, with $f = 0$ outside $[a, b]$. So

$$f = \sum_{i=1}^{n} a_i \chi_{E_i} \tag{3.18}$$

with $\bigcup_{i=1}^{n} E_i = [a, b]$. Let $\epsilon' = \epsilon/nM$ where $M = \sup f$ on $[a, b]$, and M may obviously be supposed positive. For each of the measurable sets E_i there exist open intervals $I_1, \ldots, I_k$ such that, if $G = \bigcup_{r=1}^{k} I_r$, then $m(E_i \,\Delta\, G) < \epsilon'$. But χ_G is a step function such that $\int |\chi_{E_i} - \chi_G| \, dx = m(E_i \,\Delta\, G) < \epsilon'$. Construct such step functions h_i, say, for each E_i appearing in (3.18). Then

$$\int_a^b |f - \sum_{i=1}^{n} a_i h_i| \, dx < \sum_{i=1}^{n} a_i \epsilon' \leqslant nM\epsilon' = \epsilon.$$

But $\sum_{i=1}^{n} a_i h_i$ is a step function.

(ii) From (i) there exists a step function h vanishing outside a finite interval (note that this interval need not be identical with $[a, b]$), such that

$$\int_a^b |f - h| \, dx < \epsilon/2.$$

The proof is completed by constructing a continuous function g such that $\int |h - g| \, dx < \epsilon/2$ and such that $g(x) = 0$ whenever $h(x) = 0$. Let $h = \sum_{i=1}^{n} a_i \chi_{E_i}$ where E_i is the finite interval (c_i, d_i), $i = 1, \ldots, n$. As in (i), it is sufficient to show that each χ_{E_i} may be approximated in the sense of (3.17). We may suppose that $\epsilon < 2(d_i - c_i)$ and define g by: $g = 1$ on $(c_i + \epsilon/4, d_i - \epsilon/4)$, $g = 0$ on $\mathbf{C}(c_i, d_i)$. Extend g by linearity to $(c_i, c_i + \epsilon/4)$ and $(d_i - \epsilon/4, d_i)$, as in Fig. 3.1, to get a continuous function. Clearly $\int |\chi_{E_i} - g| \, dx < \epsilon/2$, and (ii) follows.

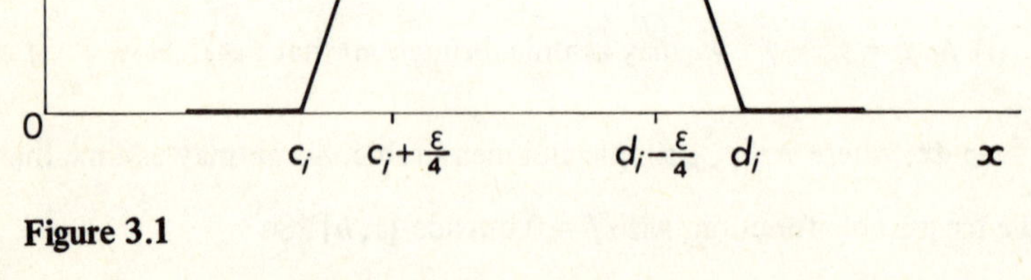

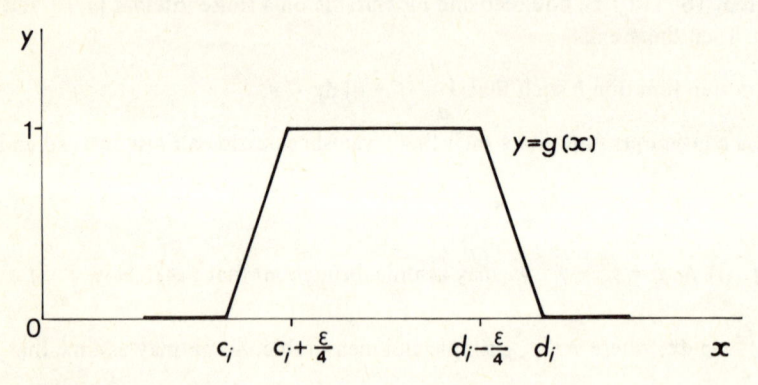

Figure 3.1

Corollary: The results of Theorem 16 hold if f is integrable over $[a, b]$, using Exercise 4, p. 60, since, as in the proof, we may assume $f \geqslant 0$.

Example 24: Let f be a bounded measurable function defined on the finite interval (a, b). Show that $\lim_{\beta \to \infty} \int_a^b f(x) \sin \beta x \, dx = 0$.

Solution: By Theorem 16, $\forall\, \epsilon > 0$, $\exists\, h = \sum_{i=1}^{n} \xi_i\, \chi_{(a_i, b_i)}$, say, with $\int_a^b |f-h|\, dx < \epsilon$. Then

$$\left|\int_a^b f \sin \beta x\, dx\right| \leqslant \int_a^b |(f-h) \sin \beta x|\, dx + \left|\int_a^b h \sin \beta x\, dx\right|$$

$$< \epsilon + \left|\int_a^b h \sin \beta x\, dx\right| .$$

Now $\left|\int_a^b \chi_{(a_i,b_i)} \sin \beta x\, dx\right| = \left|1/\beta \int_{\beta a_i}^{\beta b_i} \sin y\, dy\right| \leqslant 2/\beta < \epsilon/nM$ for $\beta > \beta_0$, say, where $M = \max\,[\xi_i, i = 1, \ldots, n]$. So $\left|\int_a^b f \sin \beta x\, dx\right| < 2\epsilon$, for $\beta > \beta_0$.

Example 25: Show that if $f \in L(a+h, b+h)$ and $f_h(x) \equiv f(x+h)$, then $f_h \in L(a, b)$ and $\int_{a+h}^{b+h} f\, dx = \int_a^b f_h\, dx$.

Solution: Clearly $(f_h)^+ = (f^+)_h$, $(f_h)^- = (f^-)_h$, so it is sufficient to prove the result for $f \geqslant 0$. By the corollary to Theorem 5, p. 58, there exists a sequence of measurable simple functions $\{\varphi_n\}$ such that $\varphi_n \leqslant f$ and $\int \varphi_n\, dx \uparrow \int f\, dx$. But then $(\varphi_n)_h \uparrow f_h$, and so by monotone convergence

$$\int_{a+h}^{b+h} f\, dx = \lim \int_{a+h}^{b+h} \varphi_n\, dx = \lim \int_a^b (\varphi_n)_h\, dx = \int_a^b f_h\, dx.$$

Exercises

41. Let S be a measurable set, $m(S) < \infty$. Show that

$$\lim \int_S \frac{dx}{2 - \sin nx} = m(S)/\sqrt{3}.$$

42. Let f be an integrable function. Show that $\forall\, \epsilon > 0$, $\exists\, g$ continuous, such that $g = 0$ outside a finite interval and such that $\int |f-g|\, dx < \epsilon$.
43. Let f be an integrable function. Show that $\lim_{h \to 0} \int |f(x+h) - f(x)|\, dx = 0$.
44. (Riemann–Lebesgue Lemma). Let f be integrable, φ bounded and measurable and suppose that there exists β such that $\varphi(x+\beta) = -\varphi(x)$, $\forall\, x \in \mathrm{R}$. Then $\lim_{k \to \infty} \int f(x)\, \varphi(kx)\, dx = 0$.
45. Show that the function $x^{-1} \sin x$ is Riemann integrable on $(-\infty, \infty)$ but that its Lebesgue integral does not exist.

Miscellaneous Exercises

46. Let f be a finite-valued non-negative function such that $m[x: f(x) > n] > 0$,

for each integer n. Show that there exists an integrable function g such that fg is not integrable.

47. Let $f_n(x)$ denote the distance from x to the nearest number of the form $k \cdot 10^{-n}$ where k is an integer, and let $f(x) = \sum_{n=1}^{\infty} f_n(x)$. Show that $\int_0^1 f\,dx = 1/36$.

48. Let $f(x) = 0$, x rational, $f(x) = 1/a$ if x is irrational and a is the first non-zero integer in the decimal representation of x. Show that f is measurable and find $\int_0^1 f\,dx$.

49. (i) Let $f_n(x) = \dfrac{nx - 1}{(x \log n + 1)(1 + nx^2 \log n)}$. Show that $\lim_{n\to\infty} f_n(x) = 0$ $(0 < x \leqslant 1)$, but that $\lim \int_0^1 f_n(x)\,dx = 1/2$.

(ii) Let $h_n(x) = nx^n$. Show that $\lim_{n\to\infty} h_n(x) = 0$ $(0 \leqslant x < 1)$, but that $\lim \int_0^1 h_n(x)\,dx = 1$.

50. Let $F_n(x) = \dfrac{n}{\pi} \displaystyle\int \frac{f(t)\,dt}{1 + n^2(t-x)^2}$. Show that $\lim_{n\to\infty} F_n(x) = f(x)$, if f is continuous at x. (It may be assumed that the integrals displayed exist.)

51. For each t, let $f(x, t)$ be an integrable function of x. Let $\partial f/\partial t$ exist for each x and satisfy $|\partial f/\partial t| \leqslant \varphi(x)$, an integrable function. Show that

$$\frac{d}{dt} \int f(x, t)\,dx = \int \frac{\partial f}{\partial t}\,dx.$$

52. Let $x^{\gamma} f(x)$ be integrable over $(0, \infty)$ for $\gamma = \alpha$, $\gamma = \beta$, where $\alpha < \beta$. Show that for each $\gamma \in (\alpha, \beta)$, $\int_0^{\infty} x^{\gamma} f(x)\,dx$ exists and is a continuous function of γ.

53. Let $\{f_n\}$ be a sequence of measurable functions such that, for each n, $|f_n| \leqslant g$, an integrable function. Show that

$$\int \liminf f_n\,dx \leqslant \liminf \int f_n\,dx \leqslant \limsup \int f_n\,dx \leqslant \int \limsup f_n\,dx.$$

Give an example where all the inequalities are strict.

CHAPTER 4

Differentiation

Differentiation and integration are closely connected. It is important to examine questions such as whether a Lebesgue integral may be differentiated with respect to the upper limit to obtain, in some sense, the integrand. In order to do this we first examine differentiation carefully. As one result we find in Section 4.4 that monotone functions are differentiable a.e., as are the functions of bounded variation which we consider in Section 4.3. The importance of these functions in connection with measure and integration will be more evident later, in Chapter 9.

4.1 THE FOUR DERIVATES

The condition of differentiability at each point is too restrictive for many purposes, and in this chapter we obtain properties of functions under slightly weaker conditions. For this purpose it is useful to have quantities related to derivatives which are defined even at points where the function is not differentiable.

Definition 1: If f is an extended real-valued function, finite at x and defined in an open interval containing x, then the following four quantities, not necessarily finite, are called respectively the **upper right derivate**, the **lower right derivate**, the **upper left derivate** and the **lower left derivate**:

$$D^+f(x) = \limsup_{h\to 0+} \frac{f(x+h)-f(x)}{x}, \qquad D_+f(x) = \liminf_{h\to 0+} \frac{f(x+h)-f(x)}{h},$$

$$D^-f(x) = \limsup_{h\to 0-} \frac{f(x+h)-f(x)}{h}, \qquad D_-f(x) = \liminf_{h\to 0-} \frac{f(x+h)-f(x)}{h}.$$

Clearly $D^+f(x) \geqslant D_+f(x)$ and $D^-f(x) \geqslant D_-f(x)$. The function f is differentiable at x if, and only if, the four derivates have a finite common value which we then write as usual $f'(x)$.

Example 1: Let $f(x) = |x|$; then at $x = 0, D^+ = D_+ = 1, D^- = D_- = -1$.

Example 2: $D^+(-f) = -D_+(f)$, $D^-(-f) = -D_-(f)$ follow from the corresponding properties of lim sup and lim inf, p. 18.

Example 3: Evaluate at $x = 0$ the four derivates of the continuous function

$$f(x) = \begin{cases} ax \sin^2 \dfrac{1}{x} + bx \cos^2 \dfrac{1}{x}, & x > 0 \\ 0, & x = 0 \\ a'x \sin^2 \dfrac{1}{x} + b'x \cos^2 \dfrac{1}{x}, & x < 0 \end{cases}$$

where $a < b, a' < b'$.

Solution: We have $D^+ f = \limsup_{h \to 0+} (a \sin^2 \frac{1}{h} + b \cos^2 \frac{1}{h})$. But $\sin 1/x$ and $\cos 1/x$ take all their values in any interval $1/(2n+2)\pi < x \leqslant 1/2n\pi$, and so in intervals arbitrarily close to 0. So $D^+ f = \sup_\theta (a \sin^2\theta + b \cos^2\theta) = b$. Similarly $D_+ f = \inf_\theta (a \sin^2\theta + b \cos^2\theta) = a, D_- f = a', D^- f = b'$.

Example 4: Let $L(x)$ be Lebesgue's function as in Chapter 1, p. 25. Show that for $x \in P$, $D^+ L = \infty$ or $D^- L = \infty$ if x is respectively, a right-hand or a left-hand end-point of a deleted interval $I_{n,r}$. Show that if $x \in P$ is not an end-point of an $I_{n,r}$, then $D^+ L = D^- L = \infty$.

Solution: $L_n(x) = 1 - L_n(1-x)$ for each n, so $L(x) = 1 - L(1-x)$. So

$$\frac{L(x+h) - L(x)}{h} = \frac{L(1-x-h) - L(1-x)}{-h}$$

and taking suprema as $h \to 0+$ we get

$$D^+ L(x) = D^- L(1-x)\,. \tag{4.1}$$

Let x be the right-hand end-point of $I_{n,r}$ so $x = \sum_{k=1}^{n} 2.3^{-k}\, \epsilon_k$. Take $m > n$ and let

$$x_m = x + \frac{1}{3^m} = x + \sum_{k=m+1}^{\infty} \frac{2}{3^k}.$$

So
$$\frac{L(x_m) - L(x)}{x_m - x} = \frac{\sum_{m+1}^{\infty} 1/2^k}{1/3^m} = \left(\frac{3}{2}\right)^m \to \infty \text{ as } m \to \infty.$$

So $D^+ L(x) = \infty$. But $1 - x$ is a left-hand end-point if x is a right-hand end-point so (4.1) gives the first result.

If $x \in P$, but is not an end-point, then $x = \sum_{k=1}^{\infty} 2.3^{-k}\, \epsilon_k$ where an infinite sequence of ϵ_k's equal 1. Let $x_n = \sum_{k=1}^{n} 2.3^{-k}\epsilon_k$, so $\{x_n\}$ is a sequence of right-hand end-points, and $x_n \uparrow x$. Then

$$\frac{L(x)-L(x_n)}{x-x_n} = \frac{\sum_{n+1}^{\infty} \epsilon_k/2^k}{2\sum_{n+1}^{\infty} \epsilon_k/3^k}.$$

Suppose ϵ_N is the first non-zero ϵ_k with $k \geqslant n+1$. Then the right-hand side

$$\geqslant \frac{1/2^N}{2\sum_{N}^{\infty} 1/3^k} = \frac{3^{N-1}}{2^N} \to \infty \text{ as } n \to \infty.$$

So $D^-L(x) = \infty$. But $x \in P$ if, and only if, $1-x \in P$, so using (4.1) again, we get $D^+L(x) = \infty$.

Exercises

1. Let f be defined by: $f(x) = x \sin(1/x)$ for $x \neq 0, f(0) = 0$. Find the four derivates at $x = 0$.
2. Let f be defined on $[0,1]$ by $f(x) = 0$ if $x \in \mathbb{Q}, f(x) = 1, x \notin \mathbb{Q}$. Find the four derivates at any x.
3. Show that the derivates of a continuous function are measurable.
4. Show that if $f'(x)$ exists then $D^+(f+g)(x) = f'(x) + D^+g(x)$, and similarly for the other derivates.
5. Give an example where $D^+(f+g) \neq D^+f + D^+g$.
6. In the notation of Example 4 show that for each right-hand end-point of an $I_{n,r}$, $D_+L = \infty$, and at each left-hand end-point $D_-L = \infty$.

4.2 CONTINUOUS NON-DIFFERENTIABLE FUNCTIONS

We now give two examples. The first is of a continuous function nowhere differentiable and the second of a continuous function non-differentiable on a given set of measure zero. We will not use the examples explicitly below, but the fact that continuous functions can be nowhere differentiable gives an extra significance to the results of Section 4.4 where we find conditions under which a function is differentiable, at least almost everywhere.

Example 5: Let $f_n(x)$ denote the distance from the real number x to the nearest number of the form $m/10^n$ where m, n are non-negative integers and $x \in (0,1)$. Show that $f = \sum_{n=1}^{\infty} f_n$ is continuous and is differentiable nowhere on $(0,1)$.

Solution: f_n has a 'saw-tooth' graph with zeros at $k/10^n, k = 1, 2, \ldots$ It is continuous and $\max f_n = 2^{-1} \cdot 10^{-n}$. So Σf_n is uniformly convergent and so f is continuous.

Let $x = 0 \cdot x_1 x_2 \ldots$ be the decimal expansion of x where we use the terminating expansion in cases of ambiguity, so $x = \sum_{m=1}^{\infty} x_m/10^m$. Let k be some fixed integer. If $x_k = 4$ or 9 write $x' = x - 1/10^k$. If $x_k \neq 4$ or 9 write $x' = x + 1/10^k$. Then for $n \geqslant k, f_n(x) = f_n(x')$. If $n < k$, then

$$f_n(x) - f_n(x') = \pm (x - x'),$$

since the choice of x' ensures that $(x, f_{k-1}(x))$ and $(x', f_{k-1}(x'))$ lie on the same monotone segment of the graph of f_{k-1} and therefore of each f_n for $n < k$. So

$$\begin{aligned} f(x) - f(x') &= \sum_{n=1}^{\infty} (f_n(x) - f_n(x')) \\ &= \Sigma \pm (x - x') \ (k-1 \text{ terms}) \\ &= p(x - x'), \text{ say.} \end{aligned}$$

For any combination of ± 1 it is easily seen that p is even if $k - 1$ is even, odd if $k - 1$ is odd. But $\dfrac{f(x) - f(x')}{x - x'} = p$ and on letting $k \to \infty$ we have $x \to x'$, but p does not tend to a limit.

Example 6: Let E be a set of measure zero. Show that there exists a function defined on R, which is continuous and increasing everywhere and for which each derivate is infinite at each point of E.

Solution: Let $\{U_n\}$ be a sequence of open sets such that $m(U_n) < 1/2^n$ and $E \subseteq U_n$ for each n (see Theorem 10, p. 36). Write $f_n(x) = m((-\infty, x) \cap U_n)$ and $f(x) = \sum_{n=1}^{\infty} f_n(x)$. Then f has the desired properties. Indeed each f_n is continuous as $|f_n(x) - f_n(y)| \leqslant |x - y|$. Also $\max f_n(x) \leqslant 1/2^n$, so Σf_n is uniformly convergent and f is continuous. Each f_n is an increasing function, so f is increasing. Let $x \in E$ and let $\delta x > 0$. Suppose $(x, x + \delta x) \subseteq U_n$ for N integers $n_1, \ldots, n_N$. Then $f(x + \delta x) - f(x) \geqslant \sum_{i=1}^{N} (f_{n_i}(x + \delta x) - f_{n_i}(x)) \geqslant N\delta x$, where N depends on δx, and since $x \in \bigcap_{n=1}^{\infty} U_n$, $N \to \infty$ as $\delta x \to 0+$. So $D^+f(x) = D_+f(x) = \infty$. Similarly $D^-f(x) = D_-f(x) = \infty$.

Exercise

7. Let E and f be as in Example 6 and let g be a continuous function on [0,1]

such that $g'(x) > 0$ for $x \notin E$ and $D_+g(x) > -\infty$, $D_-g(x) > -\infty$ for all x. Show that $g(1) + f(1) \geqslant g(0) + f(0)$.

4.3 FUNCTIONS OF BOUNDED VARIATION

We wish now to examine those functions which do not behave too erratically over an interval. Our definition will provide functions differentiable a.e., as will be shown in the next section. We suppose that the function f is defined and finite-valued on the finite interval $[a, b]$. Let $a = x_0 < x_1 < \ldots < x_k = b$ be a partition of $[a, b]$. Write

$$p = \sum_{i=1}^{k} (f(x_i) - f(x_{i-1}))^+, \quad n = \sum_{i=1}^{k} (f(x_i) - f(x_{i-1}))^-,$$

$$t = p + n = \sum_{i=1}^{k} |f(x_i) - f(x_{i-1})|,$$

where as usual we use the notation $A^+ = \max(A, 0)$, $A^- = \max(-A, 0)$. So $t, p, n \geqslant 0$ and $f(b) - f(a) = p - n$.

Definition 2: $P = \sup p$, $N = \sup n$, $T = \sup t$ where the suprema are taken over all partitions of $[a, b]$ are respectively the **positive, negative,** and **total variations** of f on $[a, b]$. If we wish to emphasize the dependence on f we will denote the variations by T_f etc., and if we wish to indicate the interval, by $T_f[a, b]$, etc. Note that T, P, N are non-negative.

Definition 3: If $T_f[a, b] < \infty$, f is said to be of **bounded variation** on $[a, b]$; we denote the class of functions with this property by $BV[a, b]$. A function is said to belong to $BV(-\infty, \infty)$ if it belongs to $BV[a, b]$ for all finite a and b, and we then put $T_f(-\infty, \infty) = \sup_{a,b} T_f[a, b]$.

Theorem 1: Let $f \in BV[a, b]$; then $f(b) - f(a) = P - N$ and $T = P + N$, all variations being on the finite interval $[a, b]$.

Proof: For any partition, $f(b) - f(a) = p - n$. So $p = n + f(b) - f(a) \leqslant N + f(b) - f(a)$. Taking the supremum over all partitions gives $P \leqslant N + f(b) - f(a)$. Similarly $n = p + f(a) - f(b)$ gives $N \leqslant P + f(a) - f(b)$. But then $P - N \leqslant f(b) - f(a) \leqslant P - N$, giving the first result. Also $T \geqslant p + n = 2p - f(b) + f(a) = 2p + N - P$. Taking the supremum gives $T \geqslant P + N$. But $t = n + p \leqslant N + P$ which similarly gives $T \leqslant N + P$ and the second result. □

Example 7: If $a < c < b$, then $T_f[a,b] = T_f[a,c] + T_f[c,b]$, with corresponding results for P and N.

Solution: We prove the result for T; the results for P and N follow similarly. Consider any partition of $[a, b]$ and, in an obvious notation, let $t[a, b]$ be the corresponding sum. Add the point c to the partition. Then t increases to t', say, and

$$t[a, b] \leqslant t'[a, c] + t'[c, b] \leqslant T[a, c] + T[c, b].$$

So we have $T[a, b] \leqslant T[a, c] + T[c, b]$. Now take any partition of $[a, c]$ and $[c, b]$ giving sums $t[a, c]$ and $t[c, b]$. These partitions give a partition of $[a, b]$ and we see that $t[a, c] + t[c, b] \leqslant T[a, b]$. Taking suprema over all such pairs of partitions gives $T[a, c] + T[c, b] \leqslant T[a, b]$, and the result follows.

As a corollary we have that the variations are increasing functions of the right-hand end-points of the intervals. The next theorem characterizes the functions of bounded variation.

Theorem 2: A function $f \in BV[a, b]$ if, and only if, f is the difference of two finite-valued monotone increasing functions on $[a, b]$, where a and b are finite.

Proof: Suppose that f is of bounded variation. Write $g(x) = P_f[a, x] + f(a)$ and $h(x) = N_f[a, x]$. Then g and h are monotone increasing functions by Example 7, and $0 \leqslant P_f[a, x] \leqslant T_f[a, x] \leqslant T_f[a, b]$. So g, and similarly h, is finite; but by Theorem 1, $f = g - h$ on $[a, b]$.

Conversely, let $f = g - h$ where g and h are finite-valued monotone increasing functions, then for any partition $a = x_0 < x_1 < \ldots < x_n = b$ we have

$$\begin{aligned}\Sigma |f(x_i) - f(x_{i-1})| &\leqslant \Sigma\, (g(x_i) - g(x_{i-1})) + \Sigma\, (h(x_i) - h(x_{i-1}))\\ &\leqslant g(b) - g(a) + h(b) - h(a).\end{aligned}$$

So $T_f[a\ b] < \infty$ as required. □

Theorem 3: Let f be a finite-valued monotone increasing function on $[a, b]$; then f is continuous except on a set of points which is at most countable.

Proof: For each $x \in [a, b]$ write $\delta f(x) = \inf\limits_{h>0} f(x + h) - \sup\limits_{h>0} f(x - h)$, where we may suppose that f is constant on $[a - 1, a]$ and on $[b, b + 1]$, so that $\delta f(x)$ is the 'jump' of the function at x. Clearly $\delta f \geqslant 0$ and f is continuous at x if, and only if, $\delta f(x) = 0$. Also the set $E_n = [x\colon \delta f(x) > 1/n]$ can contain at most $n(f(b) - f(a))$ points. But the set of points at which f is discontinuous is just $\bigcup\limits_{n=1}^{\infty} E_n$, a set which is at most countable. □

Corollary: If $f \in BV[a, b]$, then f is continuous except on a set which is at most countable. In particular, f is measurable.

Notation: We will write $l(\pi)$ for the length of the polygon π.

Theorem 4: $f \in BV[a, b]$, where a and b are finite, if, and only if, the graph of f is a rectifiable curve.

Proof: Let $a = x_0 < x_1 < \ldots < x_k = b$ be a partition of $[a, b]$ and let π be the polygon with vertices $(x_i, f(x_i))$, $i = 0, 1, \ldots, k$. Then

$$|f(x_i) - f(x_{i-1})| \leqslant \sqrt{((x_i - x_{i-1})^2 + (f(x_i) - f(x_{i-1}))^2)}$$
$$\leqslant x_i - x_{i-1} + |f(x_i) - f(x_{i-1})|.$$

So adding for $i = 1, 2, \ldots, k$ gives $t \leqslant l(\pi) \leqslant b - a + t$. Taking suprema over all partitions we get that $T_f[a, b]$ is finite if, and only if, $\sup_\pi l(\pi)$ is finite, which is the result. □

Example 8: Let $f \in BV[a, b]$ and let $f = f_1 - f_2$ where f_1 and f_2 are monotone increasing functions; show that $T_{f_1} \geqslant P_f$ and $T_{f_2} \geqslant N_f$, so that the decomposition of Theorem 2, p. 82, into monotone functions g and h, was such that $T_g + T_h$ had the minimum possible value.

Solution: Consider any partition $a = x_0 < x_1 < \ldots < x_n = b$. Then

$$(f_1(x_i) - f_1(x_{i-1})) - (f_2(x_i) - f_2(x_{i-1})) =$$
$$= f(x_i) - f(x_{i-1}) = (f(x_i) - f(x_{i-1}))^+ - (f(x_i) - f(x_{i-1}))^-. \qquad (4.2)$$

But $A = A^+ - A^- = B - C$ where $B, C \geqslant 0$ implies $A^+ \leqslant B$, $A^- \leqslant C$. Applying this to (4.2) and adding we get

$$\sum_{i=1}^{k} (f_1(x_i) - f_1(x_{i-1})) \geqslant \sum_{i=1}^{k} (f(x_i) - f(x_{i-1}))^+.$$

Taking suprema over all partitions gives the result for f_1. The result for f_2 follows similarly.

Since g is monotone increasing, $T_g[a, b] = g(b) - g(a) = P_g[a, b]$. So we always have $T_{f_1} \geqslant T_g$ and similarly $T_{f_2} \geqslant T_h$, giving the last result.

Example 9: $BV[a, b]$ is a vector spacc over the real numbers.

Solution: Let $f, g \in BV[a, b]$. For any partition, $t_{f+g} \leqslant t_f + t_g \leqslant T_f + T_g$, so $f + g \in BV[a, b]$. If $c \in \mathsf{R}$, $t_{cf} = |c| t_f \leqslant |c| T_f$, so $cf \in BV[a, b]$.

Exercises

8. Define f on $[0,1]$ by $f(0) = 0$, $f(x) = \sin(\pi/x)$ for $x > 0$. Show that $f \notin BV[0,1]$.
9. Define g on $[0,1]$ by $g(0) = 0$, $g(x) = x \sin(\pi/x)$ for $x > 0$. Show that g is continuous but that $g \notin BV[0,1]$.
10. Show that if f' exists and is bounded on $[a, b]$, then $f \in BV[a, b]$.
11. Show that if $f \in BV[a, b]$, then f is bounded on $[a, b]$.
12. Show that if $f, g \in BV[a, b]$, then $fg \in BV[a, b]$.
13. Show that if $f \in BV[a, b]$ and $x \in (a, b)$, then the limits $f(x-)$ and $f(x+)$ exist.
14. Define f on $[0,1]$ by $f(0) = 0$, $f(x) = x^p \sin 1/x$ for $x > 0$, where $p \geqslant 2$. Show that $f \in BV[0,1]$.

4.4 LEBESGUE'S DIFFERENTIATION THEOREM

We show in Theorem 8 that a function of bounded variation is differentiable a.e. First we obtain some preliminary results.

Theorem 5: Let $\mathcal{G}$ be a finite collection of intervals $[I_k]$. Then there exists a sub-collection $\mathcal{G}_0$ of disjoint intervals of $\mathcal{G}$, $\mathcal{G}_0 = [I_{k_i}]$ say, such that $m(\cup I_{k_i}) \geqslant 1/3m(\cup I_k)$.

Proof: Let $I_{k_1} \in \mathcal{G}$ be an interval of maximal length. Remove from $\mathcal{G}$ any intervals meeting I_{k_1}. The measure of the union of these intervals (including I_{k_1}) is not greater than $3l(I_{k_1})$, as $l(I_{k_1})$ is maximal. This leaves a smaller class $\mathcal{G}_1$ from which I_{k_2} is similarly chosen and the measure of the union of the intervals meeting I_{k_2} is not greater than $3l(I_{k_2})$, etc. Continue until $\mathcal{G}$ is exhausted to get intervals $I_{k_1}, I_{k_2}, \ldots, I_{k_n}$, which are disjoint from the construction. Every interval of $\mathcal{G}$ meets some I_{k_i}, so

$$m(\cup I_k) \leqslant \sum_{i=1}^{n} 3l(I_{k_i}) = 3m\left(\bigcup_{i=1}^{n} I_{k_i}\right). \;\square$$

Theorem 6: If $[I_\alpha]$ is a collection of open intervals such that $m(\cup I_\alpha) < \infty$, then there exists a finite sub-collection $I_1, \ldots, I_n$ of these intervals such that

$$m\left(\bigcup_{k=1}^{n} I_k\right) \geqslant \tfrac{1}{2} m(\cup I_\alpha).$$

Proof: By Lindelöf's Theorem, p. 23, we may choose a countable subcollection $[I_k]$ of the $[I_\alpha]$ with the same union. Then

$$\lim m\left(\bigcup_{k=1}^{n} I_k\right) = m(\cup I_\alpha) < \infty,$$

so n exists with the desired property. $\square$

Notation. If $c < d$ and f is any function, write $f(c, d)$ for $(f(d) - f(c))/(d - c)$.

Theorem 7: (i) Let $\pi(x)$ be linear on $[a, b]$, $\pi(a) \leqslant \pi(b)$. Let q be a polygon, with the same end-points as π, of which n sides, the total length of whose projections on the x-axis is d, have a slope less than $-\xi$ ($\xi > 0$). Then $l(q) > l(\pi) + d(\sqrt{(1 + \xi^2)} - 1)$.

(ii) If π and q are as in (i) but with $\pi(a) \geqslant \pi(b)$ and with n sides of q having a slope greater than ξ, then the same conclusion holds.

Proof: (i) Starting with q, replace adjacent sides, where necessary, by moving them parallel to themselves until, after a finite number of steps, there is obtained a new polygon q_1 with sides congruent to those of q and whose first n sides have slope $< -\xi$. As each replacement leaves the length unchanged, $l(q) = l(q_1)$. Clearly $q_1(a, a + d) < -\xi$.

In Fig. 4.1 B is the point $(a + d, \pi(a))$; C is the point $(a + d, q_1(a + d))$. Now $AC = AB \sec \angle BAC > AB\sqrt{(1 + \xi^2)}$. So $l(\pi) = AD < AB + BD < AB + CD < AB + CD + AC - AB\sqrt{(1 + \xi^2)}$. But $AB = d$ and $CD + AC \leqslant l(q_1) = l(q)$. So $l(\pi) < l(q) - d(\sqrt{(1 + \xi^2)} - 1)$.

To obtain (ii) replace π by $-\pi$, q by $-q$ and apply (i). □

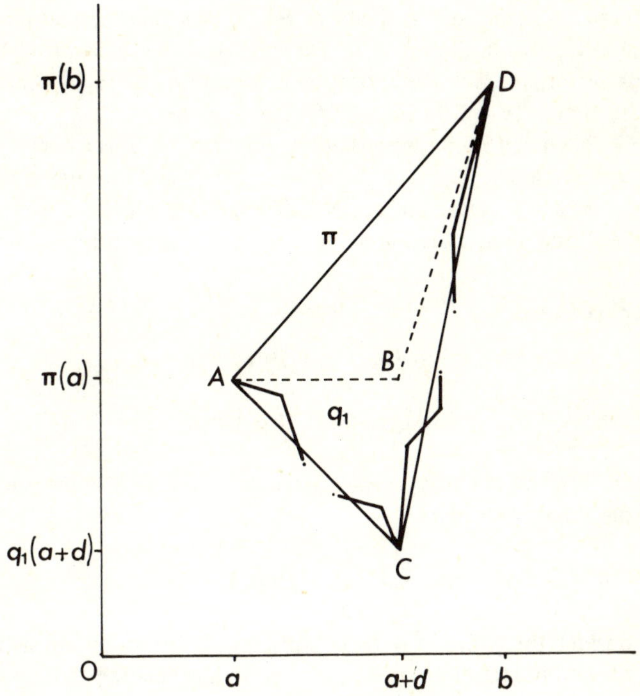

Figure 4.1

Theorem 8 (Lebesgue's Differentiation Theorem): If $f \in BV[a, b]$ where a and b are finite, then we have: (i) f is differentiable a.e., (ii) the derivative is finite a.e.

Proof (cf. [1]): (i) It is sufficient to show $D^+f \leqslant D_-f$ a.e., since $-f \in BV[a, b]$ so by Example 2, p. 78, $D_+f \geqslant D^-f$ a.e. This gives, a.e., $D^+f \geqslant D_+f \geqslant D^-f \geqslant D_-f \geqslant D^+f$ and equality a.e. follows. So we suppose that $D^+f > D_-f$ on a set of positive measure and obtain a contradiction. For by the Corollary to Theorem 3, p. 82; f is continuous a.e. and so as in Exercise 3 the derivates are measurable. Also there exists $\xi > 0$ and a set $F \subseteq [a, b]$ with $m(F) > 0$ and such that $D^+f - D_-f > 2\xi$ on F. But

$$[x: D^+f(x) - D_-f(x) > 2\xi] = \bigcup_{n=1}^{\infty} [x: D^+f(x) > r_n + \xi, D_-f(x) < r_n - \xi]$$

where $\{r_n\}$ is an enumeration of the rationals, so at least one set of this union has positive measure. We can therefore find numbers ξ, η with $\xi > 0$, and a set E in $[a, b]$ with $m(E) > 0$ and on which f is continuous, such that $D^+f > \eta + \xi$, $D_-f < \eta - \xi$ on E. Now $f - \eta x \in BV[a, b]$ and $D^+(f - \eta x) > D_-(f - \eta x)$ if, and only if, $D^+f > D_-f$, as in Exercise 4, above. So we may suppose that $\eta = 0$.

Let π be any polygon drawn, as in Theorem 4, p. 82, to approximate f, and let P be the set of points of the corresponding partition of $[a, b]$. Let $x \in E - P$ and suppose that $\pi'(x) < 0$. Since $D^+f(x) > \xi$, there exists $b_x > x$ such that $f(x, b_x) > \xi$. Then as f is continuous at x and hence $f(x, \beta)$ is a continuous function of x, we can find $a_x < x$ such that $f(a_x, b_x) > \xi$, and clearly we may choose a_x and b_x so that π is linear on (a_x, b_x). Similarly, if $\pi'(x) \geqslant 0$ we use the fact that $D_-f < -\xi$ and choose an interval (a_x, b_x) on which π is linear and $f(a_x, b_x) < -\xi$.

Then $\bigcup_x (a_x, b_x) \supseteq E - P$, so by Theorem 6, p. 84, there exists a finite subcollection of these intervals, say $I_1, \ldots, I_n$, such that

$$m\left(\bigcup_{k=1}^{n} I_k\right) > \tfrac{1}{2} m\left(\bigcup_x (a_x, b_x)\right) \geqslant \tfrac{1}{2} m(E - P) = \tfrac{1}{2} m(E).$$

By Theorem 5, p. 84, we may extract a subcollection of disjoint intervals I_{k_1}, $\ldots, I_{k_r}$ from these, such that

$$m\left(\bigcup_{i=1}^{r} I_{k_i}\right) \geqslant \tfrac{1}{3} m\left(\bigcup_{k=1}^{n} I_k\right) > \tfrac{1}{6} m(E).$$

We now consider the polygon q determined by the partition consisting of the set P and the end-points of $I_{k_1}, \ldots, I_{k_r}$. Applying Theorem 7 to each interval on which π is linear and adding we get

$$l(q) > l(\pi) + \sum_{i=1}^{r} l(I_{k_i})\left(\sqrt{(1 + \xi^2)} - 1\right) > l(\pi) + \tfrac{1}{6} m(E)\left(\sqrt{(1 + \xi^2)} - 1\right).$$

But ξ is independent of π so, since $l(\pi)$ can always be increased by a constant amount, $\sup l(\pi) = \infty$, taking the supremum over all polygons π approximating f. Hence by Theorem 4, p. 82, $f \notin BV[a, b]$ and this contradiction gives (i).

(ii) Suppose this result is false. Then, replacing f by $-f$ if necessary, we may suppose that there exists a set E on which f is continuous, $E \subseteq [a, b]$, $m(E) > 0$ and $D^+f = \infty$ on E. Then for any $M > 0$ choose, as in (i), a collection of intervals $[(a_x, b_x)]$ covering E such that $f(a_x, b_x) > M$. Pick the disjoint intervals $I_{k_1}, \ldots,$ I_{k_r}, as before, such that $\sum_{k=1}^{r} l(I_{k_i}) > \tfrac{1}{6} m(E)$. Let q be the polygon, approximating

f, determined by the end-points of the intervals I_{k_i}. The length of q in the interval I_{k_i} is greater than $l(I_{k_i})\surd(1+M^2)$ since the slope of f is greater than M. So

$$l(q) > \sum_{i=1}^{r} l(I_{k_i})\surd(1+M^2) > \tfrac{1}{6}m(E)\surd(1+M^2).$$

But M is abritrary and E is independent of M so taking the supremum over all approximating polygons π we get $\sup l(\pi) = \infty$, and (ii) follows. □

Note: Since any infinite interval may be written as the union of a sequence of finite intervals, this result extends to any interval.

Exercises

15. Show that the continuous function f of Example 5, p. 79, is not of bounded variation.
16. Construct a monotone function with a discontinuity at each rational in $[0,1]$.

4.5 DIFFERENTIATION AND INTEGRATION

Definition 4: Let $f \in L(a, b)$, then $F(x) = \int_a^x f(t)\,dt$ is the **indefinite integral** of f. So $F(a) = 0$. If $b = \infty$ and $a = -\infty$, $F(b) = \int f\,dt$.

We know from Chapter 3, Theorem 12, that if f is continuous, then F is differentiable and $F' = f$. We wish to improve on this result and consider the following questions (in the same notation):

(i) For which functions $f \in L(a, b)$ does F' exist or exist a.e.?
(ii) When is $F' = f$ a.e.?

Question (i) is answered by the corollary to Theorem 9, question (ii) by Theorem 12. The more difficult question: for which measurable functions F does there exist an integrable function f such that F is the indefinite integral of f, is answered in Chapter 9.

Theorem 9: If $f \in L(a, b)$, then: (i) $F(x) = \int_a^x f(t)\,dt$ is a continuous function on $[a, b]$, (ii) $F \in BV[a, b]$.

Proof: (i) Let $x_0 \in [a, b]$, then if $x > x_0$ we have $\int_{x_0}^x f\,dt = \int_a^b \chi_{[x_0, x]} f\,dt \to 0$ as $x \to x_0$, by Example 15, p. 64, since $f \in L(a, b)$. Similarly if $x < x_0$; so F is continuous at x_0. Similarly for continuity at b, if b is finite.

(ii) Let $a = x_0 < x_1 < \ldots < x_k = b$ be a partition of $[a, b]$. Then

$$\sum_{i=1}^{k} |F(x_i)-F(x_{i-1})| = \sum_{i=1}^{k} \left| \int_{x_{i-1}}^{x_i} f\,\mathrm{d}t \right| \leqslant \sum_{i=1}^{k} \int_{x_{i-1}}^{x_i} |f|\,\mathrm{d}t = \int_a^b |f|\,\mathrm{d}t < \infty.$$

So $F \in BV[a,b]$. □

Corollary: The indefinite integral of an integrable function is differentiable a.e.

Theorem 10: If f is a finite-valued monotone increasing function defined on the finite interval $[a,b]$, then f' is measurable and $\int_a^b f'\,\mathrm{d}x \leqslant f(b)-f(a)$.

Proof: By Theorem 8, p. 85, f' exists a.e. Define $g(x)$ to be $f'(x)$ when f' exists, $g=0$ otherwise. For $x \in (a,b)$ let $g_n(x) = n(f(x+1/n)-f(x))$, where we may suppose for convenience that $f=f(b)$ on $(b, b+1)$. Then each g_n is defined on (a,b) and is non-negative and measurable. Also, $g(x) = \lim_{n\to\infty} g_n(x)$, a.e., so clearly g is non-negative and measurable. By Fatou's Lemma

$$\begin{aligned}\int_a^b g\,\mathrm{d}x &\leqslant \liminf \int_a^b g_n\,\mathrm{d}x \\ &\leqslant \liminf\left(n\int_a^b f(x+(1/n))\,\mathrm{d}x - n\int_a^b f(x)\,\mathrm{d}x\right) \\ &\leqslant \liminf\left(n\int_{a+1/n}^{b+1/n} f(x)\,\mathrm{d}x - n\int_a^b f(x)\,\mathrm{d}x\right),\end{aligned}$$

using Example 25, p. 75, so

$$\int_a^b g\,\mathrm{d}x \leqslant \liminf\left(f(b) - n\int_a^{a+1/n} f\,\mathrm{d}x\right) \leqslant f(b)-f(a),$$

as $f(x) \geqslant f(a)$ in $(a, a+1/n)$. But $f'=g$ a.e., so the result follows. □

We cannot hope to get equality in Theorem 10 without further restrictions, for let f be any monotone increasing step function such that $f(b) > f(a)$, then $0 = \int_a^b f'\,\mathrm{d}x < f(b)-f(a)$.

Theorem 11: If $f \in L(a,b)$ and $\int_a^x f\,\mathrm{d}t = 0$ for all $x \in (a,b)$ then $f=0$ a.e. in (a,b).

Proof: Let $(\xi,\eta) \subseteq (a,b)$ so $\int_\xi^\eta f\,\mathrm{d}x = \int_a^\eta f\,\mathrm{d}x - \int_a^\xi f\,\mathrm{d}x$. So the integral of f is zero over any open interval and so over any open set $O \subseteq (a,b)$. Suppose the result of the theorem if false. Then we may suppose that there exists a set $E \subset (a,b)$, $m(E) > 0$, such that $f > 0$ on E; or else consider $-f$. Now by Theorem

10, p. 36, we can find a closed set $F \subset E$ such that $m(F) > 0$. Let $0 = (a, b) - F$. The clearly $\int_F f\,dx = \int_a^b f\,dx - \int_0 f\,dx = 0$. But $f > 0$ on F, giving a contradiction. □

Theorem 12: Let $[a, b]$ be a finite interval and let $f \in L(a, b)$ with indefinite integral F, then $F' = f$ a.e. in $[a, b]$.

Proof: Suppose first that f is bounded, $|f| \leqslant K$, say. Define f to be zero on $(b, b + 1]$ so that F is constant on $[b, b + 1]$, and define $G_n(x) = n(F(x + 1/n) - F(x))$. Then $G_n(x) = n \int_x^{x+1/n} f\,dt$, so $|G_n| \leqslant K$. But by the corollary to Theorem 9, p. 87, F' exists a.e. Hence $\lim G_n = F'$ a.e. and $|F'| \leqslant K$ a.e. So by Theorem 10, p. 63, for each $c \in (a, b)$,

$$\int_a^c F'\,dx = \lim \int_a^c G_n\,dx = \lim \left(n \int_a^c F(x + 1/n)\,dx - n \int_a^c F(x)\,dx \right)$$

$$= \lim \left(n \int_{a+1/n}^{c+1/n} F\,dx - n \int_a^c F\,dx \right),$$

by Example 25, p. 75. So

$$\int_a^c F'\,dx = \lim \left(n \int_c^{c+1/n} F\,dx - n \int_a^{a+1/n} F\,dx \right) = F(c) - F(a),$$

since F is continuous by Theorem 9. Hence $\int_a^c (F' - f)\,dx = 0$ for each $c \in [a, b]$, so $F' = f$ a.e. by Theorem 11.

Now consider the general case. Since we may consider f^+ and f^- separately, we may suppose $f \geqslant 0$. Let $f_n(x) = \min(f(x), n)$ and write

$$H_n(x) = \int_a^x (f - f_n)\,dt.$$

Then, for each n, H_n is a non-negative monotone increasing function. By the corollary to Theorem 9, H_n' exists a.e. in $[a, b]$ and when defined, $H_n' \geqslant 0$. Since f_n is bounded

$$\frac{d}{dx} \int_a^x f_n\,dt = f_n(x) \text{ a.e. in } [a, b].$$

So, a.e. in $[a, b]$ and for each n,

$$F'(x) = \frac{d}{dx} H_n(x) + \frac{d}{dx} \int_a^x f_n\,dt \geqslant f_n(x).$$

So $F' \geqslant f$ a.e. in $[a, b]$. Hence

$$\int_a^b F'\,\mathrm{d}x \geqslant \int_a^b f\,\mathrm{d}x = F(b) - F(a).$$

But by Theorem 10, applied to F, $\int_a^b F'\,\mathrm{d}x \leqslant F(b) - F(a)$. So $\int_a^b F'\,\mathrm{d}x = \int_a^b f\,\mathrm{d}x$. But $F' \geqslant f$ a.e. and so $F' = f$ a.e. in $[a, b]$. □

Corollary: Let f be integrable with indefinite integral F, then $F' = f$ a.e.

Exercises

17. Show that if f is a finite-valued step function on $[a, b]$ with indefinite integral F, then whenever F' exists, $F' = f$.
18. Give an example where $F' = f$ does not hold even when F' exists.
19. Show that Theorem 10 need not hold if f is not monotone increasing.
20. Find analogues of Theorem 10 for the cases (i) f monotone decreasing on $[a, b]$, (ii) $f \in BV[a, b]$.
21. Show that Theorem 9 does not characterize indefinite integrals, that is: there exist continuous functions of bounded variation which are not indefinite integrals.
22. Let X be a measurable set of positive finite measure. Show that

$$\lim_{h\to 0+} \frac{m(X \cap (x-h, x+h))}{2h} = \chi_X \text{ a.e.}$$

4.6 THE LEBESGUE SET

In this section we obtain, in Theorem 14, an interesting property of integrable functions which although it will not be used in this book is of importance in applications (see [3]).

Theorem 13: If $f \in L(a, b)$ where (a, b) is a finite interval, then there exists a set $E \subseteq (a, b)$ such that $m([a, b] - E) = 0$ and

$$\lim_{h\to 0} \frac{1}{h} \int_x^{x+h} |f(t) - \xi|\,\mathrm{d}t = |f(x) - \xi|$$

for all real ξ and all $x \in E$.

Proof: The last theorem gives the result immediately for a single value of ξ, but then considering all values of ξ would give an uncountable number of exceptional sets of measure zero. To avoid the difficulty we suppose that $\{\beta_n\}$ is any sequence dense in R, for example: the rationals in some order. For each n, g_n is defined by

$$g_n(t) = |f(t) - \beta_n|, \tag{4.3}$$

so that $g_n \in L(a, b)$. By Theorem 12, there exists $E_n \subseteq (a, b)$ such that $m((a, b) - E_n) = 0$ and

$$\lim_{h \to 0} \frac{1}{h} \int_x^{x+h} g_n(t)\, dt = g_n(x), \text{ for } x \in E_n.$$

Let $E = \bigcap_{n=1}^{\infty} E_n$, then $m((a, b) - E) \leqslant \sum_{n=1}^{\infty} m((a, b) - E_n) = 0$. For $\epsilon > 0$ and $\xi \in \mathrm{R}$ choose n so that $|\beta_n - \xi| < \epsilon/3$. Then for all $t \in [a, b]$

$$\big||f(t) - \xi| - |f(t) - \beta_n|\big| \leqslant |\beta_n - \xi| < \epsilon/3.$$

So $\left|\frac{1}{h} \int_x^{x+h} |f(t) - \xi|\, dt - \frac{1}{h} \int_x^{x+h} |f(t) - \beta_n|\, dt\right| \leqslant \frac{1}{h} \int_x^{x+h} \frac{\epsilon}{3}\, dt = \frac{\epsilon}{3}.$

Hence, from (4.3), we have

$$\left|\frac{1}{h} \int_x^{x+h} |f(t) - \xi|\, dt - |f(x) - \xi|\right| \leqslant \left|\frac{1}{h} \int_x^{x+h} |f(t) - \xi|\, dt - \right.$$

$$\left.\frac{1}{h} \int_x^{x+h} |f(t) - \beta_n|\, dt\right| + \left|\frac{1}{h} \int_x^{x+h} g_n(t)\, dt - g_n(x)\right| + |\beta_n - \xi| <$$

$$< \frac{\epsilon}{3} + \frac{\epsilon}{3} + \frac{\epsilon}{3} = \epsilon$$

for $x \in E$ and $|h| < \delta(\epsilon, n)$. But n depends only on ϵ and ξ, so the result follows. □

Theorem 14: Let $f \in L(a, b)$ where (a, b) is a finite interval; then

$$\lim_{h \to 0} \frac{1}{h} \int_0^h |f(x + t) - f(x)|\, dt = 0 \text{ a.e. in } [a, b].$$

Proof: By Example 25, p. 75, this is equivalent to

$$\lim_{h \to 0} \frac{1}{h} \int_x^{x+h} |f(t) - f(x)|\, dt = 0 \text{ a.e. in } [a, b].$$

But the last theorem, with $\xi = f(x)$, gives the result if $x \in E$ where $m([a, b] - E) = 0$, as required. □

Definition 5: If $f \in L(a, b)$, the set of points $x \in [a, b]$ such that

$$\lim_{h \to 0} \frac{1}{h} \int_0^h |f(x + t) - f(x)|\, dt = 0$$

is called the **Lebesgue set** of f.

Example 10: Show that the Lebesgue set of a function $f \in L(a, b)$ contains any point at which f is continuous.

Solution: Let f be continuous at x. Then $\forall\, \epsilon > 0$, $\exists\, \delta > 0$ such that $|f(x+t) - f(x)| < \epsilon$ for $|t| < \delta$. So for $0 < |h| < \delta$

$$\frac{1}{h} \int_0^h |f(x+t) - f(x)| \, \mathrm{d}t < \epsilon.$$

So the limit as $h \to 0$ exists and equals zero.

CHAPTER 5

Abstract Measure Spaces

In this chapter definitions of *measurable* as applied to sets and functions are provided for abstract spaces and we present in this general setting the main results of Chapters 2 and 3. We show in Sections 5.1–5.3 how a measure on a *ring* of sets can be extended to one on a generated *σ-ring*. The work of Chapter 2 where we went from a measure on finite unions of intervals to Lebesgue measure is an example of such an extension. The use of a ring of sets, which generalizes the notion of an algebra of sets, is necessary if the work of Chapter 2 is to be fitted into the general theory. The theory will be essential later in Chapters 9 and 10. The notions of measure and integration on abstract spaces arise also in applications, especially in the theory of probability which may be regarded as concerned with special results on classes of measurable functions on spaces of total measure one.

5.1 MEASURES AND OUTER MEASURES

We consider general spaces and generalize many of the results of Chapter 2. Those results, for example Theorem 10, p. 36 of Chapter 2, which depend on the idea of open sets are more difficult to extend and will not be examined in the general case.

Definition 1: A class of sets $\mathcal{R}$, of some fixed space is called a **ring** if whenever $E \in \mathcal{R}$ and $F \in \mathcal{R}$ then $E \cup F$ and $E - F$ belong to $\mathcal{R}$.

Example 1: The class of finite unions of intervals of the form $[a, b)$ forms a ring.

Definition 2: A ring is called a **σ-ring** if it is closed under the formation of countable unions.

Example 2: Show that every algebra is a ring and every σ-algebra a σ-ring but that the converse is not true.

Solution: The first part follows from Definitions 3 and 4, p. 30, as $E - F = \mathbf{C}(\mathbf{C}E \cup F)$. For the second, consider the σ-ring of all subsets of [0,1] which are at most countable.

If $\cup A_\alpha \in \mathcal{S}$ where the A_α are the sets of the σ-ring $\mathcal{S}$, then $\mathcal{S}$ may be regarded as a σ-algebra on the space $\cup A_\alpha$. The σ-ring considered in Example 2 shows that this need not occur.

Theorem 1: There exist a smallest ring and a smallest σ-ring containing a given class of subsets of a space; we refer to these as the *generated ring* and the *generated σ-ring* respectively.

Proof: The proof of Theorem 7, p. 32, with the appropriate replacements for 'algebra' and 'σ-algebra' applies. □

Notation: We will write $\mathcal{S}(\mathcal{R})$ for the σ-ring $\mathcal{S}$ generated by the ring $\mathcal{R}$; we write $\mathcal{H}(\mathcal{R})$ for the class consisting of $\mathcal{S}(\mathcal{R})$ together with all subsets of the sets of $\mathcal{S}(\mathcal{R})$. A class of sets with this property, namely that every subset of one of its members belongs to the class, is said to be **hereditary**.

Clearly $\mathcal{H}(\mathcal{R})$ is a σ-ring and is the smallest hereditary σ-ring containing $\mathcal{R}$. Indeed $\mathcal{H}(\mathcal{R}) = \mathcal{H}(\mathcal{S}(\mathcal{R})) = \mathcal{H}(\mathcal{H}(\mathcal{R}))$, the proof following as in Theorem 7, p. 32, as the intersection of hereditary σ-rings is again an hereditary σ-ring.

Definition 3: A set function μ defined on a ring $\mathcal{R}$ is a **measure** if (i) μ is non-negative, (ii) $\mu(\emptyset) = 0$, (iii) for any sequence $\{A_n\}$ of disjoint sets of $\mathcal{R}$ such that $\bigcup_{n=1}^{\infty} A_n \in \mathcal{R}$, we have

$$\mu\left(\bigcup_{n=1}^{\infty} A_n\right) = \sum_{n=1}^{\infty} \mu(A_n).$$

If $\mathcal{R}$ is a σ-ring, the condition $\bigcup_{n=1}^{\infty} A_n \in \mathcal{R}$ is clearly redundant.

Definition 4: A measure μ on $\mathcal{R}$ is **complete** if whenever $E \in \mathcal{R}$, $F \subseteq E$ and $\mu(E) = 0$, then $F \in \mathcal{R}$.

Definition 5: A measure μ on $\mathcal{R}$ is **σ-finite** if, for every set $E \in \mathcal{R}$, we have $E = \bigcup_{n=1}^{\infty} E_n$ for some sequence $\{E_n\}$ such that $E_n \in \mathcal{R}$ and $\mu(E_n) < \infty$ for each n.

Example 3: Show that Lebesgue measure m defined on $\mathcal{M}$, the class of measurable sets of R, is σ-finite and complete.

Solution: $\mathcal{M}$ is a σ-algebra (Theorem 4, p. 30) and so is a ring, on which m is defined. Take $E_n = E \cap (-n, n)$ to get σ-finiteness. Completeness follows from Example 4, p. 30.

Definition 6: If $\mathcal{R}$ is a ring, a set function μ^* defined on the class $\mathcal{H}(\mathcal{R})$ is an **outer measure** if (i) μ^* is non-negative, (ii) if $A \subseteq B$, then $\mu^*(A) \leqslant \mu^*(B)$, (iii) $\mu^*(\emptyset) = 0$, (iv) for any sequence $\{A_n\}$ of sets of $\mathcal{H}(\mathcal{R})$,

$$\mu^*\left(\bigcup_{n=1}^{\infty} A_n\right) \leqslant \sum_{n=1}^{\infty} \mu^*(A_n),$$

that is, μ^* is countably subadditive.

Example 4: Lebesgue outer measure m^* as defined in Chapter 2 is an outer measure in the sense of Definition 6.

We easily obtain from Definition 3 that a measure is finitely additive and from Definition 6 that an outer measure is finitely subadditive.

Example 5: Show that if $A, B \in \mathcal{R}$ and $A \subseteq B$ then $\mu(A) \leqslant \mu(B)$.

Solution: $B = A \cup (B - A)$ and as the measure μ is finitely additive the result follows.

Exercises

1. Describe the ring generated by the finite open intervals.
2. Let $\mathcal{S}$ be the class of subsets of R such that $E \in \mathcal{S}$ if either E or $\mathbf{C}E$ is at most countable. Show that $\mathcal{S}$ is a σ-ring.
3. Let A, B be subsets of a set C, let $A, B, C \in \mathcal{R}$ and let μ be a measure on $\mathcal{R}$. Show that if $\mu(A) = \mu(C) < \infty$, then $\mu(A \cap B) = \mu(B)$.
4. Show that if μ is a non-negative set function on a ring and is countably additive and is finite on some set, then μ is a measure.
5. Let μ be a measure on a ring $\mathcal{R}$, then ρ defined by $\rho(A, B) = \mu(A \,\Delta\, B)$ is a pseudometric on $\mathcal{R}$.

5.2 EXTENSION OF A MEASURE

In this section we generalize the procedure by which the outer measure m^* and the measure m were obtained in Chapter 2.

Theorem 2: Let $\{A_i\}$ be a sequence in a ring $\mathcal{R}$, then there is a sequence $\{B_i\}$ of disjoint sets of $\mathcal{R}$ such that $B_i \subseteq A_i$ for each i and $\bigcup_{i=1}^{N} A_i = \bigcup_{i=1}^{N} B_i$ for each N, so that $\bigcup_{i=1}^{\infty} A_i = \bigcup_{i=1}^{\infty} B_i$.

Proof: Define $\{B_i\}$ inductively by $B_1 = A_1, B_n = A_n - \bigcup_{i=1}^{n-1} B_i$ for $n > 1$. Clearly $B_i \in \mathcal{R}$ and $B_i \subseteq A_i$ for each i. Also, as B_n and $\bigcup_{i=1}^{n-1} B_i$ are disjoint we have $B_n \cap B_m = \emptyset$ for $n > m$. Finally, we have $B_1 = A_1$ and if $\bigcup_{i=1}^{k} B_i = \bigcup_{i=1}^{k} A_i$, it follows that

$$B_{k+1} \cup \left(\bigcup_{i=1}^{k} B_i \right) = \left(A_{k+1} - \bigcup_{i=1}^{k} B_i \right) \cup \bigcup_{i=1}^{k} B_i$$

$$= A_{k+1} \cup \bigcup_{i=1}^{k} B_i = A_{k+1} \cup \bigcup_{i=1}^{k} A_i$$

as required. □

Example 6: Show that $\mathcal{H}(\mathcal{R}) = [E: E \subseteq \bigcup_{n=1}^{\infty} E_n, E_n \in \mathcal{R}]$.

Solution: It is easily checked that the right-hand side defines a class of sets which is hereditary, contains $\mathcal{R}$, and is a σ-ring. So it contains $\mathcal{H}(\mathcal{R})$. But if $E_n \in \mathcal{R}$ for each n, we have $\bigcup_{n=1}^{\infty} E_n \in \mathcal{S}(\mathcal{R})$ and so each subset belongs to $\mathcal{H}(\mathcal{R})$. So we get equality.

The result of Example 6 ensures that the function μ^* appearing in the next theorem is well defined. The definition of μ^* extends that of Lebesgue outer measure m^* rather than the quite different method used to define the Hausdorff outer measure in Section 2.6.

Theorem 3: If μ is a measure on a ring $\mathcal{R}$ and if the set function μ^* is defined on $\mathcal{H}(\mathcal{R})$ by

$$\mu^*(E) = \inf \left[\sum_{n=1}^{\infty} \mu(E_n): E_n \in \mathcal{R}, n = 1, 2, \ldots, E \subseteq \bigcup_{n=1}^{\infty} E_n \right], \tag{5.1}$$

then (i) for $E \in \mathcal{R}$, $\mu^*(E) = \mu(E)$, (ii) μ^* is an outer measure on $\mathcal{H}(\mathcal{R})$.

Proof: (i) If $E \in \mathcal{R}$, (5.1) gives $\mu^*(E) \leqslant \mu(E)$. Suppose that $E \in \mathcal{R}$ and $E \subseteq \bigcup_{n=1}^{\infty} E_n$ where $E_n \in \mathcal{R}$. By Theorem 2 we may replace the sequence $\{E_i \cap E\}$ by a sequence $\{F_i\}$ of disjoint sets of $\mathcal{R}$, such that $F_i \subseteq E_i \cap E$ and $\bigcup_{i=1}^{\infty} F_i = E$. Then, by Example 5, $\mu(F_i) \leqslant \mu(E_i)$ for each i. So

$$\mu(E) = \mu\left(\bigcup_{i=1}^{\infty} F_i \right) = \sum_{i=1}^{\infty} \mu(F_i) \leqslant \sum_{i=1}^{\infty} \mu(E_i).$$

It follows that $\mu(E) \leqslant \mu^*(E)$ and so (i) is true.

To prove (ii): $\mu^*(\emptyset) = \mu(\emptyset)$ by (i); the only other property of an outer measure which is not immediate, namely countable subadditivity, is shown as for m^* in Chapter 2. We suppose that $\{E_i\}$ is a sequence of sets in $\mathcal{H}(\mathcal{R})$. From the definition of μ^*, for each $\epsilon > 0$, we can find for each i a sequence $\{E_{i,j}\}$ of sets of $\mathcal{R}$ such that $E_i \subseteq \bigcup_{j=1}^{\infty} E_{i,j}$ and $\sum_{j=1}^{\infty} \mu(E_{i,j}) \leqslant \mu^*(E_i) + \epsilon/2^i$. The sets $E_{i,j}$ form a countable class covering $\bigcup_{i=1}^{\infty} E_i$, so

$$\mu^*\left(\bigcup_{i=1}^{\infty} E_i\right) \leqslant \sum_{i=1}^{\infty} \sum_{j=1}^{\infty} \mu(E_{i,j}) \leqslant \sum_{i=1}^{\infty} \mu^*(E_i) + \epsilon.$$

But ϵ is arbitrary, so the result follows. □

We define measurability as in Definition 2, p. 30.

Definition 7: Let μ^* be an outer measure on $\mathcal{H}(\mathcal{R})$. Then $E \in \mathcal{H}(\mathcal{R})$ is μ^*-**measurable** if for each $A \in \mathcal{H}(\mathcal{R})$

$$\mu^*(A) = \mu^*(A \cap E) + \mu^*(A \cap \mathbf{C}E). \tag{5.2}$$

Theorem 4: Let μ^* be an outer measure on $\mathcal{H}(\mathcal{R})$ and let $\mathcal{S}^*$ denote the class of μ^*-measurable sets. Then $\mathcal{S}^*$ is a σ-ring and μ^* restricted to $\mathcal{S}^*$ is a complete measure.

Proof: That $\mathcal{S}^*$ is closed under countable unions follows precisely as in Theorem 4, p. 30. It remains to be shown that if $E, F \in \mathcal{S}^*$ then $E - F \in \mathcal{S}^*$. Let $A \in \mathcal{H}(\mathcal{R})$ and write A as the union of the four disjoint sets $A_1 = A - (E \cup F)$, $A_2 = A \cap E \cap F$, $A_3 = A \cap (F - E)$, $A_4 = A \cap (E - F)$. Since F is measurable, (5.2) gives

$$\mu^*(A) = \mu^*(A_1 \cup A_4) + \mu^*(A_2 \cup A_3). \tag{5.3}$$

Replacing A in (5.2) by $A_1 \cup A_4$ and using the fact that E is measurable gives

$$\mu^*(A_1 \cup A_4) = \mu^*(A_1) + \mu^*(A_4). \tag{5.4}$$

Replacing A in (5.2) by $A_1 \cup A_2 \cup A_3$ and using the fact that F is measurable gives

$$\mu^*(A_1 \cup A_2 \cup A_3) = \mu^*(A_1) + \mu^*(A_2 \cup A_3). \tag{5.5}$$

Then (5.3), (5.4) and (5.5) give

$$\mu^*(A) = \mu^*(A_4) + \mu^*(A_1 \cup A_2 \cup A_3),$$

which is the condition for $E - F$ to be measurable.

Suppose that $\{E_i\}$ is a sequence of disjoint sets in $\mathcal{S}^*$. Then exactly as in Theorem 5, p. 31, we have $\mu^*\left(\bigcup_{i=1}^{\infty} E_1\right) = \sum_{i=1}^{\infty} \mu^*(E_i)$. So μ^* is a measure on the σ-ring $\mathcal{S}^*$.

Also every set $E \in \mathcal{H}(\mathcal{R})$ such that $\mu^*(E) = 0$ is μ^*-measurable, for if $A \in \mathcal{H}(\mathcal{R})$,

$$\begin{aligned} \mu^*(A) &\leqslant \mu^*(A \cap E) + \mu^*(A \cap \mathbf{C}E) \\ &\leqslant \mu^*(E) + \mu^*(A) = \mu^*(A). \end{aligned}$$

So equality holds and E is μ^*-measurable. In particular if $E \in \mathcal{S}^*$ and $\mu^*(E) = 0$ and $F \subseteq E$ then it follows that $F \in \mathcal{S}^*$, so μ^* is a complete measure on $\mathcal{S}^*$. □

Theorem 4 has been proved for an arbitrary outer measure μ^* on $\mathcal{H}(\mathcal{R})$. If μ^* has been obtained from a measure μ on $\mathcal{R}$ as in Theorem 3, p. 96, we will denote the measure obtained by restricting μ^* to $\mathcal{S}^*$, by $\bar{\mu}$. Theorem 3(i) shows that $\bar{\mu}$ is an extension of μ.

Theorem 5: Let μ^* be the outer measure on $\mathcal{H}(\mathcal{R})$ defined by μ on $\mathcal{R}$, then $\mathcal{S}^*$ contains $\mathcal{S}(\mathcal{R})$, the σ-ring generated by $\mathcal{R}$.

Proof: Since $\mathcal{S}^*$ is a σ-ring it is sufficient to show that $\mathcal{R} \subseteq \mathcal{S}^*$. If $E \in \mathcal{R}$, $A \in \mathcal{H}(\mathcal{R})$ and $\epsilon > 0$, then by the definition of μ^* in (5.1) there exists a sequence $\{E_n\}$ of sets of $\mathcal{R}$ such that $A \subseteq \bigcup_{n=1}^{\infty} E_n$ and

$$\mu^*(A) + \epsilon \geqslant \sum_{n=1}^{\infty} \mu(E_n) = \sum_{n=1}^{\infty} \mu(E_n \cap E) + \sum_{n=1}^{\infty} \mu(E_n \cap \complement E)$$

as μ is a measure. So

$$\mu^*(A) + \epsilon \geqslant \mu^*(A \cap E) + \mu^*(A \cap \complement E).$$

But ϵ is arbitrary so

$$\mu^*(A) \geqslant \mu^*(A \cap E) + \mu^*(A \cap \complement E).$$

The opposite inequality is obvious, so $E \in \mathcal{S}^*$, giving the result. □

Example 7: Show that if μ is a σ-finite measure on $\mathcal{R}$, then the extension $\bar{\mu}$ of μ to $\mathcal{S}^*$ is also σ-finite.

Solution: Let $E \in \mathcal{S}^*$. Then by the definition of $\bar{\mu}$ there is a sequence $\{E_n\}$ of sets of $\mathcal{R}$ such that $\bar{\mu}(E) \leqslant \sum_{n=1}^{\infty} \mu(E_n)$. But each E_n is, by hypothesis, the union of a sequence $\{E_{n,i}, i = 1, 2, \ldots\}$ of sets of $\mathcal{R}$ such that $\mu(E_{n,i}) < \infty$ for each n and i. So

$$\bar{\mu}(E) \leqslant \sum_{n=1}^{\infty} \sum_{i=1}^{\infty} \mu(E_{n,i}),$$

and so E is the union of a countable collection of sets of finite $\bar{\mu}$-measure.

Example 8: In Chapter 2, when Lebesgue measure was constructed, $\mathcal{R}$ was the ring of finite unions of intervals $[a, b)$, $\mathcal{S}(\mathcal{R})$ was the σ-algebra of Borel sets, $\mathcal{S}^*$ the σ-algebra of (Lebesgue) measurable sets and $\mathcal{S}^*$ was greater than $\mathcal{S}(\mathcal{R})$. μ on $\mathcal{R}$ was given by $\mu\left(\bigcup_{n=1}^{N} I_n\right) = \sum_{n=1}^{N} l(I_n)$, where $I_1, \ldots, I_N$ were disjoint intervals; $\bar{\mu}$ was denoted by m.

Solution: It is easily seen that μ as defined is in fact a measure on $\mathcal{R}$ (see Chapter 9, p. 155, for details). That $\mathcal{S}^* \supset \mathcal{S}(\mathcal{R})$ follows from Theorem 18, p. 43. The remaining statements follow from the definitions.

Exercise

6. Show that the Hausdorff measures of Section 2.6 are complete.

5.3 UNIQUENESS OF THE EXTENSION

Using the definition of μ^* given in Theorem 3, p. 96, we have extended the original measure μ on $\mathcal{R}$ to a complete measure $\bar{\mu}$ on $\mathcal{S}^*$, a σ-ring containing $\mathcal{R}$. The same procedure may be applied to $\bar{\mu}$ on $\mathcal{S}^*$, but, in fact, the same measure and σ-ring are obtained, that is: $(\bar{\bar{\mu}}) = \bar{\mu}$ and $(\mathcal{S}^*)^* = \mathcal{S}^*$. This follows from the next theorem.

Theorem 6: The outer measure μ^* on $\mathcal{H}(\mathcal{R})$ defined by μ on $\mathcal{R}$ as in Theorem 3, and the corresponding outer measure defined by $\bar{\mu}$ on $\mathcal{S}(\mathcal{R})$ and $\bar{\mu}$ on $\mathcal{S}^*$ are the same.

Proof: We first observe that the outer measure β^* defined by a measure β on a σ-ring $\mathcal{T}$ satisfies, for $E \in \mathcal{H}(\mathcal{T})$

$$\beta^*(E) = \inf[\beta(F): E \subseteq F \in \mathcal{T}]. \tag{5.6}$$

This is the case since

$$\beta^*(E) = \inf\left[\sum_{n=1}^{\infty} \beta(E_n): E \subseteq \bigcup_{n=1}^{\infty} E_n, E_n \in \mathcal{T}\right],$$

and replacing the sets E_n by disjoint sets $F_n \in \mathcal{T}$, such that $F_n \subseteq E_n$ and $\bigcup_{n=1}^{\infty} E_n = \bigcup_{n=1}^{\infty} F_n$, we get

$$\sum_{n=1}^{\infty} \beta(E_n) \geqslant \sum_{n=1}^{\infty} \beta(F_n) = \beta\left(\bigcup_{n=1}^{\infty} F_n\right) \geqslant \beta^*(E)$$

so (5.6) follows.

Since $\mathcal{H}(\mathcal{R}) = \mathcal{H}(\mathcal{S}(\mathcal{R})) = \mathcal{H}(\mathcal{S}^*)$, the outer measures to be considered have the same domain of definition. As $\mu = \bar{\mu}$ on $\mathcal{R}$,

$$\begin{aligned}
\mu^*(E) &= \inf\left[\sum_{n=1}^{\infty} \mu(E_n): E \subseteq \bigcup_{n=1}^{\infty} E_n, E_n \in \mathcal{R}\right] \\
&\geqslant \inf\left[\sum_{n=1}^{\infty} \bar{\mu}(F_n): E \subseteq \bigcup_{n=1}^{\infty} F_n, F_n \in \mathcal{S}(\mathcal{R})\right] \\
&= \inf[\bar{\mu}(F): E \subseteq F \in \mathcal{S}(\mathcal{R})] \text{ by (5.6)} \\
&\geqslant \inf[\bar{\mu}(F): E \subseteq F \in \mathcal{S}^*] \text{ as } \mathcal{S}^* \supseteq \mathcal{S}(\mathcal{R}) \\
&\geqslant \mu^*(E).
\end{aligned}$$

So equality holds throughout and so by (5.6) the outer measures are equal. □

Corollary: Since the outer measure on $\mathcal{H}(\mathcal{R})$ determines the measurable sets and their measures, the measure and measurable sets obtained by extending, as in Theorem 3, μ on $\mathcal{R}$, $\bar{\mu}$ on $\mathcal{S}(\mathcal{R})$ and $\bar{\mu}$ on $\mathcal{S}^*$ are the same, namely $\bar{\mu}$ on $\mathcal{S}^*$.

Without some restrictions on $\bar{\mu}$ its extension to $\mathcal{S}(\mathcal{R})$ need not be unique, but we have:

Theorem 7: If μ is a σ-finite measure on a ring $\mathcal{R}$, then it has a unique extension to the σ-ring $\mathcal{S}(\mathcal{R})$.

Proof: By Theorem 3, p. 96, $\bar{\mu}$ on $\mathcal{S}(\mathcal{R})$ is an extension of μ. Suppose that ν is a measure on $\mathcal{S}(\mathcal{R})$ such that $\mu = \nu$ on $\mathcal{R}$; we wish to show that $\bar{\mu} = \nu$ on $\mathcal{S}(\mathcal{R})$. If $E \in \mathcal{S}(\mathcal{R})$ and $\epsilon > 0$, $\exists \{E_n\}, E_n \in \mathcal{R}, E \subseteq \bigcup_{n=1}^{\infty} E_n$ such that $\bar{\mu}(E) + \epsilon \geqslant \sum_{n=1}^{\infty} \mu(E_n)$. But $A = \bigcup_{n=1}^{\infty} E_n$ may, by Theorem 2, p. 95, may be written as the union of disjoint sets $F_n, F_n \subseteq E_n, F_n \in \mathcal{R}$; so we get

$$\bar{\mu}(E) + \epsilon \geqslant \sum_{n=1}^{\infty} \mu(F_n) = \sum_{n=1}^{\infty} \nu(F_n) = \nu(A) \geqslant \nu(E).$$

So $\bar{\mu}(E) \geqslant \nu(E)$.

Suppose that $E \in \mathcal{S}(\mathcal{R})$, $\bar{\mu}(E) < \infty$ and $\epsilon > 0$, then as above there exists $A \supseteq E$ such that $\bar{\mu}(A) < \bar{\mu}(E) + \epsilon$ where $A = \bigcup_{n=1}^{\infty} F_n$, the sets F_n being disjoint sets of $\mathcal{R}$, so that $\bar{\mu}(A) = \nu(A)$. So

$$\bar{\mu}(E) \leqslant \bar{\mu}(A) = \nu(E) + \nu(A - E).$$

But, by the first part, $\nu(A - E) \leqslant \bar{\mu}(A - E)$, also since $\bar{\mu}(E) < \infty$ we have $\bar{\mu}(A - E) < \epsilon$. So $\bar{\mu}(E) \leqslant \nu(E) + \epsilon$. Hence $\bar{\mu}(E) = \nu(E)$ if $\bar{\mu}(E) < \infty$. But by Example 6, and as μ is σ-finite, for each $E \in \mathcal{S}(\mathcal{R})$ we have $E \subseteq \bigcup_{n=1}^{\infty} E_n$ where, for each $n, E_n \in \mathcal{R}$ and $\mu(E_n) < \infty$. Then we may write $E = \bigcup_{n=1}^{\infty} F_n$ where the F_n are disjoint sets of $\mathcal{R}$ and $\mu(F_n) < \infty$. So

$$\bar{\mu}(E) = \sum_{n=1}^{\infty} \mu(F_n) = \sum_{n=1}^{\infty} \nu(F_n) = \nu(E). \square$$

Exercise

7. Give an example of a non-unique extension of a non σ-finite measure.

5.4 COMPLETION OF A MEASURE

We show in the next theorem how a measure which is not complete may be extended to one which is by adjoining to the original ring the subsets of the sets of measure zero. This could, for instance, be used to construct Lebesgue measure given the measure m on the Borel sets. It will also be relevant in the study of product measures in Chapter 10 where non-complete measures arise in a natural way.

Theorem 8: If μ is a measure on a σ-ring $\mathcal{S}$, then the class $\overline{\mathcal{S}}$ of sets of the form $E \Delta N$ for any sets E, N such that $E \in \mathcal{S}$ while N is contained in some set in $\mathcal{S}$ of zero measure, is a σ-ring, and the set function $\bar{\mu}$ defined by $\bar{\mu}(E \Delta N) = \mu(E)$ is a complete measure on $\overline{\mathcal{S}}$.

Proof: It is convenient to have two different descriptions of the sets of $\overline{\mathcal{S}}$ so we prove the set-theoretic identity

$$E \Delta N = (E - M) \cup (M \cap (E \Delta N)) \tag{5.7}$$

for any sets E, M, N such that $M \supseteq N$. Let $x \in E \Delta N$, then if $x \in M$ we have $x \in M \cap (E \Delta N)$, while if $x \in \mathbf{C}M$ we have $x \in \mathbf{C}N$ so $x \in E - N$ and hence $x \in E - M$. To get the opposite inclusion in (5.7), suppose that x belongs to the right-hand side. If $x \in M \cap (E \Delta N)$, then $x \in E \Delta N$; if $x \in E - M$, we have $x \in E - N \subseteq E \Delta N$.

Let $D \in \overline{\mathcal{S}}, D = E \Delta N$, as above, with $N \subseteq M \in \mathcal{S}$ where $\mu(M) = 0$. Then, by (5.7) $D = F \cup A$ where $F \cap A = \emptyset$ and $F \in \mathcal{S}$ and $A \subseteq M \in \mathcal{S}$ with $\mu(M) = 0$, and since for F, A disjoint we have $F \cup A = F \Delta A$ the two characterizations of the sets of $\overline{\mathcal{S}}$ are equivalent. Now if $D_i \in \overline{\mathcal{S}}, i = 1, 2, \ldots$, on writing $D_i = F_i \cup A_i$ we see that $\bigcup_{i=1}^{\infty} D_i \in \overline{\mathcal{S}}$. If $D_1 = E_1 \Delta N_1$ and $D_2 - E_2 \Delta N_2$ belong to $\mathcal{S}$ we have, using Example 1, p. 16,

$$D_1 \Delta D_2 = (E_1 \Delta E_2) \Delta (N_1 \Delta N_2).$$

So $D_1 \Delta D_2 \in \overline{\mathcal{S}}$, and so $D_1 - D_2 = (D_1 \cup D_2) \Delta D_2 \in \overline{\mathcal{S}}$. So $\overline{\mathcal{S}}$ is a σ-ring.

Also $D_1 \Delta D_2 = \emptyset$ only if $E_1 \Delta E_2 = N_1 \Delta N_2$. So, if $E_1 \Delta N_1 = E_2 \Delta N_2$, we have $\mu(E_1 \Delta E_2) = 0$ and hence $\mu(E_1) = \mu(E_2)$. So $\bar{\mu}$ is unambiguously defined. Also $\bar{\mu}$ is a measure; for clearly $\bar{\mu}(\emptyset) = 0$, and if $\{D_i\}$ is a sequence of disjoint sets of $\overline{\mathcal{S}}, D_i = F_i \cup A_i$, say, in the notation used above, so that $F_i \cap A_j = \emptyset$ for all i and j, then

$$\begin{aligned}\bar{\mu}(\cup D_i) = \bar{\mu}(\cup F_i \cup \cup A_i) = \bar{\mu}(\cup F_i \Delta \cup A_i) = \mu(\cup F_i) = \Sigma\, \mu(F_i) = \\ = \Sigma\, \bar{\mu}(F_i \cup A_i) = \Sigma\, \bar{\mu}(D_i).\end{aligned}$$

So $\bar{\mu}$ is countably additive.

Finally μ is complete, for let $D \subset D_0 \in \overline{\mathcal{S}}$ where $\bar{\mu}(D_0) = 0$. So $D_0 = E_0 \Delta N_0$ where $N_0 \subseteq M_0, E_0, M_0 \in \mathcal{S}, \mu(E_0) = \mu(M_0) = 0$, and so $D_0 \subseteq M_0' = E_0 \cup M_0 \in \mathcal{S}$ and $\mu(M_0') = 0$. Then $D = E \Delta N$ with $E = \emptyset, N = D \subseteq E_0 \cup M_0$ and so $D \in \overline{\mathcal{S}}$. □

Example 9: Show that the extension $\bar{\mu}$ of Theorem 8 is unique in the sense that if μ' is a complete measure on a σ-ring $\mathcal{S}' \supseteq \mathcal{S}$ and $\mu' = \mu$ on $\mathcal{S}$ then $\mu' = \bar{\mu}$ on $\bar{\mathcal{S}}$.

Solution: Since μ' is complete it is easily seen that $\mathcal{S}' \supseteq \bar{\mathcal{S}}$. For $D \in \bar{\mathcal{S}}$ we have as above $D = F \cup A$; F, A disjoint sets with $F \in \mathcal{S}$, $A \subseteq M \in \mathcal{S}$ with $\mu(M) = 0$. So

$$\mu'(D) = \mu'(F) + \mu'(A) = \mu(F) = \bar{\mu}(D).$$

We call $\bar{\mu}$ on $\bar{\mathcal{S}}$ the **completion** of μ on $\mathcal{S}$

Theorem 9: The completion of a σ-finite measure is σ-finite.

Proof: Let $D \in \bar{\mathcal{S}}$. As in Theorem 8, $D = F \cup A$ where $F \in \mathcal{S}$ and $\bar{\mu}(A) = 0$. So $F = \bigcup_{i=1}^{\infty} F_i$ where $\mu(F_i) < \infty$, and hence $D = A \cup \bigcup_{i=1}^{\infty} F_i$ is a countable union of sets of finite $\bar{\mu}$-measure. □

Exercise

8. Let $\bar{\mu}$ on $\bar{\mathcal{S}}$ be the completion of μ on $\mathcal{S}$. Show that, if $D \in \bar{\mathcal{S}}$, there exists $B \in \mathcal{S}$ such that $\bar{\mu}(D \,\Delta\, B) = 0$.

5.5 MEASURE SPACES

In Chapter 2 we started with the ring of finite unions of intervals of the form $[a, b)$ and obtained the σ-ring of measurable sets. In that case the σ-ring obtained was a σ-algebra. Since a σ-algebra is the most frequently occurring case we restrict ourselves to these in what follows. Some definitions are somewhat simpler for σ-algebras, for example that of measurability of functions. For an account covering the general case see for example [5].

Definition 8: A pair $[\![X, \mathcal{S}]\!]$ where $\mathcal{S}$ is a σ-algebra of subsets of a space X, is called a **measurable space**. The sets of $\mathcal{S}$ are called measurable sets.

Definition 9: A triple $[\![X, \mathcal{S}, \mu]\!]$ is called a **measure space** if $[\![X, \mathcal{S}]\!]$ is a measurable space and μ is a measure on $\mathcal{S}$.

Example 10: $[\![\mathsf{R}, \mathcal{M}, m]\!]$ and $[\![\mathsf{R}, \mathcal{B}, m]\!]$ are measure spaces, where $\mathcal{B}$ denotes the Borel sets, and where in the second example m is restricted to $\mathcal{B}$.

In the latter case m is called Borel measure on the real line.

Example 11: Let $[\![X, \mathcal{S}]\!]$ be a measurable space and let $Y \in \mathcal{S}$. Then if $\mathcal{S}' = [B \cap Y : B \in \mathcal{S}]$ we have that $[\![Y, \mathcal{S}']\!]$ is a measurable space.

In the remainder of this chapter, unless stated otherwise, we will deal with a fixed measure space $[\![X, \mathcal{S}, \mu]\!]$. Many of the definitions and results of Chapter 2 apply in general, with only changes of notation. We quote these for reference.

Theorem 10: Let $\{E_i\}$ be a sequence of measurable sets. We have

(i) if $E_1 \subseteq E_2 \subseteq \ldots$, then $\mu\left(\bigcup_{n=1}^{\infty} E_n\right) = \lim \mu(E_n)$.

(ii) if $E_1 \supseteq E_2 \supseteq \ldots$ and $\mu(E_1) < \infty$, then $\mu\left(\bigcap_{n=1}^{\infty} E_n\right) = \lim \mu(E_n)$.

Proof: See Theorem 9, p. 33. □

Definition 10: Let f be an extended real-valued function defined on X. Then f is said to be **measurable** if $\forall\, \alpha$, $[x: f(x) > \alpha] \in \mathcal{S}$.

Measurability of functions is usually associated with a measure though, strictly, only X and $\mathcal{S}$ are involved.

Example 12: Let $[\![X, \mathcal{S}]\!]$ be a measurable space and let $X = \bigcup_{n=1}^{\infty} X_n$ where, for each n, $X_n \in \mathcal{S}$ and $X_n \cap X_m = \emptyset$ for $n \neq m$. Write $\mathcal{S}_n = [B \cap X_n : B \in \mathcal{S}]$. Show that f is measurable with respect to $[\![X, \mathcal{S}]\!]$ only if, for each n, its restriction f_n to X_n is measurable with respect to $[\![X_n, \mathcal{S}_n]\!]$, and conversely if, for each n, the functions f_n are measurable with respect to $[\![X_n, \mathcal{S}_n]\!]$ and f is defined by $f(x) = f_n(x)$ when $x \in X_n$, then f is measurable with respect to $[\![X, \mathcal{S}]\!]$.

Solution: For each α, $[x: f_n(x) > \alpha] = [x: f(x) > \alpha] \cap X_n$ so f_n is measurable with respect to the measurable space $[\![X_n, \mathcal{S}_n]\!]$. The converse follows from:

$$[x: f(x) > \alpha] = \bigcup_{n=1}^{\infty} [x: f_n(x) > \alpha].$$

Theorem 11: The measurability of f is equivalent to

(i) $\forall\, \alpha$, $[f(x) \geqslant \alpha] \in \mathcal{S}$,
(ii) $\forall\, \alpha$, $[x: f(x) < \alpha] \in \mathcal{S}$,
(iii) $\forall\, \alpha$, $[x: f(x) \leqslant \alpha] \in \mathcal{S}$.

Proof: See Theorem 12, p. 38. □

Example 13: (i) If f is measurable, then $[x: f(x) = \alpha]$ is measurable for each extended real number α; (ii) the constant functions are measurable; (iii) the characteristic function χ_A is measurable if, and only if, $A \in \mathcal{S}$; (iv) a continuous function of a measurable function is measurable (cf. Exercise 35, p. 42).

Theorem 12: If c is a real number and f, g measurable functions, then $f + c$, cf, $f + g$, $g - f$ and fg are also measurable.

Proof: See Theorem 13, p. 39. □

Theorem 13: If f_i is measurable, $i = 1, 2, \ldots$, then $\sup_{1 \leqslant i \leqslant n} f_i$, $\inf_{1 \leqslant i \leqslant n} f_i$, $\sup f_n$, $\inf f_n$, $\limsup f_n$ and $\liminf f_n$ are also measurable.

Proof: See Theorem 14, p. 39. □

Definition 11: If a property holds except on a measurable set E such that $\mu(E) = 0$, we say that it holds **almost everywhere** with respect to μ, written a.e. (μ). Reference to μ may be omitted if it is obvious which measure is being considered.

Example 14: The limit of a pointwise convergent sequence of measurable functions is measurable.

Example 15: Let $f = g$ a.e.(μ), where μ is a complete measure. Show that if f is measurable, so is g.

Solution: Write $E = [x: g(x) > \alpha]$, $E_1 = [x: f(x) > \alpha]$, $E_2 = [x: f(x) \neq g(x)]$. Then E_1 and E_2 are measurable and, as μ is complete, so is $E \cap E_2$. So $E = (E_1 - E_2) \cup (E \cap E_2)$ is measurable.

We define ess sup f, ess inf f, and **essentially bounded** as in Definitions 10, p. 40, 11, p. 41, and 12, p. 41, and the properties shown there hold in general.

Exercises

9. Let $\{a_n\}$ be a sequence of non-negative numbers and for $A \subseteq \mathsf{N}$ let $\mu(A) = \sum_{n \in A} a_n$. Show that $[\![\mathsf{N}, \mathcal{P}(\mathsf{N}), \mu]\!]$ is a measure space. Show also that the measure μ is complete, and if $a_n < \infty$, for each n, it is σ-finite.
10. Let $[\![X_n, \mathcal{S}_n]\!]$ be a sequence of measurable spaces, where the X_n are disjoint subsets of a space X. Show that $[\![Y, \mathcal{S}]\!]$ is a measure space where $Y = \bigcup_{n=1}^{\infty} X_n$ and $\mathcal{S} = \left[\bigcup_{n=1}^{\infty} E_n : E_n \in \mathcal{S}_n \text{ for each } n\right]$.
11. If $E \subset \mathsf{R}$ is a measurable set of measure zero which is not a Borel set (cf. Exercise 42, p. 45), is $\chi_E = 0$ a.e. with respect to Borel measure?
12. Show that if μ is not complete, then f measurable and $f = g$ a.e. do not imply g measurable.

For the following exercises we recall Definition 6, p. 33.

13. Let $\{E_n\}$ be a sequence of subsets of X and let $F \subseteq X$; show that
    (i) $F - \liminf E_n = \limsup (F - E_n)$,
    (ii) $F - \limsup E_n = \liminf (F - E_n)$.
14. Show that if χ^* and χ_* are respectively the characteristic functions of $\limsup E_n$ and $\liminf E_n$, then $\chi^* = \limsup \chi_{E_n}$ and $\chi_* = \liminf \chi_{E_n}$.
15. Let $E_n \in \mathcal{S}$, $n = 1, \ldots$ Show that
    (i) $\mu(\liminf E_n) \leqslant \liminf \mu(E_n)$,
    (ii) if $\mu(X) < \infty$ we have $\limsup \mu(E_n) \leqslant \mu(\limsup E_n)$, and that the condition $\mu(X) < \infty$ is necessary.

5.6 INTEGRATION WITH RESPECT TO A MEASURE

We now consider the generalization of the definitions and results of Chapter 3. Much of the work of Sections 3.1 and 3.2 holds for a general measure space. Where proofs need only a variation of the notation we refer to the version given for the real line.

Definition 12: A **measurable simple function** ϕ is one taking a finite number of non-negative values, each on a measurable set; so if $a_1, \ldots, a_n$ are the distinct values of ϕ, we have $\phi = \sum_{i=1}^{n} a_i \chi_{A_i}$ where $A_i = [x: \phi(x) = a_i]$. Then the **integral** of ϕ with respect to μ is given by

$$\int \phi \, \mathrm{d}\mu = \sum_{i=1}^{n} a_i \, \mu(A_i).$$

Definition 13: Let f be measurable, $f : X \to [0, \infty]$. Then the integral of f is $\int f \, \mathrm{d}\mu = \sup[\int \phi \, \mathrm{d}\mu : \phi \leqslant f, \phi$ a measurable simple function$]$.

Definition 14: Let $E \in \mathcal{S}$, and let f be a measurable function $f : E \to [0, \infty]$; then the integral of f over E is $\int_E f \, \mathrm{d}\mu = \int f \chi_E \, \mathrm{d}\mu$.

The remarks of Example 2, p. 55, are valid for Definitions 12 and 13. The analogues of Theorem 1, p. 56, and Theorem 2, p. 56, are true apart from the obvious changes in notation.

Theorem 14 (Fatou's Lemma): Let $\{f_n\}$ be a sequence of measurable functions, $f_n: X \to [0, \infty]$. Then $\liminf \int f_n \, \mathrm{d}\mu \leqslant \int \liminf f_n \, \mathrm{d}\mu$.

Proof: See Theorem 3, p. 57. □

Theorem 15 (Lebesgue's Monotone Convergence Theorem): Let $\{f_n\}$ be a sequence of measurable functions $f_n : X \to [0, \infty]$, such that $f_n(x)\uparrow$ for each x, and let $f = \lim f_n$. Then $\int f \, \mathrm{d}x = \lim \int f_n \, \mathrm{d}\mu$.

Proof. See Theorem 4, p. 57. □

Theorem 16: Let f be a measurable function, $f : X \to [0, \infty]$. Then there exists a sequence $\{\phi_n\}$ of measurable simple functions such that, for each x, $\phi_n(x) \uparrow f(x)$.

Proof: See Theorem 5, p. 58. □

Theorem 17: Let $\{f_n\}$ be a sequence of measurable functions, $f_n : X \to [0, \infty]$; then

$$\int \sum_{n=1}^{\infty} f_n \, \mathrm{d}\mu = \sum_{n=1}^{\infty} \int f_n \, \mathrm{d}\mu.$$

Proof: See Theorem 6, p. 58, and Theorem 7, p. 59. □

We now have a new result which shows how integrals can be used to construct new measures with a special continuity property.

Theorem 18: Let $[\![X, \mathcal{S}, \mu]\!]$ be a measure space and f a non-negative measurable function. Then $\phi(E) = \int_E f \, d\mu$ is a measure on the measurable space $[\![X, \mathcal{S}]\!]$. If, in addition, $\int f \, d\mu < \infty$ then $\forall \, \epsilon > 0$, $\exists \, \delta > 0$ such that, if $A \in \mathcal{S}$ and $\mu(A) < \delta$, then $\phi(A) < \epsilon$.

Proof: The function ϕ is countably additive since, if $\{E_n\}$ is a sequence of disjoint sets of $\mathcal{S}$,

$$\phi\left(\bigcup_{n=1}^{\infty} E_n\right) = \int \chi_{\cup E_n} f \, d\mu = \sum_{n=1}^{\infty} \int \chi_{E_n} f \, d\mu$$

by Theorem 17. The other properties being obvious, ϕ is a measure on $[\![X, \mathcal{S}]\!]$. Write $f_n = \min(f, n)$. Then f_n is measurable, $f_n \uparrow f$ and $\lim \int f_n \, d\mu = \int f \, d\mu$ by Theorem 15, p. 105. So if $\int f \, d\mu < \infty$, then $\forall \, \epsilon > 0$, $\exists \, N$ such that

$$\int f \, d\mu < \int f_N \, d\mu + \epsilon/2.$$

If $A \in \mathcal{S}$ and $\mu(A) < \epsilon/2N$ we have $\int_A f_N \, d\mu < \epsilon/2$. So take $\delta = \epsilon/2N$ to get

$$\begin{aligned} \int_A f \, d\mu &= \int_A (f - f_N) \, d\mu + \int_A f_N \, d\mu \\ &\leqslant \int (f - f_N) \, d\mu + \epsilon/2 < \epsilon. \; \square \end{aligned}$$

The positive and negative parts of f, f^+ and f^- respectively, have been defined in Definition 4, p. 61. We recall the properties listed in Theorem 8, p. 61.

Definition 15: If f is measurable and both $\int f^+ \, d\mu$ and $\int f^- \, d\mu$ are finite, then f is said to be **integrable**, and the integral of f is $\int f^+ \, d\mu - \int f^- \, d\mu$.

So f is integrable if, and only if, $|f|$ is. The notation $f \in L(X, \mu)$ is used to indicate that f belongs to the class of functions integrable with respect to μ. The notation $\int_E f \, d\mu$ means $\int f \chi_E \, d\mu$, where $f \in L(X, \mu)$ and $E \in \mathcal{S}$. If $f\chi_E$ is integrable we write $f \in L(E, \mu)$ or just $f \in L(E)$.

Definition 16: As in Definition 7, p. 61, we define $\int f \, d\mu = \int f^+ \, d\mu - \int f^- \, d\mu$ provided at least one of the integrals on the right-hand side is finite.

Theorem 19: Let f and g be integrable functions and let a and b be constants. Then $af + bg$ is integrable and $\int (af + bg) \, d\mu = a \int f \, d\mu + b \int g \, d\mu$. If $f = g$ a.e., then $\int f \, d\mu = \int g \, d\mu$.

Proof: See Theorem 9, p. 61. □

The results of this theorem extend to functions where the integrals are defined as in Definition 16; the argument is that of Example 10, p. 63.

Theorem 20: Let f be integrable, then $|\int f\, d\mu| \leqslant \int |f|\, d\mu$ with equality if, and only if, $f \geqslant 0$ a.e. or $f \leqslant 0$ a.e.

Proof: See Example 8, p. 62. □

The remarks after Theorem 9, p. 62, regarding functions defined a.e. still, of course, apply.

Theorem 21 (Lebesgue's Dominated Convergence Theorem): Let $\{f_n\}$ be a sequence of measurable functions such that $|f_n| \leqslant g$ where g is an integrable function, and $\lim f_n = f$ a.e. Then f is integrable, $\lim \int f_n\, d\mu = \int f\, d\mu$, and $\lim \int |f_n - f|\, d\mu = 0$.

Proof: See Theorem 10, p. 63, and Example 13, p. 63. □

A continuous parameter version of this theorem is obtained in Example 15, p. 64.

Theorem 22: Let $\{f_n\}$ be a sequence of integrable functions such that

$$\sum_{n=1}^{\infty} \int |f_n|\, d\mu < \infty.$$

Then $\sum_{n=1}^{\infty} f_n$ converges a.e., its sum, f, is integrable, and $\int f\, d\mu = \sum_{n=1}^{\infty} \int f_n\, d\mu$.

Proof: See Theorem 11, p. 64. □

Exercises

16. Let $E_1, \ldots, E_k$ be measurable sets and let F_j $(j = 1, \ldots, k)$ be the sets of points belonging to precisely j of the E_i. Show that

$$\sum_{i=1}^{k} \mu(E_i) = \sum_{j=1}^{k} j\mu(F_j).$$

17. Let g be a measurable function such that $g \geqslant h \in L(X, \mu)$. Then $\int g\, d\mu$ exists in the sense of Definition 16, p. 106.
18. If $f \in L(X, \mu)$ and g is a measurable function such that $|g| \leqslant k|f|$ a.e., where k is a constant, then $g \in L(X, \mu)$.
19. Let E and F be measurable sets, $f \in L(E)$ and $\mu(E \,\Delta\, F) = 0$ then $f \in L(F)$ and $\int_E f\, d\mu = \int_F f\, d\mu$.
20. (Tchebychev's inequality). Let f be a measurable function and let $A = [x: f(x) \geqslant 0]$. Then for $c > 0$, $\mu[x: f(x) > c] \leqslant c^{-1} \int_A f\, d\mu$.

21. If $f \in L(X, \mu)$, then the set $[x: f(x) \neq 0]$ has σ-finite measure.
22. Let $f_n \in L(X, \mu)$, $n = 1, 2, \ldots$ and let $g = \limsup f_n$, $h = \liminf f_n$. Then the sets $G = [x: g(x) \neq 0]$ and $H = [x: h(x) \neq 0]$ have σ-finite measure.
23. Let $[\![X, \mathcal{S}, \mu]\!]$ be a measure space with $\mathcal{S} = \mathcal{P}(X)$ and $\mu([x]) = 1$ for each $x \in X$. Show that $f \in L(X, \mu)$ iff $f = 0$ except on a sequence $\{x_i\}$, $\int f \, d\mu = \sum_{i=1}^{\infty} f(x_i)$, this sequence is absolutely convergent and the value of the integral is independent of the ordering of the sequence $\{x_i\}$.
24. Let $\{f_n\}$ be a sequence of non-negative measurable functions, let $\lim f_n = f$ and $f_n \leqslant f$ for each n. Show that $\int f \, d\mu = \lim \int f_n \, d\mu$.
25. Let $\{f_n\}$ be a sequence of measurable functions and let $\lim f_n = f$ a.e. Let $\{g_n\}$ be a sequence of integrable functions such that $|f_n| \leqslant g_n$ and $g = \lim g_n$ a.e. is integrable. Show that if $\int g \, d\mu = \lim \int g_n \, d\mu$, then we have f integrable and $\int f \, d\mu = \lim \int f_n \, d\mu$.
26. Use Fatou's Lemma, p. 105, to obtain another proof of the result of Exercise 15(i), namely that $\mu(\liminf E_n) \leqslant \liminf \mu(E_n)$.
27. Let E_n, $n = 1, 2, \ldots$, be measurable sets such that $\Sigma\, \mu(E_n) < \infty$. Show that x belongs to only finitely many E_n, for almost all $x \in X$.
28. Let f be integrable and let $E_n = [x: f(x) \geqslant n\lambda]$, where $\lambda > 0$, for $n = 1, 2, \ldots$ Show that (i) $\lim_{n \to \infty} \int_{E_n} f \, d\mu = 0$; and (ii) $\mu(E_n) = o(n^{-1})$.
29. Let f be a function integrable with respect to Lebesgue measure, f: $\mathsf{R} \to [0, \infty]$. Write $F_n = [x: f(x/n) \geqslant n]$. Show that for each x outside a set of measure zero there exists a strictly increasing sequence $\{n_i\}$ such that $x \in \mathsf{c}\bigcup_{i=1}^{\infty} F_{n_i}$.
30. Use Theorem 18 to give an alternative proof of the result of Chapter 4, p. 88: if f is measurable on $[a,b]$ and $\int_a^y f \, dx = 0$ for $a \leqslant y \leqslant b$, then $f = 0$ a.e. in (a, b).
31. Let f be measurable and let $\mu(X) < \infty$; then f is integrable iff the series $\sum_{n=1}^{\infty} \mu[x: |f| \geqslant n]$ converges. Give the corresponding statement if $\mu(X) = \infty$ or if the summation is from $n = 0$.
32. Let $\mu(X)$ be finite and f a measurable function. Show that if $\lim \int f^n \, d\mu$ exists and is finite, then it equals $\mu[x: f(x) = 1]$.
33. Let E be a measurable set and let $f \in L(E)$, $f > 0$. Show that

$$\lim \int_E f^{1/n} \, d\mu = \mu(E).$$

CHAPTER 6

Inequalities and the L^p Spaces

In this chapter we change our point of view and regard suitable classes of integrable functions as spaces in their own right. It is this spproach which distinguishes analysis in this century from that in the last. The inequalities developed in Sections 6.3, 6.4 are important in examining the properties of these spaces and are also useful as computational tools; both aspects are kept in mind in what follows. The results have applications, for instance to the theory of Fourier series and to the representations of linear functionals on spaces of functions. The application to the Fourier transform will be examined in Section 10.4, and linear functionals are considered in Chapters 8 and 9.

6.1 THE L^p SPACES

Definition 1: If $[\![X, \mathcal{S}, \mu]\!]$ is a measure space and $p > 0$, we define $L^p(X, \mu)$, or more briefly $L^p(\mu)$, to be the class of measurable functions $[f\colon \int |f|^p \, d\mu < \infty]$, with the convention that any two functions equal almost everywhere specify the same element of $L^p(\mu)$. On the real line, if $X = (a, b)$ and μ is Lebesgue measure we will write $L^p(a, b)$ for the corresponding space.

Strictly, the elements of the space $L^p(\mu)$ are not functions but classes of functions such that in each class any two functions are equal almost everywhere. For example, the zero element of $L^p(\mu)$ is $[f\colon f = 0 \text{ a.e.}]$. Since any two functions equal almost everywhere have the same integrals over each set of $\mathcal{S}$, the distinction is not important for many purposes. We will write $f \in L^p(\mu)$ as an abbreviation for: f is measurable and $\int |f|^p \, d\mu < \infty$. To ask, however, for the value of an element of $L^p(\mu)$ at a particular point is, in general, meaningless. If $p = 1$, we obtain the integrable functions which we denoted by $L(X, \mu)$ in the last chapter. We will use the alternative notation $L^1(\mu)$ if we wish to emphasize that the above convention applies.

Definition 2: Let $f \in L^p(\mu)$, then the L^p**-norm** of f, denoted by $\|f\|_p$, is given by $(\int |f|^p \, d\mu)^{1/p}$.

Clearly for f and g measurable and $f = g$ a.e. we have $\|f\|_p = \|g\|_p$, so the norm can be considered as that of an element of $L^p(\mu)$. Also $\|f\|_p = 0$ if, and only if, f is the zero element of $L^p(\mu)$, and $\|af\| = |a| \cdot \|f\|$ if a is a constant. The use of the term *norm* in Definition 2 is justified below in Theorem 8, p. 115, where the L^p-norm $(p \geqslant 1)$ is shown to have the other properties of a norm in the linear space sense.

Theorem 1: Let $f, g \in L^p(\mu)$ and let a, b be constants; then $af + bg \in L^p(\mu)$.

Proof: Clearly, if $f \in L^p(\mu)$, then $af \in L^p(\mu)$ for each constant a. Also if $f, g \in L^p(\mu)$, we have $f + g \in L^p(\mu)$ since

$$|f+g|^p \leqslant 2^p \max(|f|^p, |g|^p) \leqslant 2^p (|f|^p + |g|^p),$$

giving the result. □

If F is the element of $L^p(\mu)$ containing the function f and G that containing g, then we define $aF + bG$ as the element containing $af + bg$; this is easily seen to be independent of the particular $f \in F$ and $g \in G$. Hence Theorem 1 shows that $L^p(\mu)$ is a vector space. We may use, accordingly, the same notation for elements of $L^p(\mu)$ and for the functions of which they are made up.

Definition 3: If $[\![X, \mathcal{S}, \mu]\!]$ is a measure space, we define $L^\infty(X, \mu)$, or just $L^\infty(\mu)$, to be the class of measurable functions $[f\text{: ess sup } |f| < \infty]$, with the same convention as in Definition 1. Corresponding to Definition 2 we have the L^∞-norm: $\|f\|_\infty = \text{ess sup } |f|$.

Example 1: Show that $L^\infty(X, \mu)$ is a vector space over the real numbers.

Solution: Ess sup $|af + bg| \leqslant |a|$ ess sup $|f| + |b|$ ess sup $|g|$.

Example 2: Show that if $\mu(X) < \infty$ and $0 < p < q \leqslant \infty$, then $L^q(\mu) \subseteq L^p(\mu)$.

Solution: (i) For $q < \infty$, let $f \in L^q(\mu)$; then since $|f|^p \leqslant 1 + |f|^q$, an integrable function, we have $f \in L^p(\mu)$.

(ii) $q = \infty$: $|f|^p \leqslant (\text{ess sup } |f|)^p$ a.e., and so is integrable.

Exercises

1. Show that if $f, g \in L^1(\mu)$, then $|f^2 + g^2|^{1/2} \in L^1(\mu)$.
2. Show that if $p > 0$ and $0 < a < \infty$, then
   (i) $x^{-1/p} \in L^{p-\sigma}(0, a)$ if $0 < \sigma < p$ but not if $\sigma = 0$.
   (ii) $x^{-1/p} (\log 1/x)^{-2/p} \in L^{p+\sigma}(0, a)$ if $\sigma = 0$ but not if $\sigma > 0$.
3. Show that if $0 < a < \infty$ and $0 < p < \infty$, then $\log x^{-1} \in L^p(0, a)$.
4. Show that if $0 < a < \infty$, then $e^{1/x} \notin L^p(0, a)$ for any p $(0 < p \leqslant \infty)$.
5. Show that $x^{-1/2} (1 + |\log x|)^{-1} \in L^p(0, \infty)$ if $p = 2$, but not otherwise.
6. Let f: $X \to [0, \infty)$, f a measurable, essentially bounded function. Show that if $0 < \mu(X) < \infty$ and $I_n = (\int f^n \, d\mu)^{1/n}$, then $\lim I_n = \text{ess sup } f$.

6.2 CONVEX FUNCTIONS

In this section we examine a special class of functions with a view to applications in the following sections. We will use the convention that capital letters indicate points on the graph of a function, so that if ψ is defined on (a, b) and $t \in (a, b)$ then T is the point $(t, \psi(t))$.

Definition 4: A function ψ defined on an open interval (a, b) is **convex** if for any non-negative numbers λ, μ such that $\lambda + \mu = 1$, and x, y such that $a < x < y < b$, we have

$$\psi(\lambda x + \mu y) \leqslant \lambda\psi(x) + \mu\psi(y).$$

The end-points a, b can take the values $-\infty$, ∞ respectively. Geometrically, the definition says that the segment joining the points X and Y is never below the graph of ψ. If, for all positive numbers λ, μ such that $\lambda + \mu = 1$ we have

$$\psi(\lambda x + \mu y) < \lambda\psi(x) + \mu\psi(y),$$

ψ is said to be **strictly convex**.

We recall from Chapter 4, the notation $f(a, b)$ for the ratio $(f(b) - f(a))/(b - a)$.

Theorem 2: Let ψ be convex on (a, b) and $a < s < t < u < b$, then $\psi(s, t) \leqslant \psi(s, u) \leqslant \psi(t, u)$. If ψ is strictly convex, equality will not occur.

Proof: Consider the first inequality. By Definition 4

$$\psi(t) \leqslant \left(\frac{t-s}{u-s}\right)\psi(u) + \left(\frac{u-t}{u-s}\right)\psi(s). \tag{6.1}$$

So $(u - s)\,\psi(t) \leqslant (t - s)\,\psi(u) + (u - t)\,\psi(s)$, or

$$(u - s)(\psi(t) - \psi(s)) \leqslant (t - s)(\psi(u) - \psi(s)) \tag{6.2}$$

as required. If ψ is strictly convex, equality cannot occur in (6.1) and so not in (6.2). Similarly for the second inequality of the theorem. □

Theorem 3: A differentiable function ψ is convex on (a, b) if, and only if, ψ' is a monotone increasing function. If ψ'' exists on (a, b), then ψ is convex if, and only if, $\psi'' \geqslant 0$ on (a, b), and strictly convex if $\psi'' > 0$ on (a, b).

Proof: Suppose that ψ is differentiable and convex and let $a < s < t < u < v < b$. Then Theorem 2, applied first to s, t, u and then to t, u, v gives $\psi(s, t) \leqslant \psi(u, v)$. Let $t \to s$ and $u \to v$. Then, by Theorem 2, $\psi(s, t)$ decreases to $\psi'(s)$ and $\psi(u, v)$ increases to $\psi'(v)$. So $\psi'(v) \geqslant \psi'(s)$ for all $s < v$, and so ψ' is monotone increasing and if ψ'' exists, it is never negative.

Conversely, if $\psi'' \geqslant 0$, then ψ is convex, for otherwise there would exist s, t, u with $a < s < t < u < b$ and such that T lies above SU. Then slope $ST >$ slope TU; but slope $ST = \psi'(\alpha)$ for some $\alpha \in (s, t)$ and slope $TU = \psi'(\beta)$ for some $\beta \in (t, u)$, contradicting $\psi'' > 0$. The same argument shows that ψ convex and differentiable implies that ψ' is monotone increasing. If $\psi''>0$, ψ is strictly convex, for otherwise there would exist collinear points S, T, U on its graph and we would have $\psi'(\alpha)=\psi'(\beta)$ for appropriate α and β with $\alpha<\beta$. But then $\psi''=0$ at some point between α and β, giving a contradiction. □

Example 3: (i) e^x is strictly convex on R, (ii) x^α is convex on $(0,\infty)$ for $\alpha \geqslant 1$, (iii) $-x^\alpha$ is strictly convex on $(0,\infty)$ for $0<\alpha<1$, (iv) $x \log x$ is strictly convex on (0,1).

Example 4: That ψ'' may be zero for a strictly convex function ψ can be seen from x^4 at $x=0$.

Theorem 4: Let ψ be defined on (a, b). Then ψ is convex on (a, b) if, and only if, for each x and y such that $a<x<y<b$, the graph of ψ on (a, x) and (y, b) does not lie below the line through X and Y.

Proof: Suppose that ψ is convex and let $t \in (y, b)$. Then by Theorem 2 slope $XT \geqslant$ slope XY. So T lies above the line through X and Y. Similarly for $s \in (a, x)$.

Suppose, conversely, that the condition holds but that ψ is not convex. Then there exist x, y, z with $a<x<y<z<b$ and such that Y lies above XZ and hence X lies below the line through Y and Z. But this contradicts the condition of the theorem. □

Theorem 5: Every function convex on an open interval is continuous.

Proof: Let ψ be convex on (a, b) and let $x_0 \in (a, b)$; we wish to show that ψ is continuous at x_0. Choose s, t, u such that $a<s<x_0<t<u<b$. Let $y=f_1(x)$ be the equation of the line through S and X_0, and $y=f_2(x)$ that of the line through X_0 and U. By Theorem 4, $\psi(t) \geqslant f_1(t)$, and as ψ is convex $\psi(t) \leqslant f_2(t)$. Letting t tend to x_0, we get that $\psi(x_0+)$ exists and

$$\psi(x_0)=f_1(x_0+)=f_2(x_0+)=\psi(x_0+).$$

A similar argument shows that $\psi(x_0)=\psi(x_0-)$, and so ψ is continuous. □

If, in Definition 4, we had not specified an *open* interval, Theorem 5 would not hold, for consider ψ defined by: $\psi=0$ on $[0,1)$, $\psi(1)=1$.

Example 5: A function is sometimes said to be convex on (a, b) if, for $x, y \in (a, b)$

$$f\left(\frac{x+y}{2}\right) \leqslant \tfrac{1}{2} f(x)+\tfrac{1}{2} f(y).$$

We will call such functions mid-point convex. Show that the class of functions continuous and convex in the mid-point sense is just the class of convex functions in the sense of Definition 4.

Solution: Since every convex function is mid-point convex and continuous, we need only prove the converse. Let f be continuous and mid-point convex and suppose that f is not convex. Then there exists x, y, z such that $a < x < y < z < b$ and such that Y lies above XZ. Define k and m by $k = \inf[t: S \text{ lies above } XZ \text{ for } t < s \leqslant y]$, $m = \sup[t: S \text{ lies above } XZ \text{ for } y \leqslant s < t]$. By the continuity of f we have $k < y < m$, also that K and M lie on XZ and that S lies above KM for all $s \in (k, m)$. So if $r = \frac{1}{2}(k + m)$, R lies above KM; but f is mid-point convex, giving a contradiction.

That there exist functions mid-point convex but not continuous is shown in, for example, [6].

6.3 JENSEN'S INEQUALITY

Theorem 6 (Jensen's Inequality): Let $[\![X, \mathcal{S}, \mu]\!]$ be a measure space with $\mu(X) = 1$. If ψ is convex on (a, b) where $-\infty < a < b < \infty$, and f is a measurable function such that $a < f(x) < b$, for all x, then

$$\psi(\int f \,d\mu) \leqslant \int \psi o f \,d\mu. \tag{6.3}$$

Proof: It is clear that f is integrable; put $t = \int f \,d\mu$. So $a < t < b$ since $\mu(X) = 1$. Let $\beta = \sup[\psi(x, t): x \in (a, t)]$. Then, clearly, if $s \in (a, t)$ we have $\beta(t - s) \geqslant \psi(t) - \psi(s)$. But by Theorem 2, p. 111, for $u \in (t, b)$ we have $\beta \leqslant (\psi(u) - \psi(t))/(u - t)$; so $\beta(t - u) \geqslant \psi(t) - \psi(u)$ for $u \in [t, b)$. So for $\gamma \in (a, b)$

$$\psi(\gamma) \geqslant \psi(t) + \beta(\gamma - t). \tag{6.4}$$

Put $\gamma = f(x)$ to get, for each x,

$$(\psi \circ f)(x) \geqslant \psi(t) + \beta(f(x) - t)). \tag{6.5}$$

Now f is measurable and ψ is continuous, so $\psi \circ f$ is measurable (Exercise 35, p. 42). But the right-hand side of (6.5) is integrable, so (cf. Exercise 17, p. 107) $\int \psi \circ f \,d\mu$ exists. So integrate both sides of (6.5), and note the value of t to get the inequality (6.3). □

Example 6: Let ψ be strictly convex; then equality occurs in (6.3) if, and only if, $f = \int f \,d\mu$ a.e.

Solution: If $f = \int f \,d\mu$ a.e., then equality obviously occurs, so consider the converse. If ψ is strictly convex, equality occurs in (6.4) only if $\gamma = t$. For let $\gamma \in (a, t)$ and, taking β as before, let $\nu \in (\gamma, t)$; then $\psi(\gamma, t) < \psi(\nu, t) \leqslant \beta$, and if $\gamma \in (t, b)$, let $u \in (t, \gamma)$, so $\beta \leqslant \psi(t, u) < \psi(t, \gamma)$. So for equality, t must equal γ. Now equality occurs in (6.3) only if in (6.5) equality occurs almost everywhere, that is, only if $f(x) = t = \int f \,d\mu$ a.e.

Example 7: Let $X = [x_1, \ldots, x_n]$, $\mathcal{S} = \mathcal{P}(X)$, $\mu([x_i]) = \alpha_i \geqslant 0$; then Jensen's inequality reads:

$$\psi\left(\sum_{i=1}^{n} \alpha_i x_i\right) \leqslant \sum_{i=1}^{n} \alpha_i \psi(x_i),$$

where $\sum_{i=1}^{n} \alpha_i = 1$ and ψ is any convex function. We may describe this as the discrete case of Jensen's inequality.

Example 8: Let $g: X \to (0, \infty)$ and let $\log g$ be integrable with respect to μ, where $\mu(X) = 1$. Show that

$$\exp(\int \log g \, d\mu) \leqslant \int g \, d\mu. \tag{6.6}$$

Solution: Since $\log g$ is measurable and e^x is a continuous function of x we have g measurable. We may suppose that g is integrable, otherwise there is nothing to prove. Note that the result of Theorem 6 holds if f and $\psi \circ f$ are integrable, even if a and b are infinite. For, in the proof, t and β are finite and the argument goes through as before. Then (6.6) follows on putting $\psi(x) = e^x$ and $f = \log g$.

Example 9: Let $a > 0$, $b > 0$, $1/p + 1/q = 1$ where $p > 1$ and $q > 1$. Show that

$$a^{1/p} b^{1/q} \leqslant \frac{a}{p} + \frac{b}{q}$$

with equality if, and only if, $a = b$.

Solution: e^x is strictly convex by Theorem 3, p. 111, so $\exp(1/p \log a + 1/q \log b) \leqslant a/p + b/q$, with equality if, and only if, $\log a = \log b$, that is, $a = b$, giving the result.

Exercises

7. If ψ is strictly convex, then in the discrete case of Jensen's inequality (Example 7), equality occurs iff all numbers x_i, for which $\alpha_i \neq 0$, are equal.
8. (Arithmetic–Geometric Mean Inequality). If $a_i > 0$, $i = 1, \ldots, n$, then

$$\left(\prod_{i=1}^{n} a_i\right)^{1/n} \leqslant \frac{1}{n} \sum_{i=1}^{n} a_i,$$

with equality iff all the a_i are equal.

9. If $\alpha_i > 0$, $y_i > 0$, for $i = 1, \ldots, n$ and $\sum_{i=1}^{n} \alpha_i = 1$, then

$$\prod_{i=1}^{n} y_i^{\alpha_i} \leqslant \sum_{i=1}^{n} \alpha_i y_i,$$

with equality iff all the y_i are equal.

10. Suppose that ψ is a function on R such that $\psi \circ f$ is integrable on $[0,1]$ and

$$\psi\left(\int_0^1 f \, dx\right) \leqslant \int_0^1 \psi \circ f \, dx$$

for every bounded measurable function f. Show that ψ is convex.

6.4 THE INEQUALITIES OF HÖLDER AND MINKOWSKI

Theorem 7 (Hölder's Inequality): Let $1 < p < \infty$, $1 < q < \infty$, $1/p + 1/q = 1$ and let $f \in L^p(\mu)$, $g \in L^q(\mu)$. Then $fg \in L^1(\mu)$ and

$$\int |fg| \, d\mu \leqslant \left(\int |f|^p \, d\mu\right)^{1/p} \cdot \left(\int |g|^q \, d\mu\right)^{1/q}. \tag{6.7}$$

Proof: By Example 9, if $a > 0$, $b > 0$,

$$a^{1/p} \, b^{1/q} \leqslant \frac{a}{p} + \frac{b}{q}. \tag{6.8}$$

Now, if $\|f\|_p = 0$ or $\|g\|_q = 0$ then $fg = 0$ a.e. and (6.7) is trivial. If $\|f\|_p > 0$ and $\|g\|_q > 0$ write

$$a = \frac{|f|^p}{(\|f\|_p)^p}, \; b = \frac{|g|^q}{(\|g\|_q)^q}$$

in (6.8), to get

$$\frac{|fg|}{\|f\|_p \, \|g\|_q} \leqslant \frac{1}{p} \frac{|f|^p}{(\|f\|_p)^p} + \frac{1}{q} \frac{|g|^q}{(\|g\|_q)^q}. \tag{6.9}$$

The right-hand side is integrable, so $fg \in L^1(\mu)$. Integrate both sides to get $\|fg\|_1 \leqslant \|f\|_p \, \|g\|_q$, which is (6.7). □

We will refer to the numbers p and q related as in Theorem 7 as conjugate indices. The most important special case of Theorem 7 occurs when $p = q = 2$, and is called the Schwarz or Cauchy–Schwarz inequality.

Example 10: Let f and g be non-negative measurable functions. Show that equality occurs in Hölder's inequality if, and only if,

$$s f^p + t g^q = 0 \text{ a.e.} \tag{6.10}$$

for some constants s and t not both zero.

Solution: Suppose that equality occurs in (6.7). Then if $\|f\|_p > 0$, $\|g\|_q > 0$ we must have equality in (6.9) a.e. But in (6.8) equality implies that $a = b$ so that $f^p = \alpha g^q$ a.e. where $\alpha > 0$, giving (6.10). If, say, $\|f\|_p = 0$ then $f = 0$ a.e. and (6.10) holds. Conversely if (6.10) holds we may substitute into (6.7) to eliminate f or g, and we obtain equality.

Theorem 8 (Minkowski's Inequality): Let $p \geqslant 1$ and let $f, g \in L^p(\mu)$; then

$$\left(\int |f+g|^p \, d\mu\right)^{1/p} \leqslant \left(\int |f|^p \, d\mu\right)^{1/p} + \left(\int |g|^p \, d\mu\right)^{1/p}. \tag{6.11}$$

Proof: The case $p = 1$ is trivial. So suppose that $p > 1$ and that p and q are conjugate indices. Then

$$\begin{aligned}(\|f+g\|_p)^p &= \int |f+g|^p \, d\mu \\ &\leqslant \int |f| \cdot |f+g|^{p-1} \, d\mu + \int |g| \cdot |f+g|^{p-1} \, d\mu \\ &\leqslant \|f\|_p \, \|(f+g)^{p-1}\|_q + \|g\|_p \, \|(f+g)^{p-1}\|_q \end{aligned} \tag{6.12}$$

by Hölder's inequality. But $(p-1)q = p$, so the right-hand side of (6.12) equals $(\|f\|_p + \|g\|_p)(\|f+g\|_p)^{p/q}$. So $\|f+g\|_p \leqslant \|f\|_p + \|g\|_p$ as required. □

Example 11: Show that equality occurs in Minkowski's inequality for $p = 1$ if, and only if, we have almost everywhere either $f(x) \cdot g(x) = 0$ or $sgn\, f(x) = sgn\, g(x)$; for $p > 1$ if, and only if, $sf = tg$ a.e., where s and t are non-negative constants, not both zero.

Solution: (i) $p = 1$: we have $\int (|f| + |g| - |f+g|)\, d\mu \geqslant 0$ with equality if, and only if, $|f| + |g| = |f+g|$ a.e., so the condition is necessary and sufficient.

(ii) $p > 1$: The condition is seen to be sufficient on substitution. Conversely, for equality to occur in (6.11) we must have equality in (6.12). Then, outside a set of measure zero we have, for some a, b, c, d

$$a|f| = b|f+g|^{p-1} \text{ and } c|g| = d|f+g|^{p-1}.$$

So we always have $af = \pm\, bg$ a.e. and substituting in (6.11) shows that the signs are the same, giving the result.

If p and q are conjugate indices and $q \to 1$, then $p \to \infty$. This suggests analogues of Theorems 7 and 8 for the case $p = \infty$. We recall Definition 3, p. 110.

Theorem 9: If $f \in L^1(\mu)$ and $g \in L^\infty(\mu)$, then $fg \in L^1(\mu)$ and $\|fg\|_1 \leqslant \|f\|_1\, \|g\|_\infty$.

Proof: Since $|g| \leqslant \text{ess sup}\, |g|$ a.e. we have $|fg| \leqslant |f| \cdot \|g\|_\infty$ a.e. So fg is integrable and on integrating we get the result. □

Example 12: $\|f+g\|_\infty \leqslant \|f\|_\infty + \|g\|_\infty$.

Solution: This follows immediately from Example 17, p. 41.

Example 13: If we write $\rho(f, g) = \|f-g\|_p$, then for $p \geqslant 1$, ρ is a metric on $L^p(\mu)$ that is,

(i) $\rho(a, b) \geqslant 0$, (ii) $\rho(a, b) = 0$ if, and only if, $a = b$,
(iii) $\rho(a, b) = \rho(b, a)$, (iv) $\rho(a, b) + \rho(b, c) \geqslant \rho(a, c)$.

Solution: (ii) holds by virtue of the convention regarding elements of $L^p(\mu)$; (iv) follows immediately from Minkowski's inequality; the remainder are obvious.

Exercises

11. Give discrete analogues of Hölder's and Minkowski's inequality, as provided in Example 7 for Jensen's inequality.
12. Show that the following inequalities are inconsistent for functions $f \in L^2(0, \pi)$

$$\int_0^\pi (f(x) - \sin x)^2\, dx \leqslant 4/9, \quad \int_0^\pi (f(x) - \cos x)^2\, dx \leqslant 1/9.$$

13. Show that $\int_0^\pi x^{-1/4} \sin x\, dx \leqslant \pi^{3/4}$.

14. Show that if $f, g \in L^1(\mu)$, then (i) $\sqrt{|fg|} \in L^1(\mu)$, (ii) $|f|^p\ |g|^q \in L^1(\mu)$ if $p, q \in (0,1), p+q=1$.
15. Let $f_n \in L^2(a, b)$, $n = 1, 2, \ldots$, let $f \in L^2(a, b)$ and let $\lim \|f_n - f\|_2 = 0$. Show that

(i) $$\int_a^b f^2\, \mathrm{d}x = \lim \int_a^b f_n^2\, \mathrm{d}x,$$

(ii) if a and b are finite, then $\int_a^t f\, \mathrm{d}x = \lim \int_a^t f_n\, \mathrm{d}x, a \leqslant t \leqslant b.$

(iii) Verify (i) and (ii) for $(a, b) = (-\pi, \pi)$ if $f_n(x) = \sum_{i=1}^{n} \frac{(-1)^{r-1}}{r} \sin rx$,

$f(x) = x/2$.
16. Let $p \geqslant 1$ and let $\|f_n - f\|_p \to 0$. Show that $\|f_n\|_p \to \|f\|_p$.
17. Let $f \geqslant 0, f \in L(x, 1)$ for each $x \in (0,1]$. Suppose that $t^{p-1}(f(t))^p \in L(0,1)$, where $p > 1$. Show that $F(x) = \int_x^1 f\, \mathrm{d}t$ satisfies $F(x) = o(\log 1/x)^{1-1/p}$ as $x \to 0+$.
18. Let $p > 1, f \geqslant 0, f \in L^p(0, \infty)$ and $F(x) = \int_0^x f\, \mathrm{d}t$. Show that if p and q are conjugate indices, then $F(x) = o(x^{1/q})$ as $x \to 0$ and as $x \to \infty$.
19. Show that if $k_1, k_2, \ldots, k_n > 1$ and $\sum_{i=1}^{n} 1/k_i = 1$, then if $f_i \in L^{k_i}(\mu)$ for each i,

$$\int |f_1 f_2 \ldots f_n|\, \mathrm{d}\mu \leqslant \left(\int |f_1|^{k_1}\, \mathrm{d}\mu\right)^{1/k_1} \ldots \left(\int |f_n|^{k_n}\, \mathrm{d}\mu\right)^{1/k_n}.$$

20. Equality occurs in the inequality of the last exercise iff one of the $f_i = 0$ a.e., or for each pair i, j there exist non-zero constants c_i, c_j such that

$$c_i |f_i|^{k_i} = c_j |f_j|^{k_j}. \tag{6.13}$$

21. If $\alpha, \beta, \gamma > 0$ and $p < 1/(\alpha + \beta + \gamma)$, then

$$\int_0^2 \mathrm{d}x/(x^\alpha\ |x-1|^\beta\ |x-2|^\gamma)^p < \infty.$$

22. Extend Minkowski's inequality to n functions.
23. Let f be a non-negative measurable function and ess sup $f = M > 0$. Then, if $\mu(X) < \infty$,

$$\lim \frac{\int f^{n+1}\, \mathrm{d}\mu}{\int f^n\, \mathrm{d}\mu} = M.$$

24. The limit considered in the last exercise was that of an increasing sequence.
25. Show that if $f \in L^{p_1}(\mu)$ and $g \in L^{p_2}(\mu)$ where p_1 and p_2 are positive, then $fg \in L^p(\mu)$ for a suitable p.

26. Find the minimum of

$$P_f = \left(\int_E f \, d\mu\right) \cdot \left(\int_E 1/f \, d\mu\right)$$

taken over the set of functions measurable and positive a.e. in the measurable set E of finite measure. For which functions is the minimum value obtained? Show that P_f is not bounded above for a suitable space and set E.

27. Let f and g be integrable and essentially bounded, and let $0 < p = \text{ess inf} f \leqslant \text{ess sup} f = P$, $0 < q = \text{ess inf} g \leqslant \text{ess sup} g \leqslant Q$. Show that

$$(\int fg \, d\mu)^2 \left(\sqrt{\frac{PQ}{pq}} + \sqrt{\frac{pq}{PQ}}\right)^2 \geqslant 4 \int f^2 \, d\mu \cdot \int g^2 \, d\mu.$$

28. Let $k + m = km$ and let f and g be non-negative measurable functions. Show that if $0 < k < 1$ or $k < 0$ then

$$\int fg \, d\mu \geqslant (\int f^k \, d\mu)^{1/k} (\int g^m \, d\mu)^{1/m}.$$

29. Let $0 < p < 1$ and $f \geqslant 0, g \geqslant 0, f, g \in L^p(\mu)$. Show that

$$\|f + g\|_p \geqslant \|f\|_p + \|g\|_p.$$

6.5 COMPLETENESS OF $L^p(\mu)$

The next two results show that, for $p \geqslant 1$, $L^p(\mu)$ is a complete metric space, that is, that considered as a metric space with the metric ρ of Example 13, p. 116, every Cauchy sequence converges, in the sense of the metric, to an element of the space. Although this result refers to elements of $L^p(\mu)$, the proofs depend on the choices of functions representing elements of the space following the convention discussed in Definition 1, p. 109.

Theorem 10: If $1 \leqslant p < \infty$ and $\{f_n\}$ is a sequence in $L^p(\mu)$ such that $\|f_n - f_m\|_p \to 0$ as $n, m \to \infty$, then there exists a function f and a subsequence $\{n_i\}$ such that $\lim f_{n_i} = f$ a.e. Also $f \in L^p(\mu)$ and $\lim \|f_n - f\|_p = 0$.

Proof: For each i we can choose N, depending on i, such that $\|f_n - f_m\|_p < 2^{-i}$ for $n, m \geqslant N = N(i)$. Taking $n_1 \geqslant N(1)$, $n_2 \geqslant N(2)$, . . . , we may choose by induction an increasing subsequence $\{n_i\}$ such that

$$\|f_{n_{i+1}} - f_{n_i}\|_p < 2^{-i}. \tag{6.14}$$

Put $g_k = \sum_{i=1}^{k} |f_{n_{i+1}} - f_{n_i}|$ and $g = \sum_{i=1}^{\infty} |f_{n_{i+1}} - f_{n_i}|$. Minkowski's inequality, with (6.14), gives $\|g_k\|_p < 1$. Apply Fatou's Lemma to the sequence of non-negative functions $\{g_k^p\}$ to get $(\|g\|_p)^p = \int \lim g_k^p \, d\mu \leqslant \lim \inf \int g_k^p \, d\mu \leqslant 1$. Hence g is finite a.e. and so

$$f_{n_1} + \sum_{i=1}^{\infty} (f_{n_{i+1}} - f_{n_i})$$

is absolutely convergent a.e. Define f to be the sum of this series where it converges, otherwise define f arbitrarily. Since $f_{n_1} + \sum_{i=1}^{k-1} (f_{n_{i+1}} - f_{n_i}) = f_{n_k}$, we have $f = \lim f_{n_i}$, a.e. Each function of the sequence $\{f_{n_i}\}$ is defined only up to a set of measure zero as in Definition 1, p. 109, so f is only well defined in the same sense. But this will ensure that f defines a unique element of $L^p(\mu)$.

So we now wish to show that $f \in L^p(\mu)$ and that the whole sequence $\{f_n\}$ tends to f in the desired sense. Given $\epsilon > 0$, there exists N such that $\|f_n - f_m\| < \epsilon$ for $n, m > N$. So by Fatou's Lemma, for each $m > N$,

$$\int |f - f_m|^p \, d\mu \leqslant \liminf_{i \to \infty} \int |f_{n_i} - f_m|^p \leqslant \epsilon^p. \tag{6.15}$$

So $f - f_m$ and hence $f = (f - f_m) + f_m$ are in $L^p(\mu)$. Also, from (6.15), $\|f - f_m\|_p \leqslant \epsilon$ for $m > N$, which completes the proof. □

We consider separately the case $p = \infty$, in the next theorem.

Theorem 11: Let $\{f_n\}$ be a sequence in $L^\infty(\mu)$ such that $\|f_n - f_m\|_\infty \to 0$ as $n, m \to \infty$. Then there exists a function f such that $\lim f_n = f$ a.e., $f \in L^\infty(\mu)$ and $\lim \|f_n - f\|_\infty = 0$.

Proof: We use the fact that a function is greater than its essential supremum only on a set of measure zero. Write $A_{n,m} = [x: |f_n(x) - f_m(x)| > \|f_n - f_m\|_\infty]$ and $B_n = [x: |f_n(x)| > \|f_n\|_\infty]$. Then if $E = \bigcup_{n \neq m} A_{n,m} \cup \bigcup_{k=1}^{\infty} B_k$, we have $\mu(E) = 0$ and on $\mathbf{C}E$, $\{f_n(x)\}$ is a Cauchy sequence for each x, with limit $f(x)$, say. Define f arbitrarily on E.

Given $\epsilon > 0$, there exists N such that $\|f_n - f_m\|_\infty < \epsilon$ for $n, m > N$. So, for $x \in \mathbf{C}E$, $|f_n(x) - f_m(x)| \leqslant \|f_n - f_m\|_\infty < \epsilon$, and letting $n \to \infty$, $|f(x) - f_m(x)| \leqslant \epsilon$. So $|f| \leqslant |f_m| + \epsilon$ a.e. and hence $f \in L^\infty(\mu)$; indeed by its construction a well defined element of $L^\infty(\mu)$ is obtained as in the case $0 < p < \infty$ in Theorem 10. Also $\|f - f_m\|_\infty \leqslant \epsilon$ and the result is proved. □

Exercises

30. Let $p > 0$ and $f \in L^p(\mu)$ where $f \geqslant 0$, and let $f_n = \min(f, n)$. Show that $f_n \in L^p(\mu)$ and $\lim \|f - f_n\|_p = 0$.
31. Show that if $p \geqslant 1$, the set of bounded measurable functions is dense in $L^p(\mu)$ in the sense of the metric of Example 13, p. 116.
32. Let $f \in L^p(a, b)$ with a and b finite and $p \geqslant 1$, and let $\epsilon > 0$. Show that there exist
    (i) a step function h such that $\int_a^b |f - h|^p \, dx < \epsilon$,
    (ii) a continuous function g such that g vanishes outside a bounded interval and $\int_a^b |f - g|^p \, dx < \epsilon$.

33. Show that the sets of measurable simple functions, step functions and of continuous functions are each dense in the metric space $L^p(a, b)$, $p \geqslant 1$, where a and b are finite.
34. Show that the results of the last exercise hold for $L^p(-\infty, \infty)$ for $p \geqslant 1$.
35. Let p and q be conjugate indices and let $f \in L^p(-\infty, \infty)$, $g \in L^q(-\infty, \infty)$. Show that $F(t) = \int f(x + t) g(x)\, dx$ is a continuous function of t.
36. Let p and q be conjugate indices and let $f_n \to f$ in $L^p(\mu)$, that is, $\lim \|f_n - f\|_p = 0$, and let $g_n \to g$ in $L^q(\mu)$. Show that $f_n g_n \to fg$ in $L^1(\mu)$.
37. Let $f, f_n \in L^2(\mu)$ for each n; then we say that $f_n \to f$ *weakly* if $\lim \int (f_n - f)g\, d\mu = 0$ for each $g \in L^2(\mu)$, and $\sup \|f_n\|_2 < \infty$. Show that

    (i) if $f_n \to f$ in L^2, then $f_n \to f$ weakly, but not conversely,

    (ii) if $f_n \to f$ weakly, then $\|f\|_2 \leqslant \liminf \|f_n\|_2$,

    (iii) if $X = [a, b]$ where a and b are finite, then $f_n \to f$ weakly iff $\int_a^x f_n\, dt \to \int_a^x f\, dt$ for each $x \in [a, b]$ and $\sup \|f_n\|_2 < \infty$,

    (iv) if $f_n \to f$ weakly and $\|f_n\|_2 \to \alpha \leqslant \|f\|_2$, then $f_n \to f$ in L^2.
38. Let $f_n \to f$ in $L^p(X, \mu)$ where $\mu(X) < \infty$ and $p > 1$. Show that $f_n \to f$ in $L^{p'}(X, \mu)$, $1 \leqslant p' < p$.
39. Show that if $|f_n| \leqslant K$ a.e., and if $f_n \to f$ in $L^p(X, \mu)$ where $\mu(X) < \infty$ and $p \geqslant 1$, then $f_n \to f$ in $L^{p''}$ for $1 \leqslant p'' < \infty$.

CHAPTER 7

Convergence

We now investigate systematically some forms of convergence of measurable functions. We introduce in Section 7.1 the notion of *convergence in measure*. This is of particular relevance to the theory of probability where it is often referred to as *convergence in probability*. Those results which hold in the case where $\mu(X) < \infty$ are listed separately; this is the situation encountered in probabilty theory.

7.1 CONVERGENCE IN MEASURE

We have already met convergence a.e., convergence in L^p-spaces and uniform convergence. We define in this and in the next section other forms of convergence and give theorems relating the various forms of convergence. Then in Section 7.3 these results and results from previous chapters are collected in diagrammatic form, and in Section 7.4 we give counterexamples which show which implications are not valid.

We consider throughout a measure space $[\![X, \mathcal{S}, \mu]\!]$. Any necessary conditions such as $\mu(X) < \infty$, will be imposed where necessary.

Definition 1: Let $\{f_n\}$ be a sequence of measurable functions and f a measurable function. Then f_n tends to f **in measure** if for every positive ϵ, $\lim \mu[x: |f_n(x) - f(x)| > \epsilon] = 0$.

Theorem 1: If a sequence of measurable functions converges in measure, then the limit function is unique a.e.

Proof: Let $f_n \to f$ in measure and $f_n \to g$ in measure. Since $|f-g| \leqslant |f-f_n| + |g-f_n|$, we must have, for any $\epsilon > 0$, $[x: |f(x) - g(x)| > 2\epsilon] \subseteq [x: |f(x) - f_n(x)| > \epsilon] \cup [x: |g(x) - f_n(x)| > \epsilon]$. But the measure of the set on the right-hand side tends to zero as $n \to \infty$. So $f = g$ a.e. □

Definition 2: A sequence of functions is said to be **fundamental** with respect to a particular kind of convergence if it forms a Cauchy sequence in that sense. Thus a sequence $\{f_n\}$ is fundamental in measure if for any $\epsilon > 0$, $\lim_{m,n\to\infty} \mu[x: |f_n(x) - f_m(x)| > \epsilon] = 0$.

We now prove a 'completeness' theorem for convergence in measure.

Theorem 2: If $\{f_n\}$ is a sequence of measurable functions which is fundamental in measure, then there exists a measurable function f such that $f_n \to f$ in measure.

Proof: For every integer k we can find n_k such that for $n, m \geq n_k$,

$$\mu\left[x: |f_n(x) - f_m(x)| \geq \frac{1}{2^k}\right] < \frac{1}{2^k},$$

and we may assume that for each k, $n_{k+1} > n_k$. Let

$$E_k = \left[x: |f_{n_k}(x) - f_{n_{k+1}}(x)| \geq \frac{1}{2^k}\right].$$

Then if $x \notin \bigcup_{k=m}^{\infty} E_k$, we have for $r > s \geq m$

$$|f_{n_r}(x) - f_{n_s}(x)| \leq \sum_{i=s+1}^{r} |f_{n_i}(x) - f_{n_{i-1}}(x)| < \sum_{i=s+1}^{\infty} 1/2^i = 1/2^s. \qquad (7.1)$$

So $\{f_{n_k}(x)\}$ is a Cauchy sequence for each $x \notin \limsup E_k = \bigcap_{m=1}^{\infty} \bigcup_{k=m}^{\infty} E_k$. But, for all m $\mu(\limsup E_k) \leq \mu\left(\bigcup_{k=m}^{\infty} E_k\right) \leq \sum_{k=m}^{\infty} 1/2^k = 1/2^{m-1}$. So $\{f_{n_k}\}$ converges a.e. to some measurable function f. Also from (7.1) we have that $\{f_{n_k}\}$ is uniformly fundamental in $\complement \bigcup_{k=m}^{\infty} E_k$, for each m. So $f_{n_k} \to f$ uniformly on $\complement \bigcup_{k=m}^{\infty} E_k$, and hence, for every positive ϵ,

$$\mu[x: |f_{n_k}(x) - f(x)| > \epsilon/2] \to 0 \text{ as } k \to \infty. \qquad (7.2)$$

But $[x: |f_n(x) - f(x)| > \epsilon] \subseteq$

$$[x: |f_n(x) - f_{n_k}(x)| > \epsilon/2] \cup [x: |f(x) - f_{n_k}(x)| > \epsilon/2].$$

If n and n_k are sufficiently large, the measure of the first set on the right is arbitrarily small, as $\{f_n\}$ is fundamental in measure. But the second set has been shown to have arbitrarily small measure by (7.2) and the result follows. □

Corollary: Let $f_n \to f$ in measure where f and each f_n are measurable functions. Then there exists a subsequence $\{n_i\}$ such that $f_{n_i} \to f$ a.e.

Proof: Clearly $\{f_n\}$ is fundamental in measure, so from the proof of the theorem we can find a subsequence $\{f_{n_i}\}$ and a measurable function g such that $f_{n_i} \to g$ a.e. and in measure. But $f_{n_i} \to f$ in measure, so by Theorem 1, p. 121, $f = g$ a.e., giving the result. □

An analogue of Fatou's Lemma, p. 105, holds with convergence a.e. replaced by convergence in measure.

Theorem 3: Let $\{f_n\}$ be a sequence of non-negative measurable functions and let f be a measurable function such that $f_n \to f$ in measure; then

$$\int f \, \mathrm{d}\mu \leqslant \liminf \int f_n \, \mathrm{d}\mu.$$

Proof: Suppose that $\int f \, \mathrm{d}\mu < \infty$ and that $\int f \, \mathrm{d}\mu > \liminf \int f_n \, \mathrm{d}\mu$. Then there exist $\delta > 0$ and a sequence $\{n_i\}$ such that, for each i, $\int f_{n_i} \, \mathrm{d}\mu < \int f \, \mathrm{d}\mu - \delta$. But $f_{n_i} \to f$ in measure, so by the corollary to the last theorem we can find a subsequence $\{n_i'\}$ of $\{n_i\}$ such that $f_{n_i'} \to f$ a.e. But then by Fatou's Lemma,

$$\int f \, \mathrm{d}\mu \leqslant \liminf \int f_{n_i'} \, \mathrm{d}\mu \leqslant \int f \, \mathrm{d}\mu - \delta,$$

giving a contradiction.

Now suppose that $\int f \, \mathrm{d}\mu = \infty$ and that $\liminf \int f_n \, \mathrm{d}\mu < \infty$. Then there exist $K > 0$ and a subsequence $\{f_{n_i}\}$ such that, for each i, $\int f_{n_i} \, \mathrm{d}\mu < K$. But again we find a subsequence $\{n_i'\}$ of $\{n_i\}$ such that $f_{n_i'} \to f$ a.e. But then, by Fatou's Lemma, $\liminf \int f_{n_i'} \, \mathrm{d}\mu = \infty$, giving a contradiction. So $\liminf \int f_n \, \mathrm{d}\mu = \infty$, giving the result. □

We have a corresponding analogue of the Lebesgue dominated convergence theorem, using convergence in measure.

Theorem 4: Let $\{f_n\}$ be a sequence of measurable functions such that $|f_n| < g$, an integrable function, and let $f_n \to f$ in measure, where f is measurable. Then f is integrable, $\lim \int f_n \, \mathrm{d}\mu = \int f \, \mathrm{d}\mu$ and $\lim \int |f_n - f| \, \mathrm{d}\mu = 0$.

Proof: By the corollary to Theorem 2 there exists a subsequence $\{f_{n_i}\}$ with limit f a.e., so we have $|f| \leqslant g$ and so $f \in L^1(\mu)$. Also, for each n, $g + f_n \geqslant 0$, and $g + f_n \to g + f$ in measure follows immediately from the fact that $f_n \to f$ in measure. Then by Theorem 3

$$\int g \, \mathrm{d}\mu + \int f \, \mathrm{d}\mu \leqslant \liminf \int (g + f_n) \, \mathrm{d}\mu.$$

So $\int f \, \mathrm{d}\mu \leqslant \liminf \int f_n \, \mathrm{d}\mu$.

We have, similarly, $g - f_n \geqslant 0$ and $g - f_n \to g - f$ in measure. So

$$\int g \, \mathrm{d}\mu - \int f \, \mathrm{d}\mu \leqslant \liminf \int (g - f_n) \, \mathrm{d}\mu.$$

Hence $\int f \, \mathrm{d}\mu \geqslant \limsup \int f_n \, \mathrm{d}\mu \geqslant \liminf \int f_n \, \mathrm{d}\mu \geqslant \int f \, \mathrm{d}\mu$, giving the first result. Also, it is clear from the definition of convergence in measure that $|f_n - f| \to 0$ in measure. But $|f_n - f| \leqslant 2g$, so the second result follows from the first. □

Definition 3: Convergence in $L^p(\mu)$ is often described as convergence **in the mean of order** p, that is: $f_n \to f$ in the mean of order p $(p > 0)$ if $\lim \|f_n - f\|_p = 0$. If $p = 1$, f_n is said to converge to f **in the mean**.

Theorem 5: If $f_n \to f$ in the mean of order p $(p > 0)$, then $f_n \to f$ in measure.

Proof: Suppose not. Then there exist $\epsilon > 0$, $\delta > 0$ such that $\mu[x: |f_n - f| > \epsilon] > \delta$ for infinitely many n. But then $\|f_n - f\|_p > \epsilon\, \delta^{1/p}$ for infinitely many n, giving a contradiction. □

Example 1: Let $\mu(X)$ be finite. If $n(f) = \int \frac{|f|}{1+|f|}\, d\mu$ and $\rho(f,g) = n(f-g)$, then ρ is a metric space of functions measurable and finite-valued a.e., provided that functions equal a.e. are identified. Convergence in this metric is equivalent to convergence in measure, and the metric space defined is complete.

Solution: For all a and b we have

$$\frac{|a+b|}{1+|a+b|} \leqslant \frac{|a|}{1+|a|} + \frac{|b|}{1+|b|}.$$

Put $a = f - h$, $b = h - g$ where f, g, h are measurable functions, and integrate to get $\rho(f,g) \leqslant \rho(f,h) + \rho(h,g)$. Clearly $\rho(f,g) = \rho(g,f) \geqslant 0$, $\rho(f,f) = 0$, and $\rho(f,f) = 0$ only if $f = g$ a.e. So ρ is a metric on the space of a.e. finite-valued measurable functions, if we regard any two functions equal a.e. as corresponding to the same element of the metric space. Let f_n, f be measurable functions and write $E_\epsilon = [x: |f_n(x) - f(x)| > \epsilon]$. Then for any ϵ, $0 < \epsilon < 1$,

$$\rho(f_n, f) \geqslant \tfrac{1}{2} \int_{E_\epsilon} \epsilon\, d\mu = \frac{\epsilon}{2}\, \mu(E_\epsilon).$$

So if $\rho(f_n, f) \to 0$, $\mu(E_\epsilon) \to 0$, that is: $f_n \to f$ in measure. Conversely, let $f_n \to f$ in measure. We have

$$\rho(f_n, f) \leqslant \left(\int_{E_\epsilon} + \int_{CE_\epsilon}\right) \frac{|f_n - f|}{1 + |f_n - f|}\, d\mu \leqslant \mu(E_\epsilon) + \epsilon\mu(X).$$

But $\mu(E_\epsilon) \to 0$ as $n \to \infty$, so $\rho(f_n, f) < \epsilon(1 + \mu(X))$ for all large n, and so

$$\lim \rho(f_n, f) = 0.$$

That the metric space is complete, that is: sequences that are Cauchy with respect to ρ converge, follows from Theorem 2, p. 122, since it can be seen from the proof of that theorem that the limit function f obtained will be finite-valued a.e.

Exercises

1. Show that if $f_n \to f$ in measure and $g_n \to g$ in measure, then $f_n + g_n \to f + g$ in measure.
2. Show that if $f_n \to f$ in measure and α is any real number, then $\alpha f_n \to \alpha f$ in measure.
3. Show that if $\mu(X) < \infty$ and $f_n \to f$ in measure, then $f_n^2 \to f^2$ in measure.
4. Show that if $\mu(X) < \infty$, $f_n \to f$ in measure and $g_n \to g$ in measure, then $f_n g_n \to fg$ in measure.

5. Show that the condition $\mu(X) < \infty$ is necessary for the result of the last exercise to hold.
6. Show that if $f_n \to f$ in measure, then $|f_n| \to |f|$ in measure.
7. Let S be the set of measurable functions on $[0,1)$ and let

$$U_n = [f\colon m[x\colon f(x) > 1/n] < 1/n] .$$

Then, for each n, S is the smallest convex set containing U_n.
8. Use Example 1 to make the set of all real sequences into a metric space.
9. For $\alpha \geqslant 0$ let arc tan $\alpha \in [0, \pi/2]$. Show that if

$$n(f) = \inf_{\alpha>0} \arctan(\alpha + \mu[x\colon |f(x)| > \alpha]) \text{ and } \rho(f,g) = n(f-g),$$

then ρ is a metric on the space of a.e. finite-valued measurable functions where functions equal a.e. are regarded as identical; and show that convergence with respect to ρ is equivalent to convergence in measure.

7.2 ALMOST UNIFORM CONVERGENCE

Definition 4: Let $\{f_n\}$ be a sequence of measurable functions and let f be a measurable function; then we say that f_n tends to f **almost uniformly**, and write $f_n \to f$ a.u. if, for any $\epsilon > 0$ there exists a set E with $\mu(E) < \epsilon$ and such that on $\mathbf{C}E$, $f_n \to f$ uniformly.

Theorem 6: Uniform convergence a.e. implies almost uniform convergence.

Proof: The result is obvious from the definition. □

That the converse of Theorem 6 does not hold can be seen from the sequence $\{x^n\}$ on $[0,1]$. Since $x^n \leqslant (1-\epsilon)^n$ on $[0, 1-\epsilon]$, it converges uniformly there. But $\{x^n\}$ does not converge uniformly a.e. since it does not converge uniformly in any set containing points arbitrarily close to 1.

Theorem 7: If $f_n \to f$ a.u., then $f_n \to f$ in measure.

Proof: If f_n does not tend to f in measure, there exist positive numbers ϵ and δ such that $\mu[x\colon |f_n(x) - f(x)| > \epsilon] > \delta$ for infinitely many n. But since there exists a set E, with $\mu(E) < \delta$, such that $f_n \to f$ uniformly on $\mathbf{C}E$, we get a contradiction. □

Theorem 8: If $f_n \to f$ a.u., then $f_n \to f$ a.e.

Proof: For each integer m we can find a set E_m with $\mu(E_m) < 1/m$, and on $\mathbf{C}E_m$, $f_n \to f$ uniformly. Then if $x \in \bigcup_{m=1}^{\infty} \mathbf{C}E_m$ we have $x \in \mathbf{C}E_N$, say, so $\lim f_n(x) = f(x)$. But $\mathbf{C} \bigcup_{m=1}^{\infty} \mathbf{C}E_m = \bigcap_{m=1}^{\infty} E_m$, a set of measure zero. □

Theorem 9: Let $f_n \to f$ a.e. If (i) $\mu(X) < \infty$ or (ii) for each n, $|f_n| \leqslant g$, an integrable function, then we have $f_n \to f$ a.u.

Proof: Write $E_{k,n} = \bigcap_{m=n}^{\infty} [x: |f_m(x) - f(x)| < 1/k]$. It is sufficient to prove that, for each k, $\lim_{n\to\infty} \mu(\mathbf{C}E_{k,n}) = 0$, for then if $\epsilon > 0$, $\mu(\mathbf{C}E_{k,n_k}) < \epsilon/2^k$ for an appropriate n_k. So if $E = \bigcap_{k=1}^{\infty} E_{k,n_k}$ we have $\mu(\mathbf{C}E) < \epsilon$, and on E, $|f_m - f| < 1/k$ for $m \geqslant n_k$. So $f_m \to f$ a.u.

Clearly, $[x: \lim f_m(x) = f(x)] \subseteq \bigcup_{n=1}^{\infty} E_{k,n}$, for each k, so the complementary set, $\bigcap_{n=1}^{\infty} \mathbf{C}E_{k,n}$, has measure zero. So it is sufficient to prove that $\mu(\mathbf{C}E_{k,n}) < \infty$ for some n, and for each k. For (i) this is obvious, giving the result. For case (ii) we have $|f_m - f| \leqslant 2g$. So

$$\mathbf{C}E_{k,n} = \bigcup_{m=n}^{\infty}\left[x: |f_m(x) - f(x)| \geqslant \frac{1}{k}\right] \subseteq \left[x: g(x) \geqslant \frac{1}{2k}\right].$$

But as g is integrable this is a set of finite measure and the result follows. □

Theorem 9, for the case $\mu(X) < \infty$, is usually known as Egorov's Theorem.

Example 2: Let $\{f_n\}$ be a sequence of measurable functions such that $\lim f_n = f$ uniformly, where $f \in L^p(\mu)$, and $\mu(X) < \infty$. Then $f_n \to f$ in the mean of order p $(p > 0)$.

Solution: We have $|f_n|^p \leqslant 2^p(|f|^p + 1)$ for all large n, and so $f_n \in L^p(\mu)$. But for any $\epsilon > 0$ and for all large n, $|f_n - f| < \epsilon$ and so $\int |f_n - f|^p \, d\mu < \epsilon^p \mu(X)$, giving the result.

We now prove a result resembling that of Exercise 15, p. 117, the main difference being that in that case the functions f_n were in L^2.

Example 3: Let g be a bounded function, measurable on $[0,1]$ and let $\{f_n\}$ be a sequence of functions integrable on $[0,1]$. Show that if

(i) $\left\{\int_0^1 |f_n| \, dx\right\}$ is bounded,

(ii) $\forall\, \epsilon > 0$, $\exists\, \eta > 0$ such that for each measurable subset H of $[0,1]$ with $m(H) < \eta$, we have $\left|\int_H f_n \, dx\right| < \epsilon$ for all n,

(iii) $\lim \int^u f_n \, dx = 0$ for each $u \in [0,1]$,

then $\lim \int_0^1 g f_n \, dx = 0$.

Solution: By (iii) we have $\lim \int_I f_n = 0$ for any interval I. By Theorem 10, p. 36, if A is a measurable set and $\eta > 0$, there exist intervals $I_1, \ldots, I_N$ such that $m\left(A \,\Delta \bigcup_{k=1}^{N} I_k\right) < \eta$. Then, by (ii)

$$\left| \int_A f_n \, dx - \int_E f_n \, dx \right| < 2\epsilon$$

for all n, where $E = \bigcup_{k=1}^{N} I_k$. But $\lim \int_E f_n \, dx = 0$, so

$$\lim \int_A f_n \, dx = \lim \int \chi_A f_n \, dx = 0.$$

So $\lim \int \phi f_n \, dx = 0$ where ϕ is any measurable simple function. As we may consider g^+ and g^- separately, we may suppose that $g \geqslant 0$. Then we can find a sequence $\{\phi_n\}$ of measurable simple functions, $\phi_n \uparrow g$ on $[0,1]$. So by Egorov's Theorem, $\forall\, \epsilon > 0$, $\exists\, H$ with $m(H) < \eta$, and k such that $|\phi_k - g| < \epsilon$ on $[0,1] - H$. Now

$$\int_0^1 g f_n \, dx = \int_0^1 \phi_k f_n \, dx + \int_H (g - \phi_k) f_n \, dx + \int_{[0,1]-H} (g - \phi_k) f_n \, dx.$$

But $\left| \int_{[0,1]-H} (g - \phi_k) f_n \, dx \right| < \epsilon \int_0^1 |f_n| \, dx \leqslant M\epsilon$, say, by (i) for all n. Now choose n so that $\left| \int_0^1 \phi_k f_n \, dx \right| < \epsilon$. If $g \leqslant N$ on $[0,1]$, we have

$$\left| \int_H (g - \phi_k) f_n \, dx \right| \leqslant N \int_H |f_n| \, dx \leqslant 2N\epsilon,$$

by (ii), giving the result.

Exercises

10. Let $\{f_n\}$ be a sequence of measurable functions, $f_n \in L^1(\mu)$, let $g \in L^\infty(\mu)$, let $|f_n| \leqslant F \in L^1(\mu)$ for each n and let $\lim f_n = f$ a.e. Then $fg \in L^1(\mu)$, $\int f_n g \, d\mu \to \int fg \, d\mu$ and $f_n g \to fg$ in the mean.
11. Let the sequence of measurable functions $\{f_n\}$ be almost uniformly fundamental. Then there exists a measurable function f such that $f_n \to f$ a.u.

7.3 CONVERGENCE DIAGRAMS

In this third and in the next section we set out the relationships between six important kinds of convergence of sequences of measurable functions. In place of uniform convergence we could have chosen uniform convergence a.e., for which the same implications would have held. The kinds of convergence considered are

1. convergence a.e.
2. convergence in the mean
3. uniform convergence
4. convergence in the mean of order p ($p > 0$)
5. almost uniform convergence
6. convergence in measure.

We consider the following cases

(α) no restriction on $[\![X, \mathcal{S}, \mu]\!]$,
(β) $\mu(X) < \infty$,
(γ) the sequence $\{f_n\}$ is dominated by an integrable function, that is, for some $g \in L^1(\mu)$, $|f_n| \leqslant g$ for each n.

The relations between these kinds of convergence, in the three cases, are given in Figs 7.1 to 7.3 in which arrows denote implication. The diagrams are to be understood in the following sense: (4) ⇒ (6) means that for any given $p > 0$,

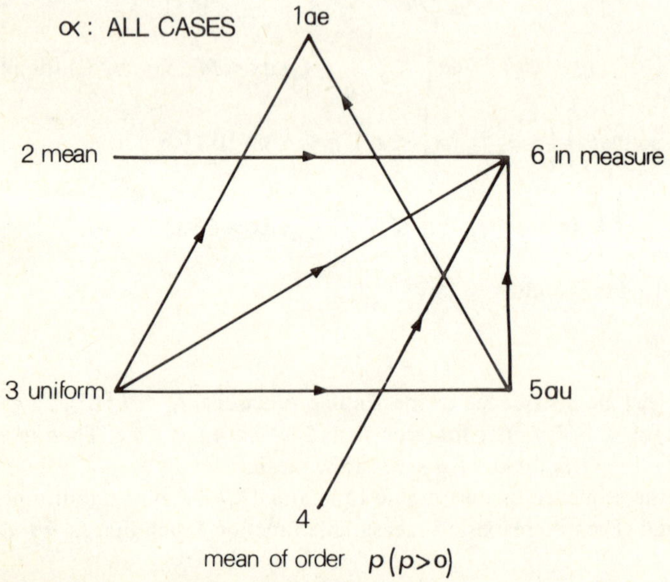

Figure 7.1

$\beta: \quad \mu(X) < \infty$

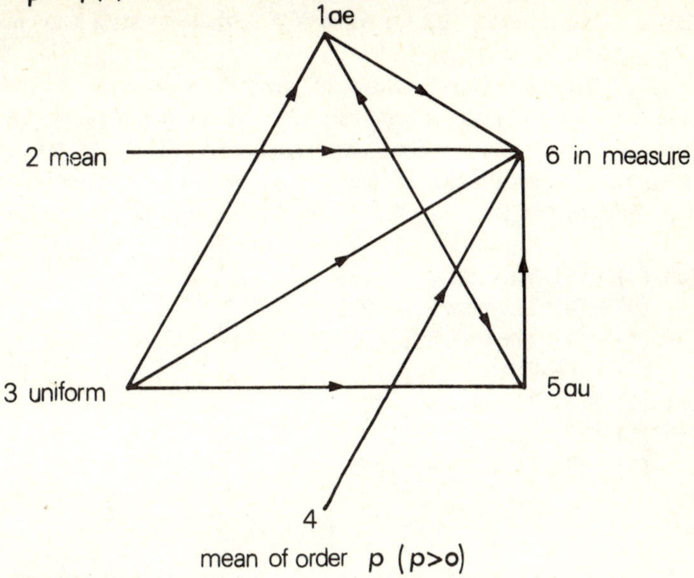

Figure 7.2

$\gamma: \quad |f_n| \leq g \in L^1(\mu)$

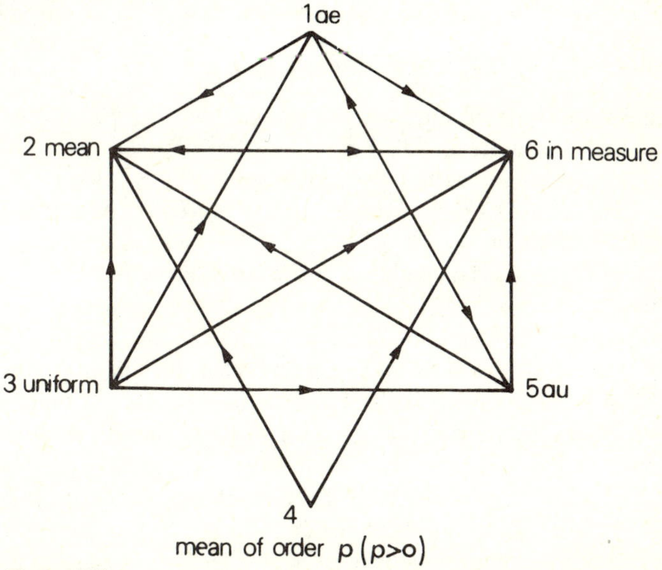

Figure 7.3

if $f_n \to f$ in $L^p(\mu)$ then $f_n \to f$ in measure; and conversely, (4) $\not\Rightarrow$ (5) means that for a suitable measure space and for some $p > 0$, there exists a sequence $\{f_n\}$ such that $f_n \to f$ in $L^p(\mu)$ but $f_n \not\to f$ a.u.

In the list of references which follows and gives the location of the proofs we refer to the case (4) $\Rightarrow$ (6); the special case (2) $\Rightarrow$ (6) is then implied. Also if we refer to the cases $A \Rightarrow B$ and $B \Rightarrow C$ we omit reference to the case $A \Rightarrow C$. Finally, if a result has been shown for case (α) we need not refer to it in cases (β) and (γ). Using these devices the following list of results is sufficient to construct the diagrams.

Case (α). (3) $\Rightarrow$ (5): Theorem 6, p. 125;
(4) $\Rightarrow$ (6): Theorem 5, p. 123;
(5) $\Rightarrow$ (1): Theorem 8, p. 125;
(5) $\Rightarrow$ (6): Theorem 7, p. 125.
Case (β). (1) $\Rightarrow$ (5): Theorem 9, p. 126.
Case (γ). (1) $\Rightarrow$ (5): Theorem 9, p. 126;
(6) $\Rightarrow$ (2): Theorem 4, p. 123.

Exercises

12. Let $[\![X, \mathcal{S}]\!]$ be a measurable space and $\{\mu_n\}$ a sequence of measures on $\mathcal{S}$ such that given $E \in \mathcal{S}$, $\mu_{n+1}(E) \geqslant \mu_n(E)$ for each n. Write $\mu(E) = \lim \mu_n(E)$. Show that

    (i) μ is a measure on $\mathcal{S}$,
    (ii) if $f \in L(X, \mu)$, then for each n, $f \in L(X, \mu_n)$ and $\int f \, d\mu = \lim \int f \, d\mu_n$.

13. In the previous exercise, let $\mu_{n+1} \geqslant \mu_n$ and $\lim \mu_n = \mu$. Show that $L(X, \mu_1) \supseteq L(X, \mu_2) \supseteq \ldots \supseteq L(X, \mu)$, but that in general $L(X, \mu) \neq \bigcap_{n=1}^{\infty} L(X, \mu_n)$.
14. If μ and ν are finite measures on the measurable space $[\![X, \mathcal{S}]\!]$ and $\mu \geqslant \nu$, then $\mu - \nu$ is a measure on $[\![X, \mathcal{S}]\!]$.
15. Let $\{\mu_n\}$ be measures on the measurable space $[\![X, \mathcal{S}]\!]$, let $\mu_n \geqslant \mu_{n+1}$ for each n, and let $\mu_1(X) < \infty$. Write $\lim \mu_n(E) = \mu(E)$ for each $E \in \mathcal{S}$. Show that μ is a measure on $\mathcal{S}$.
16. Show that the finiteness condition in the last exercise is necessary if μ is to be a measure.
17. Show that under the conditions of Exercise 15, $L(X, \mu) \supseteq \ldots \supseteq L(X, \mu_2) \supseteq L(X, \mu_1)$ and show that if $f \in L(X, \mu_n)$ for each n, then $\lim \int f \, d\mu_n = \int f \, d\mu$.
18. Let $[\![X, \mathcal{S}]\!]$ be a measurable space and $\{\mu_n\}$ a sequence of finite measures on it such that $\lim \mu_n = \mu$, uniformly on $\mathcal{S}$. Show that the set function μ is a measure.
19. Show that under the conditions of Exercise 18, if $f \in L(X, \mu_n)$ for each n, then $f \in L(X, \mu)$ and $\lim \int f \, d\mu_n = \int f \, d\mu$.

[*Note*: In Exercise 18 'uniformly' may be omitted. This is implied by the Vitali-Hahn-Saks Theorem, [17], p. 70.]

20. Show that for each of the six kinds of convergence considered in this section, if a sequence is convergent, it is fundamental.

7.4 COUNTEREXAMPLES

In the section we give a set of counterexamples to show that the set of implications given by the diagrams of Section 7.3 is the most possible. Extra examples can be found in exercises which follow. For example, 'convergence a.e. does not in general imply convergence in the mean of order p ($p > 0$)' means that there exists $p > 0$, a measure space $[\![X, \mathcal{S}, \mu]\!]$ and a sequence of functions $\{f_n\}$ with limit f a.e. such that $f_n \notin L^p(\mu)$ or $f \notin L^p(\mu)$ or, if they do, $\|f_n - f\|_p$ does not tend to zero.

As in Section 7.3, it is not necessary to give a counterexample whenever an implication is missing. Thus, if $B \Rightarrow C$, to show $A \not\Rightarrow B$ it is sufficient to show $A \not\Rightarrow C$. Also if $A \not\Rightarrow B$ in either case (β) or (γ), then clearly $A \not\Rightarrow B$ in case (α). Also, if a mode of convergence does not imply (2), that is: convergence in the mean, then it cannot imply (4), that is: convergence in the mean of order p, any $p > 0$; and conversely to show, for example, that (4) $\not\Rightarrow$ (5) it is sufficient to observe that (2) $\not\Rightarrow$ (5).

The counterexamples numbered (i) to (viii) which follow, demonstrate the following 'non-implications' between the kinds of convergence (1) to (6), from which the remaining non-implications may be deduced.

Case (α). (1) $\not\Rightarrow$ (6): (iv).

Case (β). (1) $\not\Rightarrow$ (2): (v), (2) $\not\Rightarrow$ (1): (vi), (2) $\not\Rightarrow$ (4): (vii), (3) $\not\Rightarrow$ (2): (i), (4) $\not\Rightarrow$ (2): (viii), (5) $\not\Rightarrow$ (3): (iii).

Case (γ). (2) $\not\Rightarrow$ (1): (vi), (3) $\not\Rightarrow$ (4): (ii), (5) $\not\Rightarrow$ (3): (iii).

(i) Let $X = [0,1]$, $f(0) = 0$, $f(x) = 1/x$ if $0 < x < 1$, and let $f_n(x) = f(x) + 1/n$, $n = 1, 2, \ldots$ Then $f_n \to f$ uniformly but $f \notin L^1(0,1)$.

(ii) Let $X = [0, \infty)$, and let $f_n(x) = 1/n^2$ if $0 \leqslant x \leqslant n$, $f_n(x) = 0$ if $x > n$. Then $f_n \to 0$ uniformly, and $0 \leqslant f_n \leqslant g$ where $g = 1$ on $[0,1]$, $g(x) = 1/x^2$ if $x > 1$, so that $g \in L(X)$. But $\int f_n^{1/2}\, dx = 1$, so $f_n \not\to 0$ in $L^{1/2}(X)$.

(iii) Let $X = [0,1]$, and let $f_n(x) = x^{-1/2}\, \chi_{[0,1/n]}$. Then $m[x: f_n(x) \neq 0] = 1/n \to 0$ as $n \to \infty$, so $f_n \to 0$ a.u. Also $|f_n(x)| \leqslant x^{-1/2} \in L(0,1)$; but clearly $f_n \not\to 0$ uniformly.

(iv) Let $X = [0, \infty)$, and let $f_n(x) = 1 - n(x-k)$ if $k \leqslant x \leqslant n^{-1} + k$, $f_n(x) = 0$ if $n^{-1} + k \leqslant x < k+1$, for $k = 0, 1, \ldots$ Then $\lim_{n\to\infty} f_n(x) = 0$ except for $x = 0, 1, \ldots$, but $m[x: f_n(x) > \epsilon] = \infty$ for each positive ϵ and for each n. So $f_n \to 0$ a.e. but not in measure.

(v) Let $X = [0,1]$, and let $f_n(x) = n$ if $0 \leqslant x \leqslant 1/n$, $f_n(x) = 0$ if $1/n < x \leqslant 1$. Then $f_n \to 0$ a.e., but $\int f_n\, dx = 1$ for each n. So $f_n \not\to 0$ in the mean.

(vi) Let $X = [0,1]$, and let $E_n^i = [(i-1)/n, i/n]$, $i = 1, \ldots, n$. Write χ_n^i for the characteristic function of E_n^i. The sequence $\{f_n\}$ is defined to be $\chi_1^1, \chi_2^1, \chi_2^2$,

$\chi_3^1, \chi_3^2, \chi_3^3, \ldots$, so that $f_n = \chi_k^r$ where $k(k-1)/2 < n \leqslant k(k+1)/2$, and $|f_n| \leqslant 1$, an integrable function. Also $\int_0^1 f_n\, dx = 1/k \to 0$ as $n \to \infty$, so $f_n \to 0$ in the mean. But for each $x \in [0,1]$, $\chi_k^r(x) = 1$ for each k and some r; so $f_{n_i}(x) = 1$ for an infinite subsequence $\{n_i\}$ and hence $f_n(x) \nrightarrow 0$ for any x.

(vii) Let $X = (0,1]$, and let $f(x) = x^{-1/p}$ where $p > 1$. Then $f \in L^1(0,1)$, but $f \notin L^p(0,1)$, so the sequence $f, f, \ldots$ converges in the mean but not in the mean of order p.

(viii) Let $X = (0,1]$, and let $f(x) = 1/x$. Then $f \in L^p(0,1)$ for $0 < p < 1$, but $f \notin L^1(0,1)$, so the sequence $f, f, \ldots$ converges in the mean of order p, but not in the mean.

Exercises

21. Let $f_n(x) = 1/n$, $0 \leqslant x \leqslant n$; $f_n(x) = 0, x > n$. Show that $f_n \to 0$ uniformly, but not in the mean.
22. Let $f_n(x) = n^{-1/p}, p > 0, 0 \leqslant x \leqslant n$; $f_n(x) = 0, x > n$. Show that $f_n \to 0$ uniformly but not in the mean of order p.
23. Let $f_n(x) = n\, e^{-nx}$, $0 \leqslant x \leqslant 1$; then $f_n \to 0$ a.u. on $[0,1]$, but not in the mean.
24. Let $f_n(x) = n^{1/p}\, e^{-nx}, p > 0, 0 \leqslant x \leqslant 1$; show that $f_n \to 0$ a.u. on $[0,1]$, but not in the mean of order p.
25. Show that if $\mu(X) < \infty$, then for $p > 1$ convergence in the mean of order p implies convergence in the mean.
26. Let $f_n(x) = x^{-2}$, $1 \leqslant x \leqslant n$; $f_n(x) = 0, x > n$. Show that the sequence $\{f_n\}$ is dominated by an integrable function, that $f_n \in L^{1/2}(1,\infty)$ for each n, but that $\{f_n\}$ does not converge in the mean of order 1/2.
27. Let $f_n(x) = n^{3/2}\, x\, e^{-n^2x^2}$, $0 \leqslant x \leqslant 1$. Show that $f_n \to 0$ a.e., but not in the mean of order 2.
28. Let $f_n(x) = n^{3/p}\, x^{2/p}\, e^{-n^2x^2}, p > 0, 0 \leqslant x \leqslant 1$. Show that $f_n \to 0$ a.e., but not in the mean of order p.
29. Let $f_n(x) = x^n$, $0 \leqslant x \leqslant 1$. Show that $\{f_n\}$ converges almost uniformly, but not uniformly.
30. Show that if $\mu(X) < \infty$, uniform convergence implies convergence in the mean of order p, where p is positive, provided that the functions concerned lie in $L^p(\mu)$.
31. Show that the condition $\mu(X) < \infty$ is necessary in the last exercise.

CHAPTER 8

Signed Measures and their Derivatives

We now allow measures to take negative values and then in the Hahn and Jordan decompositions show how in the study of such measures we may keep to the non-negative measures already discussed. Integrating a non-negative function over the sets of a σ-algebra produces a new measure from the original one and in the Radon-Nikodym theorem we show that any new measure continuous in a certain way can be formed in this manner. This gives rise to the derivative of one measure with respect to another and in Section 8.4 we describe the calculus of derivatives which this gives rise to and give further decomposition results. Finally, in Section 8.5 we note that for a fixed g and μ the mapping $f \to \int fg \, d\mu$ is linear and give conditions for such a linear mapping on L^p and L^1 to have this form.

8.1 SIGNED MEASURES AND THE HAHN DECOMPOSITION

We have seen in Theorem 18, p. 106, that if f is a non-negative measurable function on the measure space $[\![X, \mathcal{S}, \mu]\!]$, then the set function ϕ, defined on $\mathcal{S}$ by $\phi(E) = \int_E f \, d\mu$, is a measure. If f is any measurable function whose integral with respect to μ exists, then $\nu(E) = \int_E f \, d\mu$ is a set function on $\mathcal{S}$ which is countably additive and which behaves in most respects like a measure. This suggests extending the definition of a measure to allow negative values. This is done in Definition 1. The Jordan decomposition theorem in the next section shows that the study of these 'measures' can be reduced to that of measures in the strict sense.

Definition 1: A set function ν defined on a measurable space $[\![X, \mathcal{S}]\!]$ is said to be a **signed measure** if the values of ν are extended real numbers and

(i) ν takes at most one of the values $\infty, -\infty$,
(ii) $\nu(\emptyset) = 0$,
(iii) $\nu\left(\bigcup_{i=1}^{\infty} E_i\right) = \sum_{i=1}^{\infty} \nu(E_i)$ if $E_i \cap E_j = \emptyset$ for $i \neq j$, where if the left-hand side is

infinite, the series on the right-hand side has sum ∞ or $-\infty$ as the case may be.

Clearly, every measure is a signed measure.

Example 1: Show that if $\phi(E) = \int_E f \, d\mu$ where $\int f \, d\mu$ is defined, then ϕ is a signed measure.

Solution: We have either $\int f^+ \, d\mu < \infty$ or $\int f^- \, d\mu < \infty$ so (i) of Definition 1 follows. (ii) is trivial. Let $\{E_i\}$ be a sequence of disjoint sets of $\mathcal{S}$ and for $E \in \mathcal{S}$ write $\phi^+(E) = \int_E f^+ \, d\mu$, $\phi^-(E) = \int_E f^- \, d\mu$, so that by Theorem 18, p. 106, ϕ^+ and ϕ^- are measures. Then

$$\phi\left(\bigcup_{i=1}^{\infty} E_i\right) = \phi^+\left(\bigcup_{i=1}^{\infty} E_i\right) - \phi^-\left(\bigcup_{i=1}^{\infty} E_i\right) = \sum_{i=1}^{\infty} \phi^+(E_i) - \sum_{i=1}^{\infty} \phi^-(E_i) = \sum_{i=1}^{\infty} \phi(E_i)$$

as we cannot get $\infty - \infty$ at any stage.

Definition 2: A is a **positive set** with respect to the signed measure ν on $[\![X, \mathcal{S}]\!]$ if $A \in \mathcal{S}$ and $\nu(E) \geqslant 0$ for each measurable subset E of A. We will omit 'with respect to ν' if the signed measure is obvious from the context.

Clearly $\emptyset$ is a positive set with respect to every signed measure. Also, $\nu(A) \geqslant 0$ is necessary but not in general sufficient for A to be a positive set with respect to ν.

The next example shows a second important way of constructing a new measure from a given signed measure.

Example 2: If A is a positive set with respect to ν and if, for $E \in \mathcal{S}$, $\mu(E) = \nu(E \cap A)$, then μ is a measure.

Definition 3: A is a **negative set** with respect to ν if it is a positive set with respect to $-\nu$.

Definition 4: A is a **null set** with respect to ν, or a ν-null set, if it is both a positive and a negative set with respect to ν.

Equivalently, A is a ν-null set if $A \in \mathcal{S}$ and $\nu(E) = 0$ for all $E \in \mathcal{S}$, $E \subseteq A$.

Example 3: If A is a positive set with respect to ν, then every measurable subset of A is a positive set. The same holds for negative sets and null sets.

Theorems 1 and 2, which follow, will be used to prove the main result of the section, Theorem 3, which asserts, roughly, that X may be divided into two sets on one of which ν, and on the other $-\nu$, acts like a measure.

Theorem 1: A countable union of sets positive with respect to a signed measure ν is a positive set.

Proof: Let $\{A_n\}$ be a sequence of positive sets. Then, as in Theorem 2, p. 95, we have $\bigcup_{n=1}^{\infty} A_n = \bigcup_{n=1}^{\infty} B_n$ where the sets $B_n \in \mathcal{S}$, $B_n \subseteq A_n$ and $B_n \cap B_m = \emptyset$ if $n \neq m$. Now let $E \subsetneq \bigcup_{n=1}^{\infty} A_n$. Then $E = \bigcup_{n=1}^{\infty} (E \cap B_n)$, so $\nu(E) = \sum_{n=1}^{\infty} \nu(E \cap B_n) \geqslant 0$, as $E \cap B_n$ is a positive set for each n by Example 3. So $\bigcup_{n=1}^{\infty} A_n$ is a positive set. □

Corollary: A countable union of negative or of null sets is, respectively, a negative or a null set.

Theorem 2: Let ν be a signed measure on $[\![X, \mathcal{S}]\!]$. Let $E \in \mathcal{S}$ and $\nu(E) > 0$. Then there exists A, a set positive with respect to ν, such that $A \subseteq E$ and $\nu(A) > 0$.

Proof: If E contains no set of negative ν-measure, then E is a positive set and $A = E$ gives the result. Otherwise there exists $n \in \mathrm{N}$ such that there exists $B \in \mathcal{S}$, $B \subsetneq E$ and $\nu(B) < -1/n$. Let n_1 be the smallest such integer and E_1 a corresponding measurable subset of E with $\nu(E_1) < -1/n_1$. Let n_k be the smallest positive integer such that there is a measurable subset E_k of $E - \bigcup_{i=1}^{k-1} E_i$ with $\nu(E_k) < -1/n_k$. From the construction, $n_1 \leqslant n_2 \leqslant \ldots$ and we have a corresponding sequence $\{E_i\}$ of disjoint subsets of E. If the process stops, at n_m say, and $C = E - \bigcup_{i=1}^{m} E_i$, then C is a positive set, and $\nu(C) > 0$, for $\nu(C) = 0$ would imply that $\nu(E) = \sum_{i=1}^{m} \nu(E_i) < 0$. So C is the desired set. If the process does not stop, put $A = E - \bigcup_{k=1}^{\infty} E_k$; we wish to show that A is a positive set. We have

$$\nu(E) = \nu(A) + \nu\left(\bigcup_{k=1}^{\infty} E_k\right). \tag{8.1}$$

But ν cannot take both the values ∞, $-\infty$, $\nu(E) > 0$ and $\nu\left(\bigcup_{k=1}^{\infty} E_k\right) = \sum_{k=1}^{\infty} \nu(E_k) < 0$, so the second term on the right-hand side of (8.1) is finite. So $\sum_{k=1}^{\infty} \nu(E_k) > -\infty$; hence $\sum_{k=1}^{\infty} 1/n_k < \infty$ and, in particular, $\lim_{k\to\infty} n_k = \infty$, and $n_k > 1$ for $k > k_0$, say. So let $B \in \mathcal{S}$, $B \subseteq A$ and $k > k_0$. Then $B \subseteq E - \bigcup_{i=1}^{k} E_i$ so

$$\nu(B) \geqslant -\frac{1}{n_k - 1} \tag{8.2}$$

by the definition of n_k. But (8.2) holds for all $k > k_0$, so letting $k \to \infty$ we have $\nu(B) \geqslant 0$ and so A is a positive set. As before, $\nu(A) = 0$ would imply $\nu(E) < 0$, so $\nu(A) > 0$ as required. □

Thorem 3: Let ν be a signed measure on $[\![X, \mathcal{S}]\!]$. Then there exists a positive set A and a negative set B such that $A \cup B = X$, $A \cap B = \emptyset$. The pair A, B is said to be a Hahn decomposition of X with respect to ν. It is unique to the extent that if A_1, B_1 and A_2, B_2 are Hahn decompositions of X with respect to ν, then $A_1 \Delta A_2$ is a ν-null set.

Proof: We may suppose that $\nu < \infty$ on $\mathcal{S}$, for otherwise we consider $-\nu$, the result of the theorem for $-\nu$ implying the result for ν. Let $\lambda = \sup[\nu(C)\colon C$ a positive set$]$, so $\lambda \geqslant 0$. Let $\{A_i\}$ be a sequence of positive sets such that $\lambda = \lim \nu(A_i)$. By Theorem 1, p. 134, $A = \bigcup_{i=1}^{\infty} A_i$ is a positive set, and from the definition of λ, $\lambda \geqslant \nu(A)$. But $A - A_i \subseteq A$ and hence is a positive set. So, for each i,

$$\nu(A) = \nu(A_i) + \nu(A - A_i) \geqslant \nu(A_i).$$

So $\nu(A) \geqslant \lim \nu(A_i) = \lambda$ and hence $\nu(A) = \lambda$, that is, the value λ is achieved on a positive set. Write $B = \mathsf{C}A$. Then if B contains a set D of positive ν-measure, we have $0 < \nu(D) < \infty$. So by Theorem 2, D contains a positive set E such that $0 < \nu(E) < \infty$. But then $\nu(A \cup E) = \nu(A) + \nu(E) > \lambda$, contradicting the definition of λ. So $\nu(D) \leqslant 0$ and B is a negative set and A, B form a Hahn decomposition.

For the last part note that $A_1 - A_2 = A_1 \cap B_2$ and hence is a positive and negative set and so a null set. Similarly $A_2 - A_1$ is a null set, and so $A_1 \Delta A_2$ is null. □

Exercises

1. Let $\nu(E) = \int_E x\, e^{-x^2}\, dx$. What are the positive, negative and null sets with respect to ν? Given a Hahn decomposition of R with respect to ν.
2. Show that, if we suppose $\nu(E) < \infty$ in Theorem 2, we obtain $0 < \nu(A) < \infty$.
3. Give an example showing that a Hahn decomposition is not unique.
4. For any sequence $\{a_n\}$ and any set $E \subseteq \mathsf{N}$ let $\nu(E)$ be the sum, if it exists, of the corresponding terms of $\{a_n\}$. Which sequences correspond to signed measures on N? Show that if $\{a_n\}$ is such a sequence and $|a_n| > 0$ for each n, the Hahn decomposition of N with respect to ν is unique.
5. Let ν be a signed measure and $\{E_n\}$ a sequence of disjoint sets such that $\left|\nu\left(\bigcup_{n=1}^{\infty} E_n\right)\right| < \infty$. Show that $\sum_{n=1}^{\infty} \nu(E_n)$ is absolutely convergent.

6. Show that if ν is a signed measure, $|\nu(E)| < \infty$ and $F \subseteq E$, then $|\nu(F)| < \infty$.

7. If ν is a signed measure and $E_1 \subseteq E_2 \subseteq \ldots$, then $\nu\left(\bigcup_{i=1}^{\infty} E_i\right) = \lim \nu(E_i)$.

8. If ν is a signed measure, $E_1 \supseteq E_2 \supseteq \ldots$, and $|\nu(E_1)| < \infty$, then

$$\nu\left(\bigcap_{i=1}^{\infty} E_i\right) = \lim \nu(E_i).$$

8.2 THE JORDAN DECOMPOSITION

We now use the Hahn decomposition to obtain, in Theorem 4, a decomposition of a signed measure into the difference of measures.

Definition 5: Let ν_1, ν_2 be measures on $[\![X, \mathcal{S}]\!]$. Then ν_1 and ν_2 are said to be **mutually singular** if, for some $A \in \mathcal{S}$, $\nu_2(A) = \nu_1(\complement A) = 0$, and we then write $\nu \perp \nu_2$.

Example 4: Let μ be a measure and let the measures ν_1, ν_2 be given by $\nu_1(E) = \mu(A \cap E)$, $\nu_2(E) = \mu(B \cap E)$, where $\mu(A \cap B) = 0$ and $E, A, B \in \mathcal{S}$. Show that $\nu_1 \perp \nu_2$.

Solution: $\nu_1(B) = \mu(A \cap B) = 0, \nu_2(\complement B) = \mu(\emptyset) = 0$.

Theorem 4: Let ν be a signed measure on $[\![X, \mathcal{S}]\!]$. Then there exist measures ν^+ and ν^- on $[\![X, \mathcal{S}]\!]$ such that $\nu = \nu^+ - \nu^-$ and $\nu^+ \perp \nu^-$. The measures ν^+ and ν^- are uniquely defined by ν, and $\nu = \nu^+ - \nu^-$ is said to be the **Jordan decomposition** of ν.

Proof: Let A, B be a Hahn decomposition of X with respect to ν, and define ν^+ and ν^- by

$$\nu^+(E) = \nu(E \cap A), \nu^-(E) = -\nu(E \cap B) \tag{8.3}$$

for $E \in \mathcal{S}$. Then ν^+ and ν^- are measures by Example 2, p. 134, and $\nu^+(B) = \nu^-(A) = 0$. So $\nu^+ \perp \nu^-$. Also, for $E \in \mathcal{S}$

$$\nu(E) = \nu(E \cap A) + \nu(E \cap B) = \nu^+(E) - \nu^-(E).$$

So $\nu = \nu^+ - \nu^-$, and the proof will be complete when we show that the decomposition is unique. Let $\nu = \nu_1 - \nu_2$ be any decomposition of ν into mutually singular measures. Then we have $X = A \cup B$, where $B = \complement A$ and $\nu_1(B) = \nu_2(A) = 0$. Let $D \subseteq A$, then $\nu(D) = \nu_1(D) - \nu_2(D) = \nu_1(D) \geqslant 0$, so A is a positive set with respect to ν. Similarly B is a negative set. For each $E \in S$ we have $\nu_1(E) = \nu_1(E \cap A) = \nu(E \cap A)$ and $\nu_2(E) = -\nu(E \cap B)$, so every such decomposition of ν is obtained from a Hahn decomposition of X, as in (8.3). So it is enough to show that if A, B and A', B' are two Hahn decompositions then the measures obtained as in (8.3) are the same. We have

$$\nu(A \cup A') = \nu(A \cap A') + \nu(A \,\Delta\, A') = \nu(A \cap A')$$

by Theorem 3, p. 136. For each $E \in \mathcal{S}$, as $A \cup A'$ is a positive set we have

$$\nu(E \cap (A \cap A')) \leqslant \nu(E \cap A) \leqslant \nu(E \cap (A \cup A'))$$

and
$$\nu(E \cap (A \cap A')) \leqslant \nu(E \cap A') \leqslant \nu(E \cap (A \cup A')).$$

But the first and last terms in each of these inequalities are the same so $\nu(E \cap A) = \nu(E \cap A')$ and ν^+ defined in (8.3) is unique. But then $\nu^- = \nu^+ - \nu$ is also unique, proving the theorem. □

Notice that the Hahn decomposition is of the space and is not unique whereas the Jordan decomposition is of the signed measure and is unique.

We will henceforth use the notation ν^+ and ν^- for the measures defined by ν as in Theorem 4.

Example 5: Let $[\![X, \mathcal{S}, \mu]\!]$ be a measure space and let $\int f \, d\mu$ exist. Define ν by $\nu(E) = \int_E f \, d\mu$, for $E \in \mathcal{S}$. Find a Hahn decomposition with respect to ν and the Jordan decomposition of ν.

Solution: From Example 1, p. 134, ν is a signed measure. Let $A = [x: f(x) \geqslant 0]$, $B = [x: f(x) < 0]$. Then A, B form a Hahn decomposition, while ν^+, ν^- given by $\nu^+(E) = \int_E f^+ \, d\mu$, $\nu^-(E) = \int_E f^- \, d\mu$ form the Jordan decomposition.

Definition 6: The **total variation** of a signed measure ν is $|\nu| = \nu^+ + \nu^-$, where $\nu = \nu^+ - \nu^-$ is the Jordan decomposition of ν.

Clearly $|\nu|$ is a measure on $[\![X, \mathcal{S}]\!]$, and for each $E \in \mathcal{S}$, $|\nu(E)| \leqslant |\nu|(E)$.

Definition 7: A signed measure ν on $[\![X, \mathcal{S}]\!]$ is *σ-finite* if $X = \bigcup_{n=1}^{\infty} X_n$ where $X_n \in \mathcal{S}$ and, for each n, $|\nu(X_n)| < \infty$.

Example 6: Show that the signed measure ν is finite or σ-finite respectively if, and only if, $|\nu|$ is, or if, and only if, both ν^+ and ν^- are.

Solution: Suppose $|\nu(E)| < \infty$. Then as ν^+ and ν^- are not both infinite we have $\nu^+(E) < \infty$ and $\nu^-(E) < \infty$ and hence $|\nu|(E) < \infty$. Obviously, ν is finite if $|\nu|$ is. The corresponding results on σ-finiteness are an immediate consequence.

Exercises

9. Show that if ν_1, ν_2 and μ are measures and $\nu_1 \perp \mu, \nu_2 \perp \mu$, then $\nu_1 + \nu_2 \perp \mu$.

10. If $\nu(E) = \int_E f \, d\mu$ where $\int f \, d\mu$ exists, what is $|\nu|(E)$?

11. Show that $\nu^+ = \frac{1}{2}(\nu + |\nu|)$, $\nu^- = \frac{1}{2}(|\nu| - \nu)$, provided ν is finite-valued.

12. Show that $|\nu(E)| = \sup \sum_{i=1}^{n} |\nu(E_i)|$ where the sets E_i are disjoint and $E =$

$\bigcup_{i=1}^{n} E_i$. This result justifies the term 'total variation' in Definition 6.

13. Show that the Jordan decomposition is minimal in the sense that if ν is a signed measure and $\nu = \nu_1 - \nu_2$ where ν_1 and ν_2 are measures, then $|\nu| \leqslant \nu_1 + \nu_2$ with equality only if $\nu_1 = \nu^+$, $\nu_2 = \nu^-$.

8.3 THE RADON–NIKODYM THEOREM

We have found in Section 8.1 that $\nu(E) = \int_E f \, d\mu$ is a signed measure. It is in practice very useful to know if a measure is of this form. The essential condition is given in Definitions 8 and 9. Theorem 5 shows that if the measures or signed measures are, in addition, σ-finite, ν can be written as an integral.

Definition 8: If μ, ν are measures on the measurable space $[\![X, \mathcal{S}]\!]$ and $\nu(E) = 0$ whenever $\mu(E) = 0$, then we say that ν is **absolutely continuous** with respect to μ and we write $\nu \ll \mu$.

Definition 9: If μ, ν are signed measures on $[\![X, \mathcal{S}]\!]$ and $\nu(E) = 0$ whenever $|\mu|(E) = 0$, then ν is *absolutely continuous* with respect to μ, $\nu \ll \mu$.

Example 7: Show that the following conditions on the signed measures μ and ν on $[\![X, \mathcal{S}]\!]$ are equivalent: (i) $\nu \ll \mu$, (ii) $|\nu| \ll |\mu|$, (iii) $\nu^+ \ll \mu$ and $\nu^- \ll \mu$.

Solution: From Definition 9 we see that $\nu \ll \mu$ if, and only if, $\nu \ll |\mu|$. So we may assume that $\mu \geqslant 0$. As $|\nu| = \nu^+ + \nu^-$, we see that $|\nu| \ll \mu$ implies $\nu^+ \ll \mu$ and $\nu^- \ll \mu$, so $\nu \ll \mu$. For the opposite implications, suppose that $\nu = \nu^+ - \nu^-$ with a Hahn decomposition A, B. Then if $\nu \ll \mu$ and $\mu(E) = 0$ we have $\mu(E \cap A) = 0$ so $\nu^+(E) = 0$, and similarly $\nu^-(E) = 0$. So $|\nu|(E) = 0$.

Example 8: If μ is a measure, $\int f \, d\mu$ exists and $\nu(E) = \int_E f \, d\mu$, then $\nu \ll \mu$.

In the opposite direction we have the main theorem of this chapter.

Theorem 5 (Radon-Nikodym Theorem): If $[\![X, \mathcal{S}, \mu]\!]$ is a σ-finite measure space and ν is a σ-finite measure on $\mathcal{S}$ such that $\nu \ll \mu$, then there exists a finite-valued non-negative measurable function f on X such that for each $E \in \mathcal{S}$, $\nu(E) = \int_E f \, d\mu$. Also f is unique in the sense that if $\nu(E) = \int_E g \, d\mu$ for each $E \in \mathcal{S}$, then $f = g$ a.e. (μ).

Proof: Suppose that the result has been proved for finite measures. Then in the general case we have $X = \bigcup_{n=1}^{\infty} A_n$, $\mu(A_n) < \infty$ and $X = \bigcup_{m=1}^{\infty} B_m$, $\nu(B_m) < \infty$ and

$\{A_n\}$, $\{B_m\}$ may be supposed to be sequences of disjoint sets. So setting $X = \bigcup_{n,m=1}^{\infty} (A_n \cap B_m)$ we obtain X as the union of disjoint sets on which both μ and ν are finite, say $X = \bigcup_{n=1}^{\infty} X_n$. Let $\mathcal{S}_n = [E \cap X_n : E \in \mathcal{S}]$, a σ-algebra over X_n, and considering μ and ν restricted to $\mathcal{S}_n$ we obtain, since μ and ν are finite, a non-negative function f_n such that if $E \in \mathcal{S}_n$, $\nu(E) = \int_E f_n \, d\mu$. So if $A \in \mathcal{S}$, $A = \bigcup_{n=1}^{\infty} A_n$, say, where $A_n \in \mathcal{S}_n$, then defining $f = f_n$ on X_n gives a measurable function on X as in Example 12, p. 103, and $\nu(A) = \sum_{n=1}^{\infty} \int_{A_n} f_n \, d\mu = \int_A f \, d\mu$. So the general case follows.

So we need to show that for finite measures such a function f exists. Let $\mathcal{K}$ be the class of non-negative functions measurable with respect to μ and satisfying $\int_E f \, d\mu \leqslant \nu(E)$ for all $E \in \mathcal{S}$. Then $\mathcal{K}$ is non-empty as $0 \in \mathcal{K}$. Let $\alpha = \sup[\int f \, d\mu : f \in \mathcal{K}]$ and let $\{f_n\}$ be a sequence in $\mathcal{K}$ such that $\lim \int f_n \, d\mu = \alpha$. If B is any fixed measurable set, n a fixed positive integer and $g_m = \max(f_1, \ldots, f_m)$, then we can prove by induction that B is the union of disjoint measurable sets B_i, $i = 1, \ldots, n$, such that $g_n = f_i$ on B_i, $i = 1, \ldots, n$. For, let $n = 2$ and let $B_1 = [x : x \in B, f_1(x) \geqslant f_2(x)]$, $B_2 = B - B_1$, then $B = B_1 \cup B_2$ has the desired property. Supposing the decomposition possible for n, let $g_{n+1} = \max(f_1, \ldots, f_{n+1}) = \max(g_n, f_1)$. So $B = F_n \cup B_{n+1}$ where $g_{n+1} = f_{n+1}$ on B_{n+1}, $g_{n+1} = g_n$ on F_n and $F_n \cap B_{n+1} = \emptyset$. But then by the inductive hypothesis we have $F_n = \bigcup_{i=1}^{n} B_i$ and $g_{n+1}(x) = f_i(x)$ for $x \in B_i$, $i = 1, \ldots, n+1$.

Now, since each $f_i \in \mathcal{K}$,

$$\int_B g_n \, d\mu = \sum_{i=1}^{n} \int_{B_i} f_i \, d\mu \leqslant \sum_{i=1}^{n} \nu(B_i) = \nu(B). \tag{8.4}$$

Also we have $g_n \uparrow$, so write $f_0 = \lim g_n$. Then (8.4) and the Lebesgue Monotone Convergence Theorem imply that

$$\int_E f_0 \, d\mu = \lim \int_E g_n \, d\mu \leqslant \nu(E),$$

so $f_0 \in \mathcal{K}$. Hence

$$\alpha \geqslant \int f_0 \, d\mu \geqslant \int g_n \, d\mu \geqslant \int f_n \, d\mu,$$

so $\alpha = \int f_0 \, d\mu$. Since $\int f_0 \, d\mu \leqslant \nu(X) < \infty$, there exists a finite-valued measurable function f, also non-negative, such that $f = f_0$ a.e. (μ).

We will show now that if $\nu_0(E) = \nu(E) - \int_E f\,d\mu$, then $\nu_0(E) = 0$, for each $E \in \mathcal{S}$. By the construction of f, ν_0 is non-negative. If ν_0 is not identically zero on $\mathcal{S}$, let $C \in \mathcal{S}$ and $\nu_0(C) > 0$. Then for a suitable ϵ, $0 < \epsilon < 1$, $(\nu_0 - \epsilon\mu)(C) > 0$. But then by Theorem 2 we can find A such that $(\nu_0 - \epsilon\mu)(A) > 0$ where A is a positive set with respect to $\nu_0 - \epsilon\mu$. Also $\mu(A) > 0$, for otherwise, as $\nu \ll \mu$ we would have $\nu(A) = 0$ and hence $(\nu_0 - \epsilon\mu)(A) = 0$. So for $E \in \mathcal{S}$

$$\epsilon\mu(E \cap A) \leqslant \nu_0(E \cap A) = \nu(E \cap A) - \int_{E \cap A} f\,d\mu.$$

Hence if $g = f + \epsilon\chi_A$, for each $E \in \mathcal{S}$ we have

$$\int_E g\,d\mu = \int_E f\,d\mu + \epsilon\mu(E \cap A) \leqslant \int_{E-A} f\,d\mu + \nu(E \cap A) \leqslant \nu(E)$$

and so $g \in \mathcal{K}$. But $\int g\,d\mu = \int f\,d\mu + \epsilon\mu(A) > \alpha$, contradicting the maximality of α. So $\nu_0 = 0$ on $\mathcal{S}$, that is, $\int_E f\,d\mu = \nu(E)$.

So f has the desired properties. Let g also have these properties. Then, for $E \in \mathcal{S}$, $\int_E (f-g)\,d\mu = 0$ and taking $E = [x : f(x) > g(x)]$, we get $f \leqslant g$ a.e., and similarly $f \geqslant g$ a.e. so f is unique in the sense stated. □

Corollary 1: Theorem 5 can be extended to the case where ν is a σ-finite signed measure.

Proof: The Jordan decomposition gives $\nu = \nu^+ - \nu^-$ and by Examples 6 and 7, $\nu^+(E) = \int_E f_1\,d\mu$, $\nu^-(E) = \int_E f_2\,d\mu$ where f_1 and f_2 are finite-valued non-negative measurable functions of which at least one is integrable. So, for $E \in \mathcal{S}$, $\nu(E) = \nu^+(E) - \nu^-(E) = \int_E f\,d\mu$ where the integral of $f = f_1 - f_2$ is well-defined. □

Corollary 2: Theorem 5 can be further extended to allow μ to be a signed measure, where by $\int_E f\,d\mu$ we then mean $\int_E f^+\,d\mu - \int_E f^-\,d\mu$, provided this difference is not indeterminate. Any two such functions f and g are equal a.e. ($|\mu|$).

Proof: Let A, B be a Hahn decomposition with respect to μ, so that $\mu^+(E) = \mu(E \cap A)$, $\mu^-(E) = -\mu(E \cap B)$. Now $\nu \ll \mu^+$ by Example 7, p. 139, and μ^+ is σ-finite by Example 6, p. 138, so on applying Theorem 5 to μ^+ on A we get $\nu(E \cap A) = \int_{E \cap A} f_1\,d\mu^+$ for an appropriate function f_1 on A, and similarly $\nu(E \cap B) =$

$\int_{E\cap B} f_2 \, d\mu^-$ for f_2 defined on B. So define $f = f_1$ on A, $f = -f_2$ on B. Then by Example 12, p. 103, f is a measurable function on X, and $\nu(E) = \int_{A\cap E} f_1 \, d\mu^+ - \int_{B\cap E} (-f_2) \, d\mu^-$. As ν is a signed measure this will not be of the form $\infty - \infty$, so $\nu(E) = \int_E f \, d\mu$ is well-defined. Any two such functions, from the construction, agree except on a set of zero μ^+- and μ^--measure, giving the result. □

Exercises

14. Let μ and ν be measures and $\mu \ll \nu$. If a proposition P holds a.e. (ν) then P holds a.e. (μ). Also, if μ is a complete measure so is ν.
15. Give an example to show that in Definition 9, p. 139, the condition $|\mu|(E) = 0$ is not equivalent to $\mu(E) = 0$.
16. Give an example of measures μ and ν, on the same measurable space, such that none of the relations $\mu \ll \nu$, $\nu \ll \mu$, $\mu \perp \nu$ holds.
17. Let $0 < x_0 < 1$ and for any Lebesgue measurable set $E \subseteq [0,1]$ define $\nu(E) = \chi_E(x_0)$. Show that ν is a measure which is not absolutely continuous with respect to Lebesgue measure on $[0,1]$.
18. Show that the condition: ν σ-finite, is necessary in the Radon-Nikodym Theorem.
19. Show that if $\nu(E) = \int_E f \, d\mu$ for each $E \in \mathcal{S}$, where f is non-negative and measurable, and $f = \infty$ on a set of positive μ-measure, then ν is not σ-finite.
20. Show that the condition: μ σ-finite, is necessary in the Radon-Nikodym Theorem.
21. Show that if μ and ν are measures such that $\nu \ll \mu$ and $\nu \perp \mu$, then ν is identically zero.

8.4 SOME APPLICATIONS OF THE RADON–NIKODYM THEOREM

In this section we consider the 'calculus of derivatives' to which the Radon-Nikodym theorem gives rise and in Theorem 11 give one of the many applications of that theorem. The next section deals with a different kind of application. Our first result concerns absolute continuity and extends Theorem 18, p. 106.

Theorem 6: Let μ be a signed measure on $[\![X, \mathcal{S}]\!]$ and let ν be a finite-valued signed measure on $[\![X, \mathcal{S}]\!]$ such that $\nu \ll \mu$; then given $\epsilon > 0$ there exists $\delta > 0$ such that $|\nu|(E) < \epsilon$ whenever $|\mu|(E) < \delta$.

Proof: By Example 7, p. 139, $\nu \ll \mu$ is equivalent to $|\nu| \ll |\mu|$, and by Example 6,

p. 138, $|\nu|$ is finite-valued if, and only if, ν is, so we may suppose that μ and ν are measures. If the result is not true, then there exists a positive ϵ and a sequence $\{E_n\}$ of sets of $\mathcal{S}$ such that $\mu(E_n) < 1/2^n$, but $\nu(E_n) \geqslant \epsilon$. Then consider

$$\limsup E_n = \bigcap_{k=1}^{\infty} F_k, F_k = \bigcup_{m=k}^{\infty} E_m.$$

For each k, $\mu(\limsup E_n) \leqslant \mu(F_k) \leqslant \sum_{m=k}^{\infty} 1/2^m = 1/2^{k-1}$. So $\mu(\limsup E_n) = 0$. But, for each k, $\nu(F_k) \geqslant \epsilon$ and ν is finite, so by Theorem 10(ii), p. 103,

$$\nu(\limsup E_n) = \nu(\lim F_k) = \lim \nu(F_k) \geqslant \epsilon,$$

contradicting $\nu \ll \mu$, and the theorem is proved. □

The converse is true without any finiteness condition.

Example 9: If μ, ν are signed measures on $[\![X, \mathcal{S}]\!]$ and if $\forall\, \epsilon > 0$, $\exists\, \delta > 0$ such that whenever $|\mu|(E) < \delta$ we have $|\nu|(E) < \epsilon$, then $\nu \ll \mu$.

Solution: If $|\mu|(E) = 0$, then $|\nu|(E) < 1/n$ for all n.

Definition 10: Let μ and ν be σ-finite signed measures on $[\![X, \mathcal{S}]\!]$ and suppose that $\nu \ll \mu$. Then the **Radon-Nikodym derivative** $\mathrm{d}\nu/\mathrm{d}\mu$, of ν with respect to μ, is any measurable function f such that $\nu(E) = \int_E f\, \mathrm{d}\mu$ for each $E \in \mathcal{S}$, where if μ is a signed measure $\int f\,\mathrm{d}\mu = \int f\,\mathrm{d}\mu^+ - \int f\,\mathrm{d}\mu^-$ as in Theorem 5, Corollary 2, p. 141.

From Theorem 5, p. 139, Definition 10 specifies $\mathrm{d}\nu/\mathrm{d}\mu$ only as one of a set of functions any two of which are equal a.e. (μ); this is not important as $\mathrm{d}\nu/\mathrm{d}\mu$ will, in practice, usually appear under the integral sign. In the equations below connecting Radon-Nikodym derivatives we will indicate the measure, say μ, with respect to which the functions are equal a.e. by the notation $[\mu]$. In the case of a signed measure, the functions are equal a.e. ($|\mu|$). By $\mathrm{d}\nu/\mathrm{d}\psi \neq \mathrm{d}\phi/\mathrm{d}\lambda\,[\mu]$ we mean that either the functions differ on a set of positive μ-measure or they are not measurable with respect to the same σ-algebra.

Theorem 7: If ν_1, ν_2 are σ-finite measures on $[\![X, \mathcal{S}]\!]$ and $\nu_1 \ll \mu, \nu_2 \ll \mu$, then

$$\frac{\mathrm{d}(\nu_1 + \nu_2)}{\mathrm{d}\mu} = \frac{\mathrm{d}\nu_1}{\mathrm{d}\mu} + \frac{\mathrm{d}\nu_2}{\mathrm{d}\mu}\,[\mu]. \tag{8.5}$$

Proof: Clearly $\nu_1 + \nu_2$ is a σ-finite measure and $\nu_1 + \nu_2 \ll \mu$. For $E \in \mathcal{S}$

$$(\nu_1 + \nu_2)(E) = \nu_1(E) + \nu_2(E) = \int_E \frac{\mathrm{d}\nu_1}{\mathrm{d}\mu}\,\mathrm{d}\mu + \int_E \frac{\mathrm{d}\nu_2}{\mathrm{d}\mu}\,\mathrm{d}\mu.$$

So the uniqueness of $\mathrm{d}(\nu_1 + \nu_2)/\mathrm{d}\mu$ gives the result. □

We prove next that (8.5) holds for signed measures, using Hahn decompositions of X so that we can work with measures.

Theorem 8: If $\nu_1, \nu_2, \nu_1 + \nu_2$ and μ are σ-finite signed measures on $[\![X, \underset{\sim}{S}]\!]$ and $\nu_1 \ll \mu, \nu_2 \ll \mu$, then (8.5) holds.

Proof: $\nu_1 + \nu_2$ is a signed measure provided $\nu_1(E) + \nu_2(E)$ is never $\infty + (-\infty)$.

(i) Suppose that μ is a measure. For $i = 1, 2$ let $\nu_i = \nu_i^+ - \nu_i^-$ with Hahn decomposition A_i, B_i. Consider the four sets $A_i \cap B_j$ separately. On subsets of $A_1 \cap B_2$, for example, we have $\nu_1 + \nu_2 = \nu_1^+ - \nu_2^-$, so for $F \subseteq A_1 \cap B_2$,

$$(\nu_1 + \nu_2)(F) = \nu_1^+(F) - \nu_2^-(F)$$
$$= \int_F \left(\frac{d\nu_1^+}{d\mu} - \frac{d\nu_2^-}{d\mu}\right) d\mu = \int_F \left(\frac{d\nu_1}{d\mu} + \frac{d\nu_2}{d\mu}\right) d\mu, \tag{8.6}$$

since clearly $d\nu/d\mu = -d(-\nu)/d\mu$ $[\mu]$. But each $E \in \underset{\sim}{S}$ may be written as the union of four such sets F and adding the corresponding equations (8.6) gives the result, as the fact that ν_1, ν_2 and $\nu_1 + \nu_2$ are signed measures ensures that $\infty + (-\infty)$ will not arise.

(ii) Let μ be a signed measure and let A, B be a corresponding Hahn decomposition. Write $\underset{\sim}{S}' = [E \cap A \colon E \in \underset{\sim}{S}]$ and let μ', ν_1', ν_2' be the restriction of μ, ν_1, ν_2 to $\underset{\sim}{S}'$. Similarly, $\underset{\sim}{S}'', \mu'', \nu_1'', \nu_2''$ in the case of B. Then case (i) applied to A and to B gives

$$\begin{aligned} d(\nu_1' + \nu_2')/d\mu' &= d\nu_1'/d\mu' + d\nu_2'/d\mu' \ [\mu] \\ d(\nu_1'' + \nu_2'')/d(-\mu'') &= d\nu_1''/d(-\mu'') + d\nu_2''/d(-\mu'') \ [\mu]. \end{aligned} \tag{8.7}$$

Write $f_i = d\nu_i'/d\mu'$ on $A, f_i = -d\nu_i''/d(-\mu'')$ on B. Then for each $E \in \underset{\sim}{S}$,

$$\int_E f_i \, d\mu = \int_{E \cap A} f_i \, d\mu' - \int_{E \cap B} f_i \, d(-\mu'') = \nu_i(E \cap A) + \nu_i(E \cap B) = \nu_i(E)$$

for $i = 1, 2$ and similarly for $\nu_1 + \nu_2$. Since $(\nu_1 + \nu_2)' = \nu_1' + \nu_2'$ and $(\nu_1 + \nu_2)'' = \nu_1'' + \nu_2''$, we may subtract equations (8.7) to get the result, as again no indeterminate expressions can occur. □

Example 10: Let μ be a σ-finite measure and ν a σ-finite signed measure and let $\nu \ll \mu$; show that $d|\nu|/d\mu = |d\nu/d\mu|$ $[\mu]$.

Solution: Let $\nu = \nu^+ - \nu^-$ with a corresponding Hahn decomposition A, B. As in the last theorem, we have $|d\nu/d\mu| = d\nu^+/d\mu$ $[\mu]$ on A and $|d\nu/d\mu| = d\nu^-/d\mu$ $[\mu]$ on B. So, by Theorem 7,

$$\left|\frac{d\nu}{d\mu}\right| = \frac{d\nu^+}{d\mu} + \frac{d\nu^-}{d\mu} = \frac{d|\nu|}{d\mu} \qquad [\mu].$$

Theorem 9: Let ν be a signed measure and let μ, λ be measures on $[\![X, \underset{\sim}{S}]\!]$ such that λ, μ, ν are σ-finite, $\nu \ll \mu$ and $\mu \ll \lambda$; then

$$\frac{d\nu}{d\lambda} = \frac{d\nu}{d\mu}\frac{d\mu}{d\lambda} \ [\lambda]. \tag{8.8}$$

Proof: We may write $\nu = \nu^+ - \nu^-$ and we use the fact that $-d\nu^-/d\lambda = d(-\nu^-)/d\lambda$

$[\lambda]$, and similarly for $\mathrm{d}\nu^-/\mathrm{d}\mu$. So by Theorem 8, (8.8) need be proved for measures only. So suppose that ν is a measure and take for $\mathrm{d}\nu/\mathrm{d}\mu$ and $\mathrm{d}\mu/\mathrm{d}\lambda$ the non-negative functions, f and g respectively, provided by the Radon-Nikodym Theorem. Then we wish to show that for $F \in \mathcal{S}$, $\nu(F) = \int_F fg \, \mathrm{d}\lambda$. Let ψ be a measurable simple function, $\psi = \sum_{i=1}^{n} a_i \chi_{E_i}$; then

$$\int_F \psi \, \mathrm{d}\mu = \sum_{i=1}^{n} a_i \mu(E_i \cap F) = \sum_{i=1}^{n} a_i \int_{E_i \cap F} g \, \mathrm{d}\lambda = \int_F \psi g \, \mathrm{d}\lambda.$$

Let $\{\psi_n\}$ be a sequence of measurable simple functions such that $\psi_n \uparrow f$. Then

$$\nu(F) = \int_F f \, \mathrm{d}\mu = \lim \int_F \psi_n \, \mathrm{d}\mu = \lim \int_F \psi_n g \, \mathrm{d}\lambda = \int_F fg \, \mathrm{d}\lambda,$$

as $\psi_n g \uparrow fg$, giving the result. □

Again we may extend the result slightly to cover signed measures.

Theorem 10: Let λ, μ, ν be σ-finite signed measures on $[\![X, \mathcal{S}]\!]$ such that $\nu \ll \mu$ and $\mu \ll \lambda$; then (8.8) holds.

Proof: Let Hahn decompositions with respect to λ and to μ be given by A_1, B_1 and A_2, B_2 respectively. Consider the four sets $A_i \cap B_j$, $i, j = 1, 2$, separately. For example, on $A_1 \cap B_2$ we let $\mathcal{S}' = [E \cap A_1 \cap B_2 : E \in \mathcal{S}]$ and let λ', μ' be the restrictions of λ, μ to $\mathcal{S}'$. So λ' and $-\mu'$ are measures. Applying Theorem 9 we get, on $A_1 \cap B_2$

$$\frac{\mathrm{d}\nu}{\mathrm{d}\lambda'} = \frac{\mathrm{d}\nu}{\mathrm{d}(-\mu')} \quad \frac{\mathrm{d}(-\mu')}{\mathrm{d}\lambda'} \, [\mu].$$

As in the proof of Theorem 8, p. 144, we see that $-\mathrm{d}\nu/\mathrm{d}(-\mu')$ is the restriction of $\mathrm{d}\nu/\mathrm{d}\mu$ to $A_1 \cap B_2$ and $-\mathrm{d}(-\mu')/\mathrm{d}\lambda'$ that of $\mathrm{d}\mu/\mathrm{d}\lambda$ to $A_1 \cap B_2$. So on $A_1 \cap B_2$ we get

$$\frac{\mathrm{d}\nu}{\mathrm{d}\lambda} = \frac{\mathrm{d}\nu}{\mathrm{d}\mu}\frac{\mathrm{d}\mu}{\mathrm{d}\lambda} \, [\lambda].$$

Adding four such equations gives the result. □

Example 11: Let μ and ν be σ-finite measures on $[\![X, \mathcal{S}]\!]$ such that $\nu \ll \mu$. Show that there exists a measurable function g such that if $f \in L(X, \nu)$, then $fg \in L(X, \mu)$ and for each $E \in \mathcal{S}$,

$$\int_E f \, \mathrm{d}\nu = \int_E fg \, \mathrm{d}\mu.$$

Solution: Consider the signed measure λ defined by $\lambda(E) = \int_E f \, \mathrm{d}\nu$, for $E \in \mathcal{S}$. As $|\lambda|(X) = \int |f| \, \mathrm{d}\nu < \infty$, λ is a finite signed measure. So $\mathrm{d}\lambda/\mathrm{d}\nu = f \, [\nu]$. Let g

be a non-negative measurable function, $g = \mathrm{d}\nu/\mathrm{d}\mu\ [\mu]$; then by Theorem 9, $fg = \mathrm{d}\lambda/\mathrm{d}\mu\ [\mu]$. So

$$\int_E f\,\mathrm{d}\nu = \lambda(E) = \int_E fg\,\mathrm{d}\mu.$$

Also, $fg \in L(X, \mu)$ since using Example 10, p. 144,

$$\int |fg|\,\mathrm{d}\mu = \int \left|\frac{\mathrm{d}\lambda}{\mathrm{d}\nu}\right| \frac{\mathrm{d}\nu}{\mathrm{d}\mu}\,\mathrm{d}\mu = \int \frac{\mathrm{d}|\lambda|}{\mathrm{d}\nu}\frac{\mathrm{d}\nu}{\mathrm{d}\mu}\,\mathrm{d}\mu = |\lambda|(X) < \infty.$$

The next results show that even if the Radon-Nikodym Theorem cannot be applied to a measure, it may still be applicable to 'a part of' the measure. An alternative method of proof of the next theorem is indicated in Exercise 28.

Theorem 11 (Lebesgue Decomposition Theorem): Let $[\![X, \mathcal{S}, \mu]\!]$ be a σ-finite measure space and ν a σ-finite measure on $\mathcal{S}$. Then $\nu = \nu_0 + \nu_1$ where ν_0, ν_1 are measures on $\mathcal{S}$ such that $\nu_0 \perp \mu$ and $\nu_1 \ll \mu$. This is the **Lebesgue decomposition** of the measure ν with respect to μ and it is unique.

Proof: Clearly the measure $\lambda = \mu + \nu$ is σ-finite and $\mu \ll \lambda$. So by Theorem 5, p. 139, there exists a non-negative finite-valued measurable function f such that if $E \in \mathcal{S}$, then $\mu(E) = \int_E f\,\mathrm{d}\lambda$. Let $A = [x: f(x) > 0]$, $B = [x: f(x) = 0]$. Then $A \cup B = X, A \cap B = \emptyset$ and $\mu(B) = \int_B f\,\mathrm{d}\lambda = 0$.

Define measures ν_0, ν_1 by $\nu_0(E) = \nu(E \cap B)$, $\nu_1(E) = \nu(E \cap A)$ for each $E \in \mathcal{S}$, so that $\nu = \nu_0 + \nu_1$. Since $\nu_0(A) = 0$ we have $\nu_0 \perp \mu$. Also $\nu_1 \ll \mu$; for if $\mu(E) = 0$ we have $\int_E f\,\mathrm{d}\lambda = 0$ and so, on $E, f = 0$ a.e. (λ). But f is positive on $E \cap A$ so $\lambda(E \cap A) = 0$. From the definition of λ we have $\nu \ll \lambda$ so $\nu_1(E) = \nu(E \cap A) = 0$.

To show that the decomposition is unique we suppose that $\nu = \nu_0 + \nu_1 = \nu_0' + \nu_1'$, where $\nu_0 \perp \mu$, $\nu_0' \perp \mu$, $\nu_1 \ll \mu$, $\nu_1' \ll \mu$. So there exist sets A, B, A', B' such that $X = A \cup B = A' \cup B'$, $A \cap B = A' \cap B' = \emptyset$, and $\nu_0(B) = \mu(A) = \nu_0'(B') = \mu(A') = 0$. Let $E \in \mathcal{S}$, then

$$E = (E \cap B \cap B') \cup (E \cap A' \cap B) \cup (E \cap A \cap A') \cup (E \cap A \cap B').$$

Clearly μ is zero on the last three sets in this union and hence ν_1 and ν_1' are zero by absolute continuity. Since $\nu_1' - \nu_1 = \nu_0 - \nu_0'$ we have

$$(\nu_1' - \nu_1)(E) = (\nu_1' - \nu_1)(E \cap B \cap B') = (\nu_0 - \nu_0')(E \cap B \cap B') = 0,$$

as $\nu_0(B) = \nu_0'(B) = 0$. So $\nu_1(E) = \nu_1'(E)$, which implies $\nu_0(E) = \nu_0'(E)$ and the uniqueness of the decomposition. □

Example 12: Let $[\![X, \mathcal{S}]\!]$ be a measurable space such that the points $[x]$ of X

are measurable sets, and let ν and μ be σ-finite measures on $[\![X, \mathcal{S}]\!]$. Then $\nu = \nu_1 + \nu_2 + \nu_3$ where $\nu_1 \ll \mu$, $\nu_2 + \nu_3 \perp \mu$, $\nu_i \perp \nu_j$ for $i \neq j$ and $\nu_3([x]) = 0$ for each $x \in X$.

Solution: The last theorem gives $\nu = \nu_0 + \nu_1$ where $\nu_1 \ll \mu$ and $\nu_0 \perp \mu$. Let $D = [x: \nu_0([x]) > 0]$ and write $\nu_2(E) = \nu_0(E \cap D)$, $\nu_3(E) = \nu_0(E \cap \mathbf{C}D)$ for each $E \in \mathcal{S}$. Then $\nu_2 \perp \nu_3$ and $\nu_2 \perp \mu, \nu_3 \perp \mu$ so $\nu_2 \perp \nu_1, \nu_3 \perp \nu_1$ since $\nu_1 \ll \mu$. Clearly $\nu_3([x]) = 0$ for each x.

Exercises

22. In the special case where μ is σ-finite use Theorem 5, p. 139, to prove Theorem 6, p. 142.
23. Let $\{a_n\}$, $\{b_n\}$ be sequences of positive numbers such that $\inf a_n = 0$, $\inf b_n > 0$. Let μ, ν be the measures on $\mathcal{P}(\mathsf{N})$ such that $\mu([n]) = a_n$, $\nu([n]) = b_n$. Show that $\mu \ll \nu$ but that the result of Theorem 6 does not hold.
24. Let $\{f_k\}$ be a mean fundamental sequence in $L(X, \mu)$; show that $\forall\ \epsilon > 0$, $\exists\ \delta > 0$ such that for all k, $\int_E |f_k|\, d\mu < \epsilon$ if $\mu(E) < \delta$.
25. Show that if μ and ν are σ-finite signed measures and $\mu \ll \nu$, $\nu \ll \mu$, then
$$\frac{d\nu}{d\mu} = \left(\frac{d\mu}{d\nu}\right)^{-1} \quad [\mu].$$
26. Let $f(x) = \sqrt{(1-x)}, x \leqslant 1, f(x) = 0, x > 1$, and let $g(x) = x^2, x \geqslant 0, g(x) = 0, x < 0$. Let $\nu(E) = \int_E f\, dx$ and $\mu(E) = \int_E g\, dx$ so that ν and μ are measures on $\mathcal{M}$. Find the Lebesgue decomposition of ν with respect to μ.
27. Show that the set D occurring in Example 12 is countable.
28. Prove Theorem 11 directly, without the use of Theorem 5, using, as in Theorem 3, a sequence of sets maximizing ν to obtain the set B of the proof.

8.5 BOUNDED LINEAR FUNCTIONALS ON L^p

First we make a formal definition which was implicit in Chapter 6.

Definition 11: Let V be a real vector space. Then V is a **normed vector space** if there is a function $\|x\|$ defined for each $x \in V$ such that (i) $\forall\ x$, $\|x\| \geqslant 0$, (ii) $\|x\| = 0$ if, and only if, $x = 0$, (iii) $\|ax\| = |a| \cdot \|x\|$ for any real number a and each $x \in V$, (iv) $\|x + y\| \leqslant \|x\| + \|y\|$, $\forall x, y \in V$.

Example 13: $L^p(X, \mu)$ $(1 \leqslant p \leqslant \infty)$ is a normed vector space with norm given by $\|f\|_p$.

Solution: Only (iv) is not obvious and it is proved by Theorem 8, p. 115, for $1 \leqslant p < \infty$ and by Example 12, p. 116, for $p = \infty$.

Example 14: $C[0,1]$, consisting of the continuous real-valued functions on $[0,1]$ with norm $\|f\|_\infty = \sup[|f(x)|: 0 \leqslant x \leqslant 1]$, is a normed vector space. On this space a sequence converges with respect to the norm if it is uniformly convergent on $[0,1]$.

Definition 12: A function G on the normal linear space V to the real numbers is a **linear functional** if $\forall\, x, y \in V$ and $a, b \in \mathrm{R}$, we have $G(ax + by) = a\, G(x) + b\, G(y)$.

Definition 13: A linear functional G on the normed linear space V is **bounded** if $\exists\ K \geqslant 0$ such that

$$|G(x)| \leqslant K\, \|x\|, \ \forall\, x \in V. \tag{8.9}$$

Then the *norm* of G, denoted by $\|G\|$, is the infimum of the numbers K for which (8.9) holds. So, easily, $|G(x)| \leqslant \|G\| \cdot \|x\|$. Then dividing by $\|G\|$ we see that $\|G\| = \sup[|G(x)|: \|x\| \leqslant 1]$, and it is easy to see that apart from the trivial case when $\dim V = 0$, $\|G\| = \sup[|G(x)|: \|x\| = 1]$.

Boundedness and continuity are related as follows.

Example 15: The following are equivalent for a linear functional G: (i) G is bounded, (ii) G is continuous at 0, (iii) G is continuous at each $x \in V$.

Solution: Let $x_n \to 0$; then $|G(x_n)| \leqslant \|G\| \cdot \|x_n\| \to 0$, so (i) implies (ii). Let $x_n \to x$; then $|G(x_n) - G(x)| = |G(x_n - x)| \leqslant \|G\| \cdot \|x_n - x\| \to 0$, so (ii) implies (iii). Clearly (iii) implies (ii). If (ii) holds but G is not bounded, then $\exists\ \{x_n\}$, $\|x_n\| \leqslant 1$, but $|G(x_n)| \geqslant n$. Then if $y_n = n^{-1} x_n$, so that $\|y_n\| \to 0$, we have $|G(y_n)| \geqslant 1$, contradicting (ii).

Example 16: Define G on $L^p(\mu)$ by $G(f) = \int fg\, d\mu$ for a fixed $g \in L^q(\mu)$, p and q being conjugate indices with $p \geqslant 1$ and with $q = \infty$ in the case where $p = 1$. Then G is a bounded linear functional and $\|G\| \leqslant \|g\|_q$.

Solution: This follows from Hölders inequality, p. 115, for $1 < p < \infty$ and from Theorem 9, p. 116, for the case $p = 1$.

It will follow from the main theorems of this section that $\|G\| = \|g\|_q$ for this kind of functional. It is convenient to deal separately with the cases $1 < p < \infty$ and $p = 1$. The next theorem shows that L^q is, in a sense, the set of bounded linear functionals, or *dual space*, of L^p.

Theorem 12 (Riesz Representation Theorem for $L^p, p > 1$): Let G be a bounded linear functional on $L^p(X, \mu)$. Then there exists a unique element g of $L^q(X, \mu)$ such that

$$G(f) = \int fg\, d\mu \text{ for each } f \in L^p \tag{8.10}$$

where p, q are conjugate indices. Also

$$\|G\| = \|g\|_q. \tag{8.11}$$

Proof: Let g and g' have the desired property and let E be any set of finite measure, so that $\chi_E \in L^p$. Then $\int_E (g-g')\, d\mu = \int \chi_E (g-g')\, d\mu = 0$. So $g = g'$ a.e., since the set $[x\colon g(x) \neq g'(x)]$ has σ-finite measure. So the uniqueness is proved. That $\|G\| \leqslant \|g\|_q$ was noted in Example 16, for any g satisfying (8.10). If $\|G\| = 0$, then $G(f) = 0$ for all f, so $g \equiv 0$ satisfies (8.10) and (8.11). So suppose $\|G\| > 0$. Suppose first that $\mu(X) < \infty$. For each $E \in \mathcal{S}$ define $\lambda(E) = G(\chi_E)$; we wish to show that λ is a signed measure. Clearly $\lambda(\emptyset) = 0$. Since $\chi_{A \cup B} = \chi_A + \chi_B$ for disjoint sets A, B, λ is finitely additive. Let $E = \bigcup_{i=1}^{\infty} E_i$ and let $A_n = \bigcup_{i=1}^{n} E_i$. We have $\|\chi_{A_n} - \chi_E\|_p = (\mu(E - A_n))^{1/p} \to 0$ as $n \to \infty$. Since G is continuous by Example 15, we have $\lambda(A_n) \to \lambda(E)$, so λ is countably additive. Since G takes only finite values, λ is a signed measure. Also, if $\mu(E) = 0$, then $\|\chi_E\|_p = 0$ so $\lambda(E) = 0$, that is, $\lambda \ll \mu$. So by Theorem 5, Corollary 1, p. 141, there exists $g \in L^1(\mu)$ such that for each $E \in \mathcal{S}$

$$G(\chi_E) = \int_E g\, d\mu = \int \chi_E\, g\, d\mu.$$

We now dispense with the signed measure λ and show that g has the required properties.

By linearity we have $G(\phi) = \int \phi\, g\, d\mu$ for any measurable simple function ϕ. But each function $f \in L^\infty(\mu)$ is the uniform limit a.e. of a sequence $\{\psi_n\}$ where each ψ_n is the difference of measurable simple functions, and so $\|f - \psi_n\|_p \to 0$. So, by the continuity of G,

$$G(f) = \int fg\, d\mu \text{ for each } f \in L^\infty(\mu). \tag{8.12}$$

We now show that $\|G\| = \|g\|_q$. Let the function α on X be defined by: $\alpha = 1$ where $g > 0$, $\alpha = -1$ where $g \leqslant 0$. So α is measurable and $\alpha g = |g|$. Let $E_n = \{x\colon |g(x)| \leqslant n\}$ and put $f = \alpha\chi_{E_n}|g|^{q-1}$ where p, q are conjugate indices. Then $|f|^p = |g|^q$ on E_n, $f \in L^\infty(\mu)$ and by (8.12)

$$\int_{E_n} |g|^q\, d\mu = \int fg\, d\mu = G(f) \leqslant \|G\| \cdot \|f\|_p = \|G\| \left(\int_{E_n} |g|^q\, d\mu\right)^{1/p}. \tag{8.13}$$

So we get

$$\int \chi_{E_n} |g|^q\, d\mu \leqslant \|G\|^q. \tag{8.14}$$

For this is obvious if $\left(\int_{E_n} |g|^q\, d\mu\right)^{1/p} = 0$; otherwise divide (8.13) across by this factor and raise to the power q. Since $\chi_{E_n} \uparrow 1$, (8.14) and Theorem 15, p. 105, give $\|g\|_q \leqslant \|G\|$, and, in particular, $g \in L^q(\mu)$. So by Example 16, $\|g\|_q = \|G\|$.

So (8.10) holds for $f \in L^\infty(X, \mu)$. But the bounded functions are dense in L^p. For it is sufficient to show that every non-negative function $f \in L^p$ is the limit, in the mean of order p, of a sequence $\{f_n\}$ of bounded functions. Put $f_n = \min(f, n)$. Then $0 \leqslant (f - f_n)^p \leqslant f^p$ and $f - f_n \to 0$ a.e. So by Theorem 21, p. 107, $\|f - f_n\|_p \to 0$. Then by the continuity of G, $G(f_n) \to G(f)$. Also, by Hölders inequality, $\int f_n g \, d\mu \to \int fg \, d\mu$. So $G(f) = \int fg \, d\mu$, proving the result of the theorem for finite measure spaces.

We now extend the result to the case when $X = \bigcup_{i=1}^{\infty} X_i$, where the X_i are disjoint measurable sets of finite μ-measure. Any function f_i on X_i, measurable with respect to the σ-algebra of sets $E \cap X_i$, $E \in \mathcal{S}$, can be extended to f on X by putting $f = 0$ on $\complement X_i$. Then G has the restriction G_i on $L(X_i, \mu)$ where $G_i(f_i) = G(f)$, and we have $\|G_i\| \leqslant \|G\|$. By the first part, $G_i(f_i) = G(\chi_{X_i} f) = \int_{X_i} f g_i \, d\mu$ for each $f \in L^p(X, \mu)$, for each i, and for a suitable $g_i \in L^q(X_i, \mu)$. Extend g_i to X by putting $g_i = 0$ on $\complement X_i$ and write $g = \Sigma g_i$. By linearity, if $Y_n = \bigcup_{i=1}^{n} X_i$,

$$G(\chi_{Y_n} f) = \int_{Y_n} f(g_1 + \ldots + g_n) \, d\mu, \ \forall f \in L^p(X, \mu).$$

As in the first part, since $\mu(Y_n) < \infty$, we have $\|g_1 + \ldots + g_n\|_q \leqslant \|G\|$ for each n. So

$$(\|g\|_q)^q = \int |\Sigma g_i|^q \, d\mu = \int \lim \left| \sum_{i=1}^{n} g_i \right|^q d\mu$$

$$\leqslant \liminf \int \left| \sum_{i=1}^{n} g_i \right| d\mu \leqslant \|G\|^q$$

by Fatou's Lemma, p. 105, giving $\|g\|_q = \|G\|$ by Example 16, as before. Also $\chi_{Y_n} f \to f$ in the mean of order p so $G(\chi_{Y_n} f) \to G(f)$. But $\sum_{i=1}^{n} g_i \to g$ in the mean of order q, so by Hölders inequality $\int \chi_{Y_n} f \sum_{i=1}^{n} g_i \, d\mu \to \int fg \, d\mu$.

Now consider the general case where μ need not be σ-finite. We show that there exists a set $X_0 \in \mathcal{S}$ which is of σ-finite measure, that is, X_0 is the union of a sequence of sets of finite measure, and such that if $f = 0$ on X_0 then $G(f) = 0$. Let $\{f_n\}$ be such that $\|f_n\|_p = 1$ and $G(f_n) \geqslant \|G\| (1 - 1/n)$. By Exercise 21, p. 108, we see that $X_0 = \bigcup_{n=1}^{\infty} [x \colon f_n(x) \neq 0]$ has σ-finite measure. Let $E \in \mathcal{S}$ with $E \subseteq \complement X_0$, then $\|f_n + t\chi_E\|_p = (1 + t^p \mu(E))^{1/p}$ for $t \geqslant 0$. Also

$$G(f_n) - G(\pm t\chi_e) \leqslant |G(f \mp t\chi_E)| \leqslant \|G\| (1 + t^p \mu(E))^{1/p},$$

and it follows that

$$|G(t\chi_E)| \leqslant \|G\| \, [(1 + t^p \, \mu(E))^{1/p} - 1 + n^{-1}],$$

for every n. Let $n \to \infty$ and then divide by $t(> 0)$ to get

$$|G(\chi_E)| \leqslant \|G\| \, \frac{(1 + t^p \, \mu(E))^{1/p} - 1}{t} \, .$$

Since $p > 1$ we may apply l'Hôpital's rule as $t \to 0$ to get $G(\chi_E) = 0$. So G vanishes for simple functions and hence for measurable functions which equal zero on X_0. So by the proof for the σ-finite case we can find $g \in L^q(X_0)$ such that

$$G(\chi_{X_0} f) = \int_{X_0} fg \, \mathrm{d}\mu.$$

Define g to be zero on $\mathbf{C}X_0$ to get the required function g of the theorem. □

Theorem 13 (Riesz Representation Theorem for L^1): Let $[\![X, \mathcal{S}, \mu]\!]$ be a σ-finite measure space and let G be a bounded linear functional on $L^1(X, \mu)$. Then there exists a unique $g \in L^\infty(X, \mu)$ such that

$$G(f) = \int fg \, \mathrm{d}\mu \text{ for each } f \in L^1(\mu). \tag{8.15}$$

Also, $\|G\| = \|g\|_\infty$.

Proof. Suppose first that $[\![X, \mathcal{S}, \mu]\!]$ is a finite measure space. As in the last theorem we construct a unique g such that (8.15) holds for $f \in L^\infty(X, \mu)$. We wish to show that $g \in L^\infty$. We have

$$\left| \int_E g \, \mathrm{d}\mu \right| \leqslant \|G\| \, \|\chi_E\|_1 = \|G\| \, \mu(E), \; \forall E \in \mathcal{S}. \tag{8.16}$$

Suppose that $|g(x)| > \|G\|$ on a set A of positive measure and write $E_n = [x: |g(x)| > (1 + 1/n) \, \|G\|]$. So $A = \cup E_n$. So, for some n, $\mu(E_n) > 0$ and $|g(x)| > (1 + 1/n) \, \|G\|$ on E_n. Then $\int_{E_n} g \, \mathrm{d}\mu \geqslant \|G\| \, (1 + 1/n) \, \mu(E_n)$, contradicting (8.16) as we may suppose $\|G\| > 0$. So $\|g\|_\infty \leqslant \|G\|$, and hence $\|g\|_\infty = \|G\|$ by Example 16.

We extend (8.15), as in the last theorem, to all functions $f \in L^1(\mu)$. Extend, as before, to the σ-finite case; we now have $\|g_1 + \ldots + g_n\|_\infty \leqslant \|G\|$ for each n, so $\|g\|_\infty \leqslant \|G\|$ as required. For the last part of the σ-finite case in the last theorem, Hölders inequality is replaced by Theorem 9, p. 116. □

Exercises

29. Prove the result used in Theorem 12: if $f_n \to f$ in $L^p(X, \mu)$, and $g_n \to g$ in $L^q(X, \mu)$, where p, q are conjugate, then $f_n g_n \to fg$ in $L^1(X, \mu)$.
30. Let V be a normed vector space and let V^* be the 'dual space' of bounded

linear functionals on V with norm as given in Definition 13, p. 148. Show that V^* is a normed vector space.

31. Let V be a normed vector space. Show that the mapping $G \to G(f)$, for fixed $f \in V$, is a bounded linear functional, f^{**} say, on V^* and $\|f^{**}\| \leqslant \|f\|$.
32. In the case $V = L^p(\mu)$ $(p > 1)$, show that the correspondence $f \to f^{**}$ of the last exercise is linear and norm preserving.
Note: In this notation, Theorem 12 says that $(L^p)^*$ $(1 < p < \infty)$ may be identified with L^q, and $(L^p)^{**}$ with L^p. For more results of this kind, see, for example, [15].
33. Show that the element $g \in L^q$ of norm 1 such that $\int fg \, d\mu = \|f\|$, for a given $f \in L^p$, is unique, where p, q are conjugate indices $1 < p, q < \infty$.
34. $L^1(X, \mu)$ is a normed linear space, and if μ is σ-finite the dual space is $L^\infty(X, \mu)$.

CHAPTER 9

Lebesgue-Stieltjes Integration

It is often natural on the real line to consider integration with respect to an increasing function, particularly in probability and classical applied mathematics, and integrals are considered from this view point in this chapter. It turns out that this is the natural context in which to consider indefinite integrals and integration by parts, and in Sections 9.3 and 9.4 we extend the calculus developed in earlier chapters. Section 9.2 provides results in this direction which complement those of Chapter 2, on Hausdorff measures. In the construction of such measures we use Helly's theorem which is of independent interest, especially in probability theory. In Section 9.5 we examine the change of variable in an intergal; the change of variable for Lebesgue integrals is a special case. Finally, in Section 9.6 we provide an analogue for the representation theorems of Chapter 8 and show that every bounded linear functional on the space of functions continuous on a closed bounded interval arises from a measure.

9.1 LEBESGUE-STIELTJES MEASURE

Let g be a finite-valued left-continuous monotone increasing function on R and for an interval $[a, b)$ define $\mu([a, b)) = g(b) - g(a)$. If $a = b$ the interval is empty, so $\mu(\emptyset) = 0$. Also the values of μ are non-negative. We wish to show first that μ can be extended to be a measure on $\mathcal{R}$, the ring of finite unions of intervals of the form $[a, b)$. Then the results of Chapter 5 provide a unique extension of μ to the Borel sets. For the rest of this section μ, g and $\mathcal{R}$ will have the connotations just described and, as in Chapter 2, intervals will be of the form $[a, b)$ unless stated otherwise, and will be finite. We could instead use intervals $(a, b]$ and right-continuous functions; the proofs follow with appropriate modifications. That there is a certain lack of symmetry is seen from Exercise 7 below.

Theorem 1: If E_i, $i = 1, \ldots, n$, are disjoint intervals such that $\bigcup_{i=1}^{n} E_i \subseteq I$, where I is an interval, then $\sum_{i=1}^{n} \mu(E_i) \leqslant \mu(I)$.

Proof: Let $E_i = [a_i, b_i)$ for each i, and $I = [a, b)$. Order the intervals E_i so that $a_1 \leqslant a_2 \leqslant \ldots \leqslant a_n$. Then

$$\begin{aligned}\sum_{i=1}^{n} \mu(E_i) &= \sum_{i=1}^{n} (g(b_i) - g(a_i)) \\ &\leqslant \sum_{i=1}^{n} (g(b_i) - g(a_i)) + \sum_{i=1}^{n-1} (g(a_{i+1}) - g(b_i)) \\ &= g(b_n) - g(a_1) \leqslant g(b) - g(a) = \mu(I). \quad \square\end{aligned}$$

Theorem 2: If $[a, b] \subset \bigcup_{i=1}^{n} (a_i, b_i)$, where $a_i, b_i, i = 1, \ldots, n$, are bounded, then

$$g(b) - g(a) \leqslant \sum_{i=1}^{n} (g(b_i) - g(a_i)). \tag{9.1}$$

Proof: Write $U_i = (a_i, b_i)$ and select intervals as follows. Let $a \in U_{k_1}$, say. If $b_{k_1} \leqslant b$, let k_2 be such that $b_{k_1} \in U_{k_2}$, etc., by induction, the sequence ending when $b_{k_m} > b$. Renumbering the intervals, we have chosen $U_1, \ldots, U_m$ where $a_{i+1} < b_i < b_{i+1}, i = 1, \ldots, m-1$. So

$$\begin{aligned}g(b) - g(a) &\leqslant g(b_m) - g(a_1) \\ &= g(b_1) - g(a_1) + \sum_{i=1}^{m-1} (g(b_{i+1}) - g(b_i)) \\ &\leqslant \sum_{i=1}^{m} (g(b_i) - g(a_i)).\end{aligned}$$

But $m \leqslant n$ and (9.1) follows. $\square$

Theorem 3: Let $\{E_i\}$ be a sequence of intervals and I an interval.

(i) If $I \subseteq \bigcup_{i=1}^{\infty} E_i$, then $\mu(I) \leqslant \sum_{i=1}^{\infty} \mu(E_i)$.

(ii) If the E_i are disjoint and $I = \bigcup_{i=1}^{\infty} E_i$, then $\mu(I) = \sum_{i=1}^{\infty} \mu(E_i)$.

Proof: (i) Suppose that $I = [a, b)$ and $E_i = [a_i, b_i)$, each i. We may suppose that $b > a$ and if $\epsilon > 0$ we may choose c such that $0 < c < b - a$ and $g(b) - g(b - c) < \epsilon$. Also, for each i, choose ξ_i such that $\xi_i < a_i$ and $g(a_i) - g(\xi_i) < \epsilon/2^i$. Write $F = [a, b - c]$ and $U_i = (\xi_i, b_i)$. So $F \subset \bigcup_{i=1}^{\infty} U_i$ and by the Heine-Borel Theorem, p. 18, $F \subset \bigcup_{i=1}^{n} U_i$ for some n. Then by Theorem 2

$$g(b - c) - g(a) \leqslant \sum_{i=1}^{n} (g(b_i) - g(\xi_i))$$

$$< \sum_{i=1}^{n} (g(b_i) - g(a_i) + \epsilon/2^i) < \sum_{i=1}^{\infty} (g(b_i) - g(a_i)) + \epsilon.$$

So $g(b) - g(a) - \epsilon < \sum_{i=1}^{\infty} \mu(E_i) + \epsilon$. But ϵ is arbitrary and the result follows.

(ii) From Theorem 1, p. 153, we obtain $\sum_{i=1}^{\infty} \mu(E_i) \leqslant \mu(I)$. But then (i) gives the result. □

Theorem 4: There exists a unique measure $\bar{\mu}$ on $\mathcal{R}$ such that, if I is an interval, we have $\bar{\mu}(I) = \mu(I)$.

Proof: Each set $E \in \mathcal{R}$ can be written as $E = \bigcup_{i=1}^{n} E_i$ where the E_i are *disjoint* intervals. Define $\bar{\mu}(E) = \sum_{i=1}^{n} \mu(E_i)$. This defines $\bar{\mu}$ uniquely on $\mathcal{R}$ since if $E = \bigcup_{j=1}^{m} F_j$ is another decomposition of E into disjoint intervals, then $E = \bigcup_{i,j} (E_i \cap F_j)$, the intervals $E_i \cap F_j$ are disjoint and

$$\begin{aligned} \bar{\mu}(E) &= \sum_{i=1}^{n} \mu(E_i) = \sum_{i=1}^{n} \sum_{j=1}^{m} \mu(E_i \cap F_j) \\ &= \sum_{j=1}^{m} \sum_{i=1}^{n} \mu(E_i \cap F_j) = \sum_{j=1}^{m} \mu(F_j), \end{aligned}$$

using the additivity of μ given by Theorem 3(ii). So μ and $\bar{\mu}$ are equal for intervals; also $\bar{\mu}$ is clearly finitely additive.

Let $\{E_i\}$ be a sequence of disjoint sets of $\mathcal{R}$ such that $E = \bigcup_{i=1}^{\infty} E_i \in \mathcal{R}$. Then, for each i, E_i is a finite union of disjoint intervals

$$E_i = \bigcup_{j=1}^{m(i)} E_{i,j},$$

so $\bar{\mu}(E_i) = \sum_{j=1}^{m(i)} \mu(E_{i,j})$. If E is an interval, then Theorem 3(ii) gives

$$\bar{\mu}(E) = \mu(E) = \sum_{i=1}^{\infty} \sum_{j=1}^{m(i)} \mu(E_{i,j}) = \sum_{i=1}^{\infty} \bar{\mu}(E_i) \tag{9.2}$$

as the intervals $E_{i,j}$ are disjoint. In general, we can write $E = \bigcup_{k=1}^{m} F_k$ where the F_k are disjoint intervals. Then, as $\bar{\mu}$ is finitely additive

$$\bar{\mu}(E) = \sum_{k=1}^{m} \bar{\mu}(F_k) = \sum_{k=1}^{m} \sum_{i=1}^{\infty} \bar{\mu}(F_k \cap E_i)$$

by (9.2). So $\bar{\mu}(E) = \sum_{i=1}^{\infty} \sum_{k=1}^{m} \bar{\mu}(F_k \cap E_i) = \sum_{i=1}^{\infty} \bar{\mu}(E_i)$. So $\bar{\mu}$ is countably additive. Since $\bar{\mu}(\emptyset) = \mu(\emptyset) = 0$, $\bar{\mu}$ is a measure.

Clearly, any measure on $\mathcal{R}$ which extends μ must, from the definition of $\bar{\mu}$, equal $\bar{\mu}$ on each set of $\mathcal{R}$. So the extension is unique. □

As a result of this theorem we may drop the notation $\bar{\mu}$ and write $\mu(E)$ for any set E in $\mathcal{R}$. Theorems 1 to 4 give the unique extension from μ defined on the intervals to a measure on the ring of finite unions of intervals. In the special case $g(x) \equiv x$, the extended measure μ equals the length if the set is an interval. In general, by Theorem 4, p. 97, and Theorem 5, p. 98, the measure μ on $\mathcal{R}$ has an extension to a measure on a σ-ring $\mathcal{S}^*$ which contains the σ-ring generated by $\mathcal{R}$, that is, $\mathcal{S}^*$ contains the Borel sets. Since g is finite-valued, μ is σ-finite and hence, by Example 7, p. 98, its extension is σ-finite and by Theorem 7, p. 100, the extension $\bar{\mu}$ of μ to the Borel sets is unique. The notation is introduced in the following definition.

Definition 1: The **Lebesgue-Stieltjes measure** $\bar{\mu}_g$ induced by the monotone increasing left-continuous function g is the completion of the extension $\bar{\mu}$ of the measure μ given by Theorem 4, p. 155, to the Borel sets; $\bar{\mu}_g$ is defined on a σ-algebra $\bar{\mathcal{S}}_g$ such that $\mathcal{B} \subseteq \bar{\mathcal{S}}_g$.

In the next example we see how to go from a measure to an associated function.

Example 1: Let μ be a finite measure on $[\![\mathsf{R}, \mathcal{B}]\!]$ and let $g(x) = \mu(-\infty, x)$. Then $\mu([a, b)) = g(b) - g(a)$, g is monotone increasing and left-continuous; and on $\mathcal{B}$, $\mu = \bar{\mu}_g$. Then any function $g + K$ where K is constant, is described as a **primitive** of μ. The primitive g constructed above is characterized by the fact that $g(-\infty) = \lim_{x \to -\infty} g(x) = 0$.

The following example shows that every montone increasing function has associated left-continuous and right-continuous monotone increasing functions, and will be referred to in Section 9.4. In probability theory the right-continuous function given in Example 2(iii) is referred to as the *distribution function*. In that case $h(-\infty) = 0, h(+\infty) = 1$.

Example 2: Let f be a finite-valued monotone increasing function defined on (a, b). Show that

(i) $g(x) = f(x-)$ is left-continuous and monotone increasing on (a, b),
(ii) $h(x) = f(x+)$ is right-continuous and monotone increasing on (a, b),
(iii) if μ is a finite measure on $[\![\mathsf{R}, \mathcal{B}]\!]$, $g(x) = \mu(-\infty, x)$, and $h(x) = \mu(-\infty, x]$, then $h(x) = g(x+)$.

Solution: Given $\epsilon > 0$, there exists $\delta_0 > 0$ such that if $0 < \delta < \delta_0$, $f(x-) - f(x - 2\delta) < \epsilon$. But $f(x - 2\delta) \leqslant f((x - \delta) -)$, so $g(x) - g(x - \delta) < \epsilon$, and g is left-continuous. It is clearly monotone increasing, for if $x_1 < x_2$, $g(x_1) \leqslant$

$f(x_2-) = g(x_2)$. The proof for (ii) is similar. Note that since f is monotone increasing, g and h are well defined on (a, b). For (iii), using the finiteness of μ, $\mu(-\infty, x] = \lim_{n\to\infty} \mu(-\infty, x + 1/n) = g(x+)$.

Exercises

1. If g is continuous, $\bar{\mu}_g([x]) = 0$ for each x.
2. The set $[x: \bar{\mu}_g([x]) > 0]$ is at most countable.
3. Let h be a finite-valued right-continuous monotone increasing function on R and for any interval $(a, b]$ define $\mu((a, b]) = h(b) - h(a)$. Then μ is a measure on the ring $\mathcal{R}'$, say, of finite unions of intervals of the type $(a, b]$.
4. If $g(-\infty) = 0$ then, for each $x, g(x) = \bar{\mu}_g(-\infty, x)$.
5. The Lebesgue function L, p. 25, corresponding to the Cantor set P, induces a Lebesgue-Stieltjes measure ν on $[0,1]$ such that $\nu([0,1]) = 1, \nu \perp m$, and the ν_3 component in the decomposition of Example 12, p. 146, is zero.
6. Let $\bar{\mu}_g$ be the Lebesgue-Stieltjes measure induced by a function g constant on R except at a finite number of points. Then every subset of R is $\bar{\mu}_g$-measurable.
7. Let the function $f_n, n = 1, 2, \ldots$, be monotone increasing and let $f_n(x) \uparrow f(x)$ for each x, where f is finite-valued. Then: (i) f is monotone increasing; (ii) $\lim f_n(x-) = f(x-)$; (iii) $\lim f_n(x+) = f(x+)$ is not true in general; (iv) $\lim f_n(x+) = f(x+)$ if $f_n \to f$ uniformly; (v) if, in addition, each f_n is left-continuous, so is f.

9.2 APPLICATIONS TO HAUSDORFF MEASURES

Definition 2: The set E **supports** the measure μ if E is a measurable set and $\mu(CE) = 0$.

The following result is of interest, especially in probability theory, and will be used in Theorem 6 below.

Theorem 5: Let A be a countable set which is dense in R. Every sequence $\{g_k\}$ of primitives, such that $0 \leqslant g_k \leqslant M$ for all k, has a subsequence $\{g_{k_j}\}$ that converges to a primitive g at all points of A.

Proof: Let A be the sequence $\{a_i\}$. Then it is possible to find a subsequence $\{g_n^{(1)}\}$ of $\{g_n\}$ such that $\{g_n^{(1)}(a_1)\}$ converges. From $\{g_n^{(1)}\}$ we extract a further subsequence $\{g_n^{(2)}\}$ such that $\{g_n^{(2)}(a_2)\}$ converges. By induction we construct, for each n, a sequence $\{g_k^{(n)}, k = 1, 2, \ldots\}$ converging at a_n and contained in the previous subsequence. Now consider the 'diagonal' subsequence $g_n^{(n)}$. Except for its first $(k-1)$ terms it is contained in the k-th subsequence $g_h^{(k)}$ and so it converges at a_k. So, for each fixed a_k, $\{g_n^{(n)}(a_k)\}$ converges; write the limit at a_k as $g(a_k)$. Since $0 \leqslant g_n \leqslant M$ for each n, we have $0 \leqslant g \leqslant M$. Define $g(x) = \sup[g(a_j): a_j \leqslant x]$, extending g to R. Then g is easily seen to be monotone increasing and left-continuous and $g_n \to g$ on A. □

Corollary 1 (Helly's Theorem): Every bounded sequence $\{g_n\}$ of primitives has a subsequence $\{g_{n_k}\}$ that converges to a primitive g at all points of continuity of g.

Proof: We choose any countable dense set A as in the theorem. If x is a point of continuity of g, and $\epsilon > 0$, there exist a_i, a_j such that $a_i < x < a_j$ and $g(a_j) - g(a_i) < \epsilon$. If $\{g_{n_k}\}$ is the subsequence constructed in the theorem, we have $g_{n_k}(a_i) \leqslant g_{n_k}(x) \leqslant g_{n_k}(a_j)$. Letting $k \to \infty$, the first and last terms in this inequality have limits $g(a_i)$ and $g(a_j)$ respectively. So $\limsup g_{n_k}(x) - \liminf g_{n_k}(x) < \epsilon$ and we deduce that $\lim g_{n_k}(x)$ exists and equals $g(x)$. □

Corollary 2: Suppose that the primitives g_n are distribution functions so that $g_n(\infty) = 1$, $g_n(-\infty) = 0$. Then g is a distribution function provided that $\forall\ \epsilon$, $0 < \epsilon < 1$, $\exists\ a > 0$ such that $|g_n(a) - g_n(-a)| > 1 - \epsilon$ for all n sufficiently large.

Theorem 6: Let h be a Hausdorff measure function such that $h(2t) \leqslant 2h(t)$. Then $H(E) > 0$ if, and only if, E supports a non-zero Borel measure μ whose primitive g is continuous and such that $\omega_g(t) = O(h(t))$ as $t \to 0+$.

Proof: Suppose that such a measure μ exists. For $H(E)$ to be a positive it is necessary and sufficient that $H(E \cap [n, n+1]) > 0$ for some n, so we may suppose that $E \subseteq [0,1]$. Let $\{(u_i, v_i)\}$ be a covering of E, and suppose that $\omega_g(t) \leqslant \lambda h(t)$ for $t \leqslant 1$. Then

$$0 < \mu([0,1]) \leqslant \mu(\mathbf{C}E) + \Sigma\, \mu((u_i, v_i)) = \Sigma\, g(v_i - u_i) \leqslant \lambda\, \Sigma\, h(v_i - u_i).$$

So $H(E) > 0$, and we note that the condition $h(2t) \leqslant 2h(t)$ has not been used for the proof in this direction.

Conversely, suppose $H(E) > 0$. Then $\exists\ \rho$ such that for coverings of E by open intervals $\{I_k\}$ of length at most ρ, $\Sigma\, h(l(I_k)) > \frac{1}{2}H(E)$. For all other coverings $\Sigma\, h(l(I_k)) > h(\rho)$ so we always have

$$\inf \Sigma\, h(l(I_k)) > 0 \tag{9.3}$$

where the infimum is taken over all coverings $\{I_k\}$ of E by open intervals. Now $E = \cup E_n$, where $E_n = E \cap [n, n+1)$. So some $H(E_n) > 0$ and for that n we will find a measure μ_n as described, supported by $[n, n+1)$, and put $\mu_n = 0$ outside this interval. So without loss of generality we suppose $E \subseteq [0,1)$. For each positive integer n divide $[0,1)$ into intervals $I_{n,j} = [j/2^n, (j+1)/2^n)$, $j = 0, \ldots, 2^n - 1$. Write these intervals as I_n, for convenience, and their lengths as $\delta_n = 2^{-n}$. Give each interval I_n which meets E a uniformly distributed measure of total value $h(\delta_n)$, the other I_n intervals are given measure zero. This gives a measure ν_n on $[0,1)$. Then consider the intervals I_{n-1}. Any of these which have been given a measure at most $h(\delta_{n-1})$ have their measures unchanged. On the others the measure ν_n is multiplied by a factor which decreases the total for each interval to $h(\delta_{n-1})$. This gives a new measure $\nu_{n,1}$ on $[0,1)$. Similarly, having constructed $\nu_{n,j}$ we consider the intervals I_{n-j-1} and define the measure $\nu_{n,j+1}$. Since $h(2t) \leqslant 2h(t)$ the sequence of measures $\{\nu_{n,j}\}$ is clearly decreasing, for each n. Denote by μ_n the measure $\nu_{n,n}$. Then for each interval I_j, with $j \leqslant n$, we have

$$\mu_n(I_j) \leqslant h(\delta_j) \tag{9.4}$$

and in particular

$$\mu_n([0,1)) \leqslant h(1). \tag{9.5}$$

Each point of E belongs to at least one I_j for which $\mu_n(I_j) = h(\delta_j)$. Choose a set of such intervals which cover E and do not overlap. We get $\mu_n([0,1)) = \Sigma\, h(\delta_j)$. So by (9.3)

$$\mu_n([0,1)) \geqslant b > 0 \tag{9.6}$$

where b is independent of n.

By Theorem 5 and (9.5), $\{\mu_n\}$ has a subsequence $\{\mu_{n_k}\}$ converging to a measure μ. It is easy to see that μ is supported by E, and we may suppose that the set A in Theorem 5 is chosen so that the primitive of μ is the limit of the primitives of the measures μ_{n_k} on all the end-points of the I_j. So

$$\mu(I_j) \leqslant h(\delta_j) \tag{9.7}$$

and $\mu([0,1)) \geqslant b > 0$. This shows that μ is a positive measure. From (9.7) we can calculate the modulus of continuity of the primitive g of μ. Any two points at distance at most δ_j belong either to the same I_j or to two adjacent I_j intervals. So $\omega_g(\delta_j) \leqslant 2h(\delta_j)$. So if $0 < \delta \leqslant 1$ and $\delta \leqslant \delta_j < 2\delta$ we have $\omega_g(\delta) \leqslant \omega_g(\delta_j) \leqslant 2h(\delta_j) \leqslant 2h(2\delta) \leqslant 4h(\delta)$ and the theorem is proved. □

Corollary: The Hausdorff dimension of a compact set E is the supremum of the numbers $\beta \geqslant 0$ such that there exists a non-zero positive measure μ, supported by E, whose primitive g satisfies $\omega_g(t) = O(t^\beta)$ as $t \to 0+$.

Proof. For $h(t) = t^\alpha$, $0 < \alpha \leqslant 1$, we have $h(2t) \leqslant 2h(t)$. Now the result follows immediately from the theorem and Definition 15, p. 50. □

Example 3: A Hausdorff measure function providing a measure H such that $0 < H(E) < \infty$, where $E = P_\xi$ is given by ω_L where L is the Lebesgue function corresponding to P_ξ, and ω_L is its modulus of continuity.

Solution: L is a monotone increasing continuous function which defines a Lebesgue-Stieltjes measure $\bar{\mu}_L$ supported by P_ξ. So $H(E) > 0$ by the first part of Theorem 6. To show $H(E)$ is finite: at the j-th stage of the construction of P_ξ, let the 2^j residual intervals each have length b_j. Then by the remarks after Definition 16, p. 52, we have $\omega_L(b_j) \leqslant 2.2^{-j}$. Then P_ξ has a covering by 2^j intervals of length b_j. These intervals are closed but since ω_L is continuous, the Hausdorff measure is unaffected, as in Example 24, p. 46. So $H(E) \leqslant \liminf\,(2^j\omega_L(b_j)) \leqslant 2$.

9.3 ABSOLUTELY CONTINUOUS FUNCTIONS

Definition 3: A function f is **absolutely continuous** on $[a, b]$, $-\infty \leqslant a < b \leqslant \infty$ if, given $\epsilon > 0$, there exists $\delta > 0$ such that $\sum_{i=1}^{n} |f(x_i) - f(y_i)| < \epsilon$ whenever $\sum_{i=1}^{n} |x_i - y_i| < \delta$ for any finite set of disjoint intervals such that $(x_i, y_i) \subset [a, b]$ for each i.

The usefulness of this definition is due to the result of Theorem 8, where we show the close connection with absolute continuity of measures and with indefinite integrals. Clearly, an absolutely continuous function is continuous. That absolute continuity is a stringent condition is shown by the next example.

Example 4: Let $f(x) = \sqrt{x}$, $0 \leqslant x \leqslant 1/2$. Let $f(1) = 0$ and define f to be linear on $[1/2, 1]$. Let $f(x + k) = f(x)$ for each $k \in \mathsf{Z}$ and each x. Show that f is continuous on R but not absolutely continuous.

Solution: From its definition, f is continuous on $[0,1]$ and so on R. Given δ, $0 < \delta < 1/2$, let $x_i = i$, $y_i = i + \delta/i^2$. Then, for each n, $\sum_{i=1}^{n} |x_i - y_i| < 2\delta$ but $\sum_{i=1}^{n} |f(x_i) - f(y_i)| = \sum_{i=1}^{n} \sqrt{\delta}/i$ which tends to infinity with n. So f is not absolutely continuous.

Recall Definition 3, p. 81.

Theorem 7: Let f be absolutely continuous on $[a, b]$, where a and b are finite; then $f \in BV[a, b]$.

Proof: Let ϵ and δ be as in Definition 3, and let $a = x_0 < x_1 < \ldots < x_N$ be any partition of $[a, b]$. Introduce new points $a + i(b - a)/n, i = 0, 1, \ldots, n$, where n is such that $(b - a)/\delta < n < (b - a)/\delta + 1$, so that the new partition points are a distance less than δ apart. Let $[z_i: i = 0, 1, \ldots, M]$ be the complete set of partition points, then

$$\sum_{i=1}^{N} |f(x_i) - f(x_{i-1})| \leqslant \sum_{i=1}^{M} |f(z_i) - f(z_{i-1})|.$$

Collect the subintervals into groups beginning and ending with the added partition points; then this sum, with the z_i's renumbered, is

$$\sum_{k=1}^{n} \sum_{i=1}^{n_k} |f(z_{i,k}) - f(z_{i-1,k})| \leqslant \sum_{k=1}^{n} \epsilon = n\,\epsilon \leqslant (1 + (b - a)/\delta)\epsilon.$$

But this bound is independent of the original partition $x_0, \ldots, x_N$, so $f \in BV[a, b]$. □

Example 5: Show that if $f \in L(a, b)$, its indefinite integral F is absolutely continuous on $[a, b]$.

Solution: Let $\epsilon > 0$ be given. For any set of disjoint intervals $(x_i, y_i), i = 1, \ldots, n$, with $\bigcup_{i=1}^{n} (x_i, y_i) = E$, say, we have

$$\sum_{i=1}^{n} |F(y_i) - F(x_i)| = \sum_{i=1}^{n} \left| \int_{x_i}^{y_i} f \, dt \right| \leqslant \int_E |f| \, dt.$$

But by Theorem 18, p. 106, $\int_E |f| \, dt < \epsilon$ provided $m(E) < \delta$, for some $\delta > 0$, giving the result.

Example 6: Let f be absolutely continuous on the finite interval $[a, b]$; so $f \in BV[a, b]$. Let $f = f_1 - f_2$ be the decomposition of f into non-negative monotone increasing functions provided by Theorem 2, p. 82. Show that f_1 and f_2 are absolutely continuous on $[a, b]$.

Solution: Let ϵ and δ be as in Definition 3 and $(x_i, y_i), i = 1, \ldots, n$, be a set of disjoint subintervals of $[a, b]$ such that $\sum_{i=1}^{n} |x_i - y_i| < \delta$. We easily have, in the notation of Definition 2, p. 81, that

$$\sum_{i=1}^{n} T_f[x_i, y_i] \leqslant \epsilon.$$

But then $\sum_{i=1}^{n} |f_1(x_i) - f_1(y_i)| = \sum_{i=1}^{n} P_f[x_i, y_i] \leqslant \epsilon$, and similarly for f_2.

Example 7: Let f and g be absolutely continuous on the finite interval $[a, b]$. Show that fg is absolutely continuous on $[a, b]$.

Solution: Since f and g are continuous, they are bounded; suppose $|f|, |g| \leqslant M$ on $[a, b]$. Since $f(x)g(x) - f(y)g(y) = (f(x) - f(y))\, g(x) + f(y)(g(x) - g(y))$, we have $|f(x)g(x) - f(y)g(y)| \leqslant M|f(x) - f(y)| + M|g(x) - g(y)|$. So the absolute continuity of f and g with the ϵ and δ of Definition 3, p. 160, implies that of fg with $2M\epsilon$ and δ.

In the following theorem the absolutely continuous function g is supposed defined on R. If the function given is defined on a finite interval $[\alpha, \beta]$ only, we may extend it to R by defining it to be constant on $(-\infty, \alpha]$ and $[\beta, \infty)$.

Theorem 8: Let g be a monotone increasing and absolutely continuous function on R. Then $\mathcal{M} \subseteq \bar{\mathcal{S}}_g$, and on $\mathcal{M}$, $\bar{\mu}_g \ll m$.

Proof: Let E be a Borel set such that $m(E) = 0$. For any $\delta > 0$ there exists an open set $O \supset E$ such that $m(O) < \delta$, and if O is the union of disjoint open intervals $(a_n, b_n), n = 1, 2, \ldots$, we have $\sum_{n=1}^{\infty} (b_n - a_n) < \delta$. Let $\epsilon > 0$ be given and choose $\delta > 0$ such that $\sum_{i=1}^{n} (g(\eta_i) - g(\xi_i)) < \epsilon$ whenever the intervals $(\xi_i, \eta_i), i = 1, \ldots, n$, are disjoint and such that $\sum_{i=1}^{n} (\eta_i - \xi_i) < \delta$. Then $\sum_{i=1}^{n} (g(b_i) - g(a_i)) < \epsilon$ for each n, so that $\sum_{i=1}^{\infty} (g(b_i) - g(a_i)) \leqslant \epsilon$. But $E \subset \bigcup_{i=1}^{\infty} (a_i, b_i)$ so $\bar{\mu}_g(E) \leqslant \epsilon$. So $\bar{\mu}_g(E) = 0$. But if A is any Lebesgue measurable set with $m(A) = 0$, there exists a Borel set $E \supseteq A$ such that $m(E) = 0$. Since $\bar{\mu}_g$ is complete by Definition 1, p. 156, $A \in \bar{\mathcal{S}}_g$ and $\bar{\mu}_g(A) = 0$. □

Corollary 1: With the conditions of the theorem, $g(x) - g(a) = \int_a^x g'(t)\,dt$ for all $x \in [a, b]$, where a and b are any fixed finite numbers.

Proof: By Theorem 5, p. 139, $\bar{\mu}_g([a, x)) = \int_a^x f(t)\,dt$ where f is a non-negative measurable function, and $f \in L(a, b)$ as g is finite. So $g(x) - g(a) = \int_a^x f(t)\,dt$. But then $g' = f$ by Theorem 12, p. 89. □

Corollary 2: Again with the conditions of the theorem, $d\bar{\mu}_g/dm = g'\ [m]$.

Proof: The set function $\phi(E) = \int_E g'\,dx$ agrees with $\bar{\mu}_g$ on every finite interval $[a, x)$, by Corollary 1, and so on every Borel set. So if $F \in \mathcal{M}$, let $E \in \mathcal{B}$ be such that $m(F \,\Delta\, E) = 0$. Then $\bar{\mu}_g(F \,\Delta\, E) = 0$, so $\bar{\mu}_g(F) = \bar{\mu}_g(E) = \phi(E) = \phi(F)$ as required. □

Corollary 3: Let f be absolutely continuous on the finite interval $[a, b]$; then if $x \in [a, b]$,

$$f(x) - f(a) = \int_a^x f'(t)\,dt.$$

Proof: Let $f = f_1 - f_2$ as in Example 6, so that f_1 and f_2 are monotone increasing and absolutely continuous on $[a, b]$. Then Corollary 1, applied to f_1 and f_2 separately, gives the result. □

From this last result and from Example 5, p. 161, we see that on finite intervals a function is an indefinite integral if, and only if, it is absolutely continuous, thus answering the question raised in Chapter 4, p. 87.

Exercises

8. Show that f absolutely continuous on $[a, b]$ does not imply $f \in BV[a, b]$ if a or b is allowed to be infinite.
9. Show that if f satisfies a Lipschitz condition of order 1 on $[a, b]$, i.e. $|f(x) - f(y)| \leqslant K|x - y|$ for some K and each x and y, then f is absolutely continuous.
10. Show that if ϕ is convex on (a, b), then ϕ is absolutely continuous on each closed subinterval $[c, d] \subset (a, b)$.
11. Show that if f is non-negative and absolutely continuous on $[a, b]$, then so is f^p for $p \geqslant 1$.
12. Show that the absolute continuity of the non-negative function f on the finite interval $[a, b]$ does not imply that f^p is absolutely continuous, if $0 < p < 1$.
13. Show that the Lebesgue function L on $[0,1]$, p. 25, is not absolutely continuous.
14. Let f be a non-negative integrable function on (a, b) with indefinite integral F and let $K = [x\colon f(x) \neq 0]$. Define g by $g(x) = \text{distance}\,(x, K), x \in (a, b)$. Show that $\int_a^b g(x)\, \mathrm{d}\bar{\mu}_F(x) = 0$.
15. Show that the condition that the intervals (x_i, y_i) be disjoint, may not be omitted from Definition 2.
16. If f and g are absolutely continuous functions on R, is $f \circ g$ necessarily absolutely continuous?

9.4 INTEGRATION BY PARTS

We recall from Definition 4 of Chapter 4 the notation $F(x) = \int_a^x f\, \mathrm{d}t$ for the indefinite integral F of an integrable function f, and the result of Theorem 12, p. 89, that $F' = f$ a.e.

Theorem 9: Let $f, g \in L(a,b)$, where a and b are finite, and F, G be their indefinite integrals; then Fg and $Gf \in L(a, b)$ and for each $x \in (a, b)$

$$F(x)\, G(x) - F(a)\, G(a) = \int_a^x (fG + Fg)\, \mathrm{d}t. \tag{9.8}$$

Proof: In (a, b), outside a set E of zero measure, we have $F' = f$ and $G' = g$. Since F and G are bounded on $[a, b]$, Fg and $Gf \in L(a, b)$ and the usual proof for the derivative of a product gives $(FG)' = Fg + Gf$ on $[a, b] - E$. By Example 5, p. 161, F and G are absolutely continuous, and hence FG is absolutely continuous by Example 7, p. 161. So by Corollary 3 to Theorem 8, p. 162, FG is the indefinite integral of its derivative, giving (9.3). □

We now wish to extend Theorem 9 to the case of integrals with respect to a Lebesgue-Stieltjes measure $\bar{\mu}_g$. It is convenient to state this extension in terms of a left continuous function $g \in BV[a, b]$, and for this purpose we need the following result for which we use the notations of Definition 2, p. 81.

Theorem 10: Let $g \in BV[a, b]$ and let g be left-continuous at x, $a < x < b$. Then $g_1(y) = P_g[a, y]$ and $g_2(y) = N_g[a, y]$ are left-continuous at x.

Proof: Since $T_g[a, y] = P_g[a, y] + N_g[a, y]$ and all three functions increase with y, it is sufficient to show that $T_g[a, y]$ is left-continuous at x. Suppose it is not, then, using Example 7, p. 81, there exists $\epsilon > 0$ such that $T_g[y, x] > \epsilon$ for all $y \in [a, x)$. So we can find a partition $a = a_0 < a_1 < \ldots < a_{n_1-1} < x$ such that

$$\sum_{i=1}^{n_1-1} |g(a_i) - g(a_{i-1})| + |g(x) - g(a_{n_1-1})| > \epsilon/2.$$

But as g is left-continuous at x we can find a_{n_1}, say, such that $a_{n_1-1} < a_{n_1} < x$ and

$$\sum_{i=1}^{n_1} |g(a_i) - g(a_{i-1})| > \epsilon/2.$$

But $T_g[a_{n_1}, x] > \epsilon$, so we get similarly $a_{n_1} < a_{n_1+1} < \ldots < a_{n_2} < x$ where

$$\sum_{i=n_1+1}^{n_2} |g(a_i) - g(a_{i-1})| > \epsilon/2.$$

So we get a sequence $\{a_i\}$ such that for each integer k we get $t_g > k\epsilon/2$ for the partition $a = a_0 < a_1 < \ldots < a_{n_k} < x$. But this implies that $T_g[a, x] = \infty$, a contradiction. □

Definition 4: Let g be a left-continuous function, $g \in BV[a, b]$ and let $g = g_1 - g_2$ be the decomposition of g into positive and negative variations. Let $\bar{\mu}_g$ denote the signed measure $\bar{\mu}_{g_1} - \bar{\mu}_{g_2}$ which, by the last theorem, is well defined. We will say that f is ***integrable with respect to*** g if it is integrable with respect to $\bar{\mu}_{g_1}$ and $\bar{\mu}_{g_2}$ and we will write its integral as $\int f \, d\bar{\mu}_g = \int f \, d\bar{\mu}_{g_1} - \int f \, d\bar{\mu}_{g_2}$ or, in an alternative notation, as $\int f \, dg = \int f \, dg_1 - \int f \, dg_2$.

Clearly $\bar{\mu}_g$ is defined on a σ-algebra which contains the Borel sets. Note that if g is, initially, defined only on a finite interval $[\alpha, \beta]$ the integral $\int f \, d\bar{\mu}_g$ may still be considered since we may extend g to R by defining it to be constant on $(-\infty, \alpha]$ and $[\beta, \infty)$. In the following theorem we consider integrals of f and g over an interval $[a, b]$ which will normally be a proper subinterval of that in which the functions are defined, so that $f(b+)$ and $g(b+)$ are meaningful; if the functions are given on $[a, b]$ only, $f(b+)$ and $g(b+)$ are to be taken as $f(b)$ and $g(b)$, respectively. The integration-by-parts formula of Theorem 9 is a special case of the next result, as is indicated in Example 8 below.

Theorem 11: If f and g are left-continuous functions on the finite interval $[a, b]$ and $f, g \in BV[a, b]$, then

$$\int_{[a,b]} f(x+)\, d\bar{\mu}_g + \int_{[a,b]} g\, d\bar{\mu}_f = f(b+)\,g(b+) - f(a)\,g(a). \tag{9.9}$$

Proof: We may suppose that f and g are non-negative and monotone increasing functions, for otherwise we may decompose f and g with $f = f_1 - f_2$ and $g = g_1 - g_2$ as above, and combining the resulting equations of the form (9.9) obtain the result of the theorem. Let $\{f_n\}$ and $\{g_m\}$ be the sequences of measurable simple functions tending to f and g, respectively, given by Theorem 5, p. 58. Then f_n and g_m are of the form $\sum_{i=1}^{k} \lambda_i\, \chi_{(a_i, a_{i+1}]}$ where $\lambda_i < \lambda_{i+1}$, as f and g are monotone increasing and left-continuous. We show that (9.9) holds for f and g replaced by f_n and g_m, respectively. Since f_n and g_m are left-continuous monotone increasing Borel measurable functions, the integrals are defined. It is convenient, as f and g are left-continuous, to suppose them defined on an interval $(a - \epsilon, b + \epsilon]$, $\epsilon > 0$ and constant on the intervals $(a - \epsilon, a]$ and $[b, b + \epsilon]$. Then if $g_m = \sum_{i=0}^{s-1} \xi_i\, \chi_{(a_i, a_{i+1}]}$, we have $a_0 < a \leqslant a_1 < \ldots < a_{s-1} < b < a_s$. Clearly $\bar{\mu}_{g_m}([a_i]) = g_m(a_{i+1}) - g_m(a_i)$ for each i, and $\bar{\mu}_{g_m}([x]) = 0$ if x is not a partition point. So if h is any finite-valued Borel measurable function,

$$\int_{[a,b]} h(x+)\, d\bar{\mu}_{g_m} = \sum_{i=1}^{s-1} h(a_i+)\,(g_m(a_{i+1}) - g_m(a_i)),$$

and similarly

$$\int_{[a,b]} h\, d\bar{\mu}_{f_n} = \sum_{j=1}^{r-1} h(a_j')\,(f_n(a_{j+1}') - f_n(a_j')).$$

Let $[c_0, c_1, \ldots, c_p]$ be the union of the points of the two partitions $[a_i]$ and $[a_j']$, to which we add a and b. As we may assume the points distinct, we have $c_0 < a = c_1 < c_2 < \ldots < c_{p-1} = b < c_p$. Then

$$\begin{aligned}\int_{[a,b]} f_n(x+)\, d\bar{\mu}_{g_m} &= \sum_{i=1}^{p-1} f_n(c_i+)\,(g_m(c_{i+1}) - g_m(c_i)) \\ &= \sum_{i=1}^{p-1} f_n(c_{i+1})\,(g_m(c_{i+1}) - g_m(c_i))\end{aligned}$$

$$= -g_m(c_1)\,f_n(c_2) - \sum_{i=2}^{p-1} g_m(c_i)\,(f_n(c_{i+1}) - f_n(c_i)) + g_m(c_p)\,f_n(c_p).$$

But
$$\int_{[a,b]} g_m\, d\bar{\mu}_{f_n} = \sum_{i=1}^{p-1} g_m(c_i)\,(f_n(c_{i+1}) - f_n(c_i)),$$

so adding and observing that $g_m(c_p) f_n(c_p) = g_m(b+) f_n(b+)$ and that $c_1 = a$, we get (9.9) for f_n and g_m, that is

$$\int_{[a,b]} f_n(x+)\, d\bar{\mu}_{g_m} + \int_{[a,b]} g_m\, d\bar{\mu}_{f_n} = f_n(b+) g_m(b+) - f_n(a) g_m(a). \qquad (9.10)$$

Now, if h is any non-negative monotone increasing Borel measurable function, $\int h\, d\bar{\mu}_{fn} \to \int h\, d\bar{\mu}_f$. For, given $\epsilon' > 0$, we can find by Theorem 5, p. 58, a step function ϕ, $0 \leqslant \phi \leqslant h$ such that $|h - \phi| < \epsilon'$, uniformly on $(a - \epsilon, b + \epsilon)$. Then

$$|\int h\, d\bar{\mu}_{f_n} - \int h\, d\bar{\mu}_f| \leqslant \int (h - \phi)\, d\bar{\mu}_{f_n} + \int (h - \phi)\, d\bar{\mu}_f$$
$$+ |\int \phi\, d\bar{\mu}_{f_n} - \int \phi\, d\bar{\mu}_f|.$$

The first and second terms on the right-hand side are less than $\epsilon' \bar{\mu}_{f_n}([a, b]) + \epsilon' \bar{\mu}_f([a, b]) < K\epsilon'$, say, for $n > n_0$, and the third clearly tends to zero as $n \to \infty$. A similar result holds for $\bar{\mu}_{g_m}$ and $\bar{\mu}_g$.

Letting $n \to \infty$ in (9.10), using Exercise 7, p. 157, and the fact that $f_n \uparrow f$ uniformly on $(a - \epsilon, b + \epsilon]$ we get therefore

$$\int_{[a,b]} f(x+)\, d\bar{\mu}_{g_m} + \int_{[a,b]} g_m\, d\bar{\mu}_f = f(b+) g_m(b+) - f(a) g_m(a).$$

Then letting $m \to \infty$, we get (9.9). □

Corollary: Interchanging the roles of f and g in (9.9), adding, and noting that $f(x) = f(x-)$, $g(x) = g(x-)$, we obtain the more symmetrical form:

$$\int_{[a,b]} \tfrac{1}{2}(f(x+) + f(x-))\, d\bar{\mu}_g + \int_{[a,b]} \tfrac{1}{2}(g(x+) + g(x-))\, d\bar{\mu}_f$$
$$= f(b+) g(b+) - f(a-) g(a-).$$

This form is particularly appropriate if we define $\bar{\mu}_f$, for a monotone increasing function f, to be the measure induced by the monotone increasing function $f(x-)$ which was shown in Example 2, p. 156, to be left-continuous.

Example 8: Deduce the result of Theorem 9 from that of Theorem 11.

Solution: In (9.9) write $b = x, f = F, g = G$. Then as F is continuous, the left-hand side of (9.9) reads

$$\int_{[a,x]} F\, d\bar{\mu}_G + \int_{[a,x]} G\, d\bar{\mu}_F.$$

If G and F are monotone increasing, Corollary 2 to Theorem 8, p. 162, gives $d\bar{\mu}_G/dm = g\ [m]$, $d\bar{\mu}_F/dm = f\ [m]$. So, by Example 11, p. 145, since $\bar{\mu}_G$ and $\bar{\mu}_F$ are finite measures and F and G are bounded, the result follows. In general, $G = G_1 - G_2$, where G_1 and G_2 are the primitives

$$G_1(x) = \int_a^x g^+ \,dt, \; G_2(x) = \int_a^x g^- \,dt.$$

Then $$\frac{d\bar{\mu}_G}{dm} = \frac{d\bar{\mu}_{G_1}}{dm} - \frac{d\bar{\mu}_{G_2}}{dm} = g\ [m].$$

Similarly for F, and the obvious extension of Example 11, p. 145, to signed measures gives the result.

Exercises

17. Given that $g(x) = e^{-x^2}$, find $\int_{-\infty}^{\infty} x \,dg$ and $\int_{-\infty}^{\infty} x^2 \,dg$.
18. Use Theorem 11 to show that if $h \in L(0, \infty)$,
$$\int_0^x \left(\int_0^t h(u)du\right)dt = \int_0^x (x-u)\,h(u)\,du.$$
19. Prove Theorem 9 by using continuous functions to approximate f and g.
20. Let $h \in L(0, d)$ and write $k(x) = \int_x^d \frac{h(t)}{t}\,dt$. Show that
(i) k is well defined on $(0, d]$, (ii) $\lim_{x \to 0+} xk(x) = 0$, (iii) $k \in L(0, d)$,
(iv) $\int_0^d k\,dx = \int_0^d h\,dx$.

9.5 CHANGE OF VARIABLE

A change of variable within an integral involves a transformation of one space into another and the identification of the corresponding integrals. We give a general result in Theorem 12 and obtain in Theorems 14 and 15 the special case where all the functions are defined on the real line.

Definition 5: Let f be a measurable function from the measure space $[\![X, \mathcal{S}, \mu]\!]$ to R. On the class of subsets of R, $[E: f^{-1}(E) \in \mathcal{S}]$, define the measure μf^{-1} by $(\mu f^{-1})(E) = \mu(f^{-1}(E))$. This class is clearly a σ-algebra and we call its members the μf^{-1}-measurable sets of R.

Clearly every Borel set is μf^{-1}-measurable by an obvious extension of Example 19, p. 41.

Example 9: If μ is a complete measure on $[\![X, \mathcal{S}]\!]$ and $f: X \to \mathsf{R}$ is measurable, then μf^{-1} is a complete measure.

Solution: Let $\mu f^{-1}(E) = 0$ and $F \subseteq E$. Then $f^{-1}(F) \subseteq f^{-1}(E)$ so $f^{-1}(F) \in \mathcal{S}$.

Theorem 12: Let f be a measurable function from $[\![X, \mathcal{S}, \mu]\!]$ to R, and g a Borel measurable function on R; then

$$\int g \, \mathrm{d}\mu f^{-1} = \int g \circ f \, \mathrm{d}\mu \tag{9.11}$$

in the sense that if either exists, so does the other and the two are equal.

Proof: It is sufficient to consider the case where g is non-negative. If $F \subseteq \mathsf{R}$, then $\chi_F \circ f = \chi_{f^{-1}(F)}$ and is measurable for $F \in \mathcal{B}$, so

$$\int \chi_F \, \mathrm{d}\mu f^{-1} = (\mu f^{-1})(F) = \int \chi_{f^{-1}(F)} \, \mathrm{d}\mu = \int \chi_F \circ f \, \mathrm{d}\mu,$$

and so (9.11) holds when g is a simple function. In the general case, let $\{\phi_n\}$ be an increasing sequence of Borel measurable simple functions $\phi_n \uparrow g$. Then $\{\phi_n \circ f\}$ is an increasing sequence of measurable simple functions tending to $g \circ f$, which is therefore measurable, and the result follows on taking limits with respect to n. □

Corollary: If F is a Borel set, then $\int_F g \, \mathrm{d}\mu f^{-1} = \int_{f^{-1}(F)} g \circ f \, \mathrm{d}\mu$.

Proof: Replace g in the theorem by $g \chi_F$. The result follows since $(g \chi_F) \circ f = (g \circ f)(\chi_F \circ f) = \chi_{f^{-1}(F)} (g \circ f)$. □

Example 10: Let f be a measurable function on $[\![X, \mathcal{S}, \mu]\!]$, where $\mu(X) < \infty$, such that $\lambda < f < \gamma$, and let $e(y) = (\mu f^{-1})(\lambda, y)$. Then e is a left-continuous monotone increasing function such that for $y_1 < y_2$, $\mu f^{-1}([y_1, y_2)) = e(y_2) - e(y_1)$. So the measures μf^{-1} and $\bar{\mu}_e$ agree on the Borel sets and by the last Corollary we have for any Borel measurable function g

$$\int_F g(y) \, \mathrm{d}e(y) = \int_{f^{-1}(F)} g \circ f \, \mathrm{d}\mu.$$

In the presence of absolute continuity, the result of the last theorem has a simpler form.

Theorem 13: Let f be a measurable function on the measure space $[\![X, \mathcal{S}, \mu]\!]$, such that μf^{-1} is σ-finite and $\mu f^{-1} \ll m$ on $\mathcal{B}$. Then there exists a non-negative finite-valued Borel measurable function p such that

$$\int g \circ f \, \mathrm{d}\mu = \int g(x) p(x) \, \mathrm{d}x \tag{9.12}$$

whenever the left-hand side exists. In (9.12), $p(x) = \mathrm{d}\mu f^{-1}/\mathrm{d}m$ $[m]$, where μf^{-1} and m are considered as measures on $\mathcal{B}$.

Proof: Theorem 12 gives us that $\int g \circ f \, \mathrm{d}\mu = \int g \, \mathrm{d}\mu f^{-1}$ and that this latter integral exists. We may suppose, as in Theorem 12, that $g \geqslant 0$. Since μf^{-1} and m are σ-finite measures on $\mathcal{B}$, Example 11, p. 145, gives $\int g \, \mathrm{d}\mu f^{-1} = \int g(x) p(x) \, \mathrm{d}x$, that is: (9.12) holds whenever g is integrable with respect to μf^{-1}. Since $p = \mathrm{d}\mu f^{-1}/\mathrm{d}m$ as in the proof of Example 11, we may suppose p finite-valued and Borel measurable. Also, since μf^{-1} is σ-finite we may write $\mathsf{R} = \bigcup_{n=1}^{\infty} E_n$ where, for each n, $E_n \subseteq E_{n+1}$ and $\mu f^{-1}(E_n) < \infty$. Then if $g_n = \chi_{E_n} \min(g, n)$, g_n is integrable with respect to μf^{-1} and $g_n \uparrow g$. So (9.12) holds with g replaced by g_n. Letting $n \to \infty$ gives the result. □

Corollary: For any Borel set F, $\int_{f^{-1}(F)} g \circ f \, d\mu = \int_F g(x)\, p(x)\, dx$.

We now give a classical result on the real line, in which $p(x)$ of the last theorem equals 1.

Theorem 14: Let k be a non-negative integrable function on R, let g be a Borel measurable function integrable over $[a, b]$, let K be the indefinite integral of k and let ξ, η be such that $a = K(\xi)$, $b = K(\eta)$. Then

$$\int_a^b g(t)\, dt = \int_\xi^\eta g(K(x))\, k(x)\, dx.$$

Proof: We may define $g = 0$ on $(-\infty, a)$, (b, ∞) to get functions g, k, K defined on R, with $K(x) = \int_{-\infty}^x k\, dt$. Since k is integrable, K is continuous. In Theorem 12 let $X = \mathsf{R}$, $\mathcal{S} = \bar{\mathcal{S}}_K$, $\mu = \bar{\mu}_K$, $f = K$, so $\mu f^{-1} = \bar{\mu}_K K^{-1}$. Consider any interval $[c, d]$. Let $\gamma = \inf[u: K(u) = c]$ and $\delta = \sup[u: K(u) = d]$. So $K^{-1}[c, d] = [\gamma, \delta]$ as K is continuous and monotone increasing. Then $\bar{\mu}_K K^{-1}[c, d] = K(\delta) - K(\gamma) = d - c$, so that $\bar{\mu}_K K^{-1}$ agrees with m on Borel sets and hence on all Lebesgue measurable sets, as $\bar{\mu}_K K^{-1}$ is complete by Example 9, p. 167. So Theorem 12 gives

$$\int g(t)\, dt = \int g \circ K \, d\bar{\mu}_K = \int g(K(x))\, k(x)\, dx$$

by Example 11, p. 145, and by Corollary 2 to Theorem 8, p. 162. Hence by the Corollary to Theorem 12,

$$\int_a^b g(t)\, dt = \int_\xi^\eta g(K(x))\, k(x)\, dx.$$

Clearly any ξ and η such that $K(\xi) = a$, $K(\eta) = b$ may be chosen, since for example, any two values of ξ such that $K(\xi) = a$ are separated by an interval on which $k = 0$ a.e. □

The result of Theorem 14 holds for any integrable function g, but the change from Borel to Lebesgue measurable is non-trivial as the next proof shows.

Theorem 15: Let k be a non-negative integrable function on R, let $h \in L(a, b)$ and let K, ξ, η be as in Theorem 14. Then $h(K(x))\, k(x) \in L(\xi, \eta)$ and

$$\int_a^b h(t)\, dt = \int_\xi^\eta h(K(x))\, k(x)\, dx. \tag{9.13}$$

Proof: Since h is measurable, we can find a Borel measurable function g such that $h = g$ a.e., as in Exercise 11, p. 60, and then $\int_a^b g\, dt = \int_a^b h\, dt$. If we show that $h(K(x))\,k(x) = g(K(x))\,k(x)$ a.e., then the former is measurable and its integrability and the result of the theorem follows from Theorem 14. We need consider only points x such that $h(K(x)) \neq g(K(x))$. Let $E = [x\colon h(x) \neq g(x)]$, so that $m(E) = 0$. Write $[\xi, \eta] \cap K^{-1}(E) = \bigcup_{i=1}^{4} A_i$ where: on A_1, K' does not exist; on A_2, $K' = k = 0$; on A_3, $K' \neq k$; on A_4, $K' = k > 0$. These sets are disjoint; $m(A_1) = 0$ by Theorem 9, p. 87, and $m(A_3) = 0$ by Theorem 12, p. 89. On A_2, $h(K(x))\,k(x) = g(K(x))\,k(x)$. If we show that $m(A_4) = 0$, the result follows. Write $A_{4,n} = [x\colon K'(x) > 1/n, K(x) \in E]$. Clearly, it is sufficient to show that $m(A_{4,n}) = 0$ for each fixed integer n. Given $\epsilon > 0$, we can find an open set O, such that $E \subseteq O$ and $m(O) < \epsilon$. Since K is continuous, for each x in $A_{4,n}$ there exists $h_x > 0$ such that $(K(x), K(x + h_x)) \subseteq O$ and $K(x + h_x) - K(x) > h_x/n$. Using again the continuity of K we can find for each x a point x', $x' < x < x' + h_x$, such that $(K(x'), K(x' + h_x)) \subseteq O$ and $K(x' + h_x) - K(x') > h_x/n$. The intervals $(x', x' + h_x)$ cover $A_{4,n}$ so by Theorem 5, p. 84, and Theorem 6, p. 84, we can find a finite disjoint collection $[I_j, j = 1, \ldots, N]$ of these intervals such that $\frac{1}{6} m(A_{4,n}) \leqslant m\left(\bigcup_{j=1}^{N} I_j\right)$. But

$$m\left(\bigcup_{j=1}^{N} I_j\right) = \sum_{j=1}^{N} h_{x_j} < n \sum_{j=1}^{N} (K(x'_j + h_{x_j}) - K(x'_j)) < n\, m(O) < n\epsilon.$$

As n and ϵ are independent, $m(A_{4,n}) = 0$ and the result follows. □

Corollary: Since $\mathrm{d}\bar{\mu}_F/\mathrm{d}m = F'$ $[m]$ when F is absolutely continuous, (9.13) may be written

$$\int_a^b h(t)\,\mathrm{d}t = \int_\xi^\eta h(K(x))\,\mathrm{d}K(x).$$

With the help of the following example we obtain a version of Theorem 15 when K is monotone decreasing, instead of monotone increasing.

Example 11: Show that if $f \in L(-\infty, \infty)$, then g defined by $g(x) = f(-x)$ is an integrable function and $\int_a^b f\,\mathrm{d}x = \int_{-b}^{-a} g\,\mathrm{d}x$ for $-\infty \leqslant a < b \leqslant \infty$.

Solution. Since $(f(-x))^+ = f^+(-x)$ and $(f(-x))^- = f^-(-x)$, we may consider non-negative functions only. Clearly g is measurable and if $\phi_n \uparrow f$, where $\phi_n = \sum_{i=1}^{N} \lambda_i \chi_{E_i}$ is a measurable simple function, then if $\psi_n = \sum_{i=1}^{N} \lambda_i \chi_{-E_i}$, we have $\psi_n \uparrow g$. But $m(E) = m(-E)$ (cf. Exercise 9, p. 34). So $\int g\,\mathrm{d}x = \int f\,\mathrm{d}x$ and so

$$\int f(x)\,\chi_{[a,b]}(x)\,\mathrm{d}x = \int g(x)\,\chi_{[a,b]}(-x)\,\mathrm{d}x = \int_{-b}^{-a} g\,\mathrm{d}x.$$

Example 12: From the last example obtain the analogue of Theorem 15 for K monotone decreasing.

Solution: We suppose that $k \leqslant 0$, k integrable with indefinite integral K, $K(\xi) = a$, $K(\eta) = b$ and $h \in L(a, b)$. Let $k_1(x) = -k(-x)$ with indefinite integral K_1. So $K_1(x) = -L + K(-x)$, where $L = \int k \, dx$. Then Example 25, p. 75, gives $\int_a^b h(t)\,dt = \int_{a-L}^{b-L} h(t+L)\,dt$. Applying Theorem 15 to $h_L(t) = h(t+L)$, K_1 and k_1 we get

$$\int_{a-L}^{b-L} h(t+L)\,dt = \int_{\xi_1}^{\eta_1} h(K_1(x)+L)k_1(x)\,dx$$

where $K_1(\xi_1) = a - L, K_1(\eta_1) = b - L$. So by the last example,

$$\int_a^b h(t)\,dt = -\int_{\xi_1}^{\eta_1} h(K(-x))k(-x)\,dx$$

$$= \int_{-\eta_1}^{-\xi_1} h(K(x))k(x)\,dx = -\int_{\eta}^{\xi} h(K(x))k(x)\,dx$$

is the desired result.

Exercises

21. Let f be a finite-valued measurable function from the measure space $[\![X, \mathcal{S}, \mu]\!]$ to R and define the monotone increasing functions F and G by: $F(x) = \mu f^{-1}(-\infty, x]$, $G(x) = \mu f^{-1}(-\infty, x)$. Show that F and G have the following properties:

    (i) If $\mu(X) < \infty$, then $F(-\infty) = G(-\infty) = 0$ and the finiteness of μ may not be omitted in general,
    (ii) $F(\infty) = G(\infty) = \mu(X)$,
    (iii) $F(x-) = G(x)$, and if μ is finite $F(x) = G(x+)$,
    (iv) G is left continuous, and F is right-continuous if μ is finite but not necessarily if $\mu(X) = \infty$,
    (v) F is not necessarily left-continuous nor G right-continuous,
    (vi) $\mu f^{-1} = \bar{\mu}_G$ on $\mathcal{B}$, and if μ is complete, $\mu f^{-1} = \bar{\mu}_G$ on $\bar{\mathcal{S}}_G$.

22. Show that if μf^{-1} is σ-finite, then μ is σ-finite.
23. Show by an example that μ σ-finite does not imply μf^{-1} σ-finite even if $\mu f^{-1} \ll m$.
24. Show that if μ is σ-finite and f is one-to-one, then μf^{-1} is σ-finite.
25. Let f be a measurable function on R, and on the measurable set B such that $m(B) < \infty$ let $0 \leqslant f \leqslant M < \infty$. Show that

$$\int_B f^n \, dx = M^n \, m(B) - n \int_0^M y^{n-1} \, e(y) \, dy,$$

for $n = 1, 2, \ldots$, where $e(y) = mf^{-1}(0, y)$.

26. In the last exercise take $B = (0,1), M = \frac{1}{2}$ and let $F(t) = \sum_{n=0}^{\infty} a_n t^n$ for $|t| < 1$.

Show that

$$\int_0^1 F(f(x)) \, dx = F(\tfrac{1}{2}) - \int_0^{\frac{1}{2}} F'(y) \, e(y) \, dy.$$

9.6 RIESZ REPRESENTATION THEOREM FOR $C(I)$

We recall the notation $C(0,1)$ of Example 14, p. 148, for the real vector space of continuous functions with the supremum norm $\|f\|_\infty$. More generally, for any bounded closed set A in R, $C(A)$ will denote the class of functions continuous on A and with norm $\|f\|_\infty = \sup[|f(x)|: x \in A]$. In particular we are interested in the case when A is an interval, and consider linear functionals on the vector space $C(I)$ where I will always denote a finite closed interval, in this context.

Definition 6: A linear functional G on $C(I)$ is a **positive linear functional** if $G(f) \geqslant 0$ for any non-negative function $f \in C(I)$.

Example 13: Show that if $f, g \in C(I), f \geqslant g$, and G is a positive linear functional on $C(I)$, then $G(f) \geqslant G(g)$.

Solution: Since $f = (f-g) + g$, we have $G(f) = G(f-g) + G(g) \geqslant G(g)$.

We have defined a bounded linear functional in Definition 13, p. 148. It is clear that if μ is any measure on the Borel subsets of I then G defined by $G(f) = \int_I f \, d\mu$ is a positive linear functional on $C(I)$. If, in addition, $\mu(I) < \infty$, then G is bounded. The main result of this section is that all positive bounded linear functionals are of this form. This is then extended to all bounded linear functionals, measures being replaced by signed measures.

The construction in the L^p case suggests that the measure of a subinterval J of I may be obtained as $\mu(J) = G(\chi_J)$. But χ_J is not continuous, so in the proof of the theorem it is approximated by a piecewise-linear continuous function. Since G is positive and continuous, this allows $\mu(J)$ to be defined as a limit.

Theorem 16 (Riesz Representation Theorem for $C(I)$): Let G be a positive bounded linear functional on $C(I)$. Then there exists a unique measure μ on $\mathcal{B}$ such that

$$G(f) = \int_I f \, d\mu \tag{9.14}$$

for each $f \in C(I)$. Also, $\|G\| = \mu(I)$.

Proof: Let $I = [a, b]$ and if $t \in (a, b]$ and n is a sufficiently large positive integer, define $h_{t,n} \in C(I)$ by $h_{t,n} = 1$ on $[a, t - 1/n]$, $h_{t,n} = 0$ on $[t, b]$, $h_{t,n}$ linear on $[t - 1/n, t]$. Then the family of functions $[h_{t,n}]$ increases with respect to t and with respect to n. Also $h_{t,n} \leqslant 1$ for all t and n. So $G(h_{t,n}) \leqslant G(\chi_I)$. Set $g(t) = \lim_{n\to\infty} G(h_{t,n})$, for $t \in (a, b]$, and set $g(t) = 0$ for $t \leqslant a$, $g(t) = G(\chi_I)$ for $t > b$. Since $h_{t,n}$ increases with t, so does $G(h_{t,n})$ by the last example and therefore $g(t)$ is a monotone increasing function.

Also g is left-continuous. This is clear if $t \leqslant a$ or $t > b$. So let $t \in (a, b]$, let $\epsilon > 0$ and let $n > \max(2, \|G\|/\epsilon)$ be so large that

$$G(h_{t,n}) \leqslant g(t) \leqslant G(h_{t,n}) + \epsilon. \tag{9.15}$$

Let k_n be the function in $C(I)$, $k_n = 1$ on $[a, t - 1/n + 1/n^2]$, $k_n = 0$ on $[t - 1/n^2, b]$ and k_n linear on the intermediate interval. Then the maximum difference between $h_{t,n}$ and k_n can be seen from a diagram to occur at $t - 1/n + 1/n^2$ and at $t - 1/n^2$ when it equals $1/n$. So $\|k_n - h_{t,n}\|_\infty = 1/n$. Therefore $G(h_{t,n}) \leqslant G(k_n) + n^{-1}\|G\| < g(t - 1/n^2) + \epsilon$. So by (9.15) we have $g(t - 1/n^2) \leqslant g(t) \leqslant g(t - 1/n^2) + 2\epsilon$, and so g is left-continuous.

Let $\mu = \bar{\mu}_g$ so that $\mu([\alpha, \beta)) = g(\beta) - g(\alpha)$. Then for any measurable set $E \subseteq \mathbf{C}I$, $\mu(E) = 0$ and $\mu([a, c)) = g(c)$ for each $c \geqslant a$. Also $\|G\| = G(\chi_I)$; for obviously $G(\chi_I) \leqslant \|G\|$ and by the definition of $\|G\|$ there exists a sequence $\{f_n\}$ in $C(I)$ with $\|f_n\|_\infty = 1$ and such that $\lim G(f_n) = \|G\|$. But $f_n \leqslant \chi_I$ implies $G(f_n) \leqslant G(\chi_I)$. So $\|G\| = G(\chi_I) = g(b+) = \mu(I)$.

We now show that (9.14) holds for all $f \in C(I)$, and we may suppose that f is not identically zero. Now f is uniformly continuous on I so, given $\epsilon > 0$, there exists $\delta > 0$ such that $x, y \in I$ and $|x - y| < \delta$ imply $|f(x) - f(y)| < \epsilon$. Now let $a = t_0 < t_1 < \ldots < t_m = b$ be a partition such that $\max(t_k - t_{k-1}) < \delta/2$, and choose n so large that $n^{-1} < \min(t_k - t_{k-1})$ and that for $k = 1, \ldots, m$ we have

$$G(h_{t_k,n}) \leqslant g(t_k) \leqslant G(h_{t_k,n}) + \frac{\epsilon}{m\|f\|_\infty}. \tag{9.16}$$

This is possible by (9.15). Now consider the functions defined on I by

$$f_1(x) = f(t_1)\,h_{t_1,n}(x) + \sum_{k=2}^{m} f(t_k)(h_{t_k,n}(x) - h_{t_{k-1},n}(x))$$
$$+ f(t_m)\,(\chi_I(x) - h_{t_m,n}(x))$$

and
$$f_2(x) = \sum_{k=1}^{m} f(t_k)\,\chi_{[t_{k-1},t_k)}(x) + f(t_m)\,\chi_{[t_m]}(x).$$

So $f_1 \in C(I)$ and f_2 is a step function. Clearly

$$\sup[|f_2(x) - f(x)| : x \in I] \leqslant \epsilon. \tag{9.17}$$

Using (9.16) for $k = 1, \ldots, m$ we have

$$|G(h_{t_k,n} - h_{t_{k-1},n}) - (g(t_k) - g(t_{k-1}))| \leqslant \frac{2\epsilon}{m\,\|f\|}. \tag{9.18}$$

Since G is linear,

$$G(f_1) = f(t_1)\,G(h_{t_1,n}) + \sum_{k=2}^{m} f(t_k)\,G(h_{t_k,n} - h_{t_{k-1},n}) + f(t_m)(G(\chi_I) - G(h_{t_m,n})).$$

Using $g(t_0) = g(a) = 0$,

$$\int f_2 \,\mathrm{d}\mu = f(t_1)g(t_1) + \sum_{k=2}^{m} f(t_k)(g(t_k) - g(t_{k-1})) + f(t_m)(g(b+) - g(b)).$$

So using (9.16), (9.18) and $g(b+) = G(\chi_I)$ we have $|G(f_1) - \int f_2 \,\mathrm{d}\mu| \leqslant 2\epsilon$. So, by (9.17),

$$|\int f \,\mathrm{d}\mu - G(f_1)| \leqslant 2\epsilon + \epsilon\,\mu(I). \tag{9.19}$$

We now show that $\|f_1 - f\|_\infty \leqslant 2\epsilon$. On $[t_0, t_1 - 1/n]$ it is clear that $|f - f_1| < \epsilon$. On $[t_1 - 1/n, t_1]$, $f - f_1 = (f - f(t_2)\,h_{t_2,n}) + h_{t_1,n}\,(f(t_2) - f(t_1))$. But $h_{t_2,n} = 1$ on this interval, so $|f - f_1| < 2\epsilon$. Similarly for the succeeding intervals. So $|G(f) - G(f_1)| \leqslant 2\epsilon\;\|G\|$ and (9.19) gives $|\int f \,\mathrm{d}\mu - G(f)| \leqslant 2\epsilon + \epsilon\mu(I) + 2\epsilon\,\|G\| = \epsilon(2 + 3\,\|G\|)$ as $\mu(I) = \|G\|$. But ϵ is arbitrary and (9.14) follows.

It remains to be shown that μ is unique. Let μ_0 be any other measure with the required properties and let μ_0 have primitive g_0 with $g_0(a) = 0$. Then $g_0(t - 1/n) \leqslant \int h_{t,n} \,\mathrm{d}\mu_0 \leqslant g_0(t)$. Letting $n \to \infty$ the middle term has limit $g(t)$, and since g_0 is left continuous $g_0 = g$ on I. □

Theorem 17: Every bounded linear functional F on $C(I)$ can be written as $F = F^+ - F^-$ where F^+, F^- are positive linear functionals.

Proof: For $f \geqslant 0$ let $F^+(f) = \sup[F(g)\colon 0 \leqslant g \leqslant f]$. Then $F^+(f) \geqslant 0$ and $|F^+(f)| \leqslant \|F\| \cdot \|f\|_\infty$. Also, obviously, $F^+(cf) = cF^+(f)$ if $c > 0$. Consider now a pair f_1, f_2 of non-negative functions in $C(I)$. If $0 \leqslant g_1 \leqslant f_1$ and $0 \leqslant g_2 \leqslant f_2$ then $0 \leqslant g_1 + g_2 \leqslant f_1 + f_2$ and $F^+(f_1 + f_2) \geqslant \sup F(g_1 + g_2) = \sup F(g_1) + \sup F(g_2) = F^+(f_1) + F^+(f_2)$. Conversely, if $0 \leqslant g \leqslant f_1 + f_2$, then $0 \leqslant \min(f_1, g) \leqslant f_1$ and $0 \leqslant g - \min(f_1, g) \leqslant f_2$ so that $F^+(f_1 + f_2) = \sup F(g) \leqslant \sup F(\min(f_1, g)) + \sup F(g - \min(f_1, g)) \leqslant F^+(f_1) + F^+(f_2)$. Thus F^+ is additive on non-negative functions. But F^+ can be extended to a linear functional on $C(I)$ by $F^+(f_1 - f_2) = F^+(f_1) - F^+(f_2)$, since every function $f \in C(I)$ is the difference of non-negative functions $f_1, f_2 \in C(I)$. Then F^+ is bounded since $|F^+(f)| \leqslant F^+(f^+) + F^+(f^-) = F^+(|f|) \leqslant \|F\| \cdot \|f\|_\infty$. This extension is unique for if $f_1 - f_2 = g_1 - g_2$, then $f_1 + g_2 = f_2 + g_1$ so applying F^+ and rearranging we get $F^+(f_1) - F^+(f_2) = F^+(g_1) - F^+(g_2)$. Also the extension is a linear functional, for if $a > 0$ and $f =$

$f_1 - f_2$, then $F^+(af) = F^+(af_1) - F^+(af_2) = aF^+(f)$. If $a = -1$ and $f = f_1 - f_2$, then $F^+(-f) = F^+(f_2) - F^+(f_1) = -F^+(f)$. Also, in the same notation, $F^+(f+g) = F^+(f_1 + g_1 - (f_2 + g_2)) = F^+(f_1 + g_1) - F^+(f_2 + g_2)$. Expand and rearrange to get $F^+(f+g) = F^+(f) + F^+(g)$. So F^+ is a positive bounded linear functional.

Now let $F^-(f) = F^+(f) - F(f)$. Then F^- is a bounded linear functional, being the difference of bounded linear functionals, and F^- is positive since for $f \geqslant 0$ we have, from its definition, that $F^+(f) \geqslant F(f)$. Since $F = F^+ - F^-$ the theorem is proved. □

We now extend Theorem 16 with the aid of Theorem 17.

Theorem 18: Let G be a bounded linear functional on $C(I)$. Then there exists a finite signed measure μ on the Borel subsets of I such that $G(f) = \int_I f \, d\mu$ for each $f \in C(I)$.

Proof: By Theorem 17, $G = G^+ - G^-$ where G^+, G^- are positive bounded linear functions. Then by Theorem 18, $G^+(f) = \int_I f \, d\mu_1$, $G^-(f) = \int_I f \, d\mu_2$ so $\mu = \mu_1 - \mu_2$ gives the result. □

CHAPTER 10

Measure and Integration in a Product Space

In order to consider multiple integrals we need to deal with measure and integration on the Cartesian product of spaces. We now examine these in a fairly general framework and use the general theory developed in Chapter 5. The most important special case, that of finite-dimensional Euclidean spaces is examined in Section 10.3, where the usual procedures of integration with respect to polar coordinates is fitted into this framework, and their use for Lebesgue integrals is justified. In Section 10.4 we examine the Laplace and Fourier transforms and show how integration in a product space can be used to obtain their properties. Finally, using also the results on L^2 obtained in Chapter 6, we obtain the Parseval and Plancherel theorems for the Fourier transform.

10.1 MEASURABILITY IN A PRODUCT SPACE

In this section we assemble the basic definitions for product measures and measurability with respect to such measures and obtain some properties.

Definition 1: If X and Y are sets, their *Cartesian product* $X \times Y$ is the set of ordered pairs $[(x,y): x \in X, y \in Y]$. If X and Y are spaces, $X \times Y$ is the *product space*.

In considering measures on a product space the important sets will be those which are Cartesian products of subsets of X and Y, as described in the next definition.

Definition 2: A set in $X \times Y$ is a **rectangle** if it may be written $A \times B$ for $A \subseteq X$, $B \subseteq Y$.

To define measures on $X \times Y$ we first suppose that $[\![X, \mathcal{S}]\!]$ and $[\![Y, \mathcal{T}]\!]$ are measurable spaces; appropriate rectangles then provide the building bricks for the measurable sets in $X \times Y$.

Definition 3: **A measurable rectangle** in $X \times Y$ is any set which may be written as $A \times B$ for $A \in \mathcal{S}$, $B \in \mathcal{T}$.

Definition 4: The class of **elementary sets** $\mathcal{E}$ consists of those sets which may be written as the union of a finite number of disjoint measurable rectangles.

If we are given measures on $\mathcal{S}$ and $\mathcal{T}$ it is easy to see how we should define a measure on the measurable rectangles and hence on the sets of $\mathcal{E}$. That we may then use the results of Section 5.3 is guaranteed by the following theorem.

Theorem 1: $\mathcal{E}$ is an algebra.

Proof: Clearly $\mathcal{E}$ is closed under finite disjoint unions. It is closed also under finite intersections, for let $P = \bigcup_{i=1}^{n} U_i \in \mathcal{E}$, $Q = \bigcup_{j=1}^{m} V_j \in \mathcal{E}$, where U_i, V_j are measurable rectangles, $U_i \cap U_k = \emptyset$ for $i \neq k$, $V_j \cap V_s = \emptyset$ for $j \neq s$. Then

$$P \cap Q = \bigcup_{\substack{1 \leq i \leq n \\ 1 \leq j \leq m}} U_i \cap V_j \in \mathcal{E}$$

as the intersection of two measurable rectangles is a measurable rectangle.

If $A \times B$ is a measurable rectangle, $\complement(A \times B) = (\complement A \times Y) \cup (A \times \complement B) \in \mathcal{E}$. So if $P = \bigcup_{i=1}^{n} U_i \in \mathcal{E}$, $\complement P = \bigcap_{i=1}^{n} \complement U_i \in \mathcal{E}$. Also $\mathcal{E}$ is closed under finite unions for if $P \in \mathcal{E}$, $Q \in \mathcal{E}$, then $P \cup Q = (P - Q) \cup Q$, a disjoint union, belongs to $\mathcal{E}$. Since clearly $X \times Y \in \mathcal{E}$, $\mathcal{E}$ is an algebra. □

We may now define a class of measurable sets of $X \times Y$.

Definition 5: $\mathcal{S} \times \mathcal{T}$ denotes the σ-algebra generated by the class of measurable reactangles. Also, $[\![X \times Y, \mathcal{S} \times \mathcal{T}]\!]$ is the **product of the measurable spaces** $[\![X, \mathcal{S}]\!]$ and $[\![Y, \mathcal{T}]\!]$.

Example 1: $\mathcal{S} \times \mathcal{T} = \mathcal{S}(\mathcal{E})$, the σ-algebra generated by $\mathcal{E}$.

A more convenient characterization of $\mathcal{S} \times \mathcal{T}$ may be provided in terms of monotone classes. We formally define these in the next definition and obtain from Theorem 3 a characterization of monotone classes which provides the essential tool for integration theory in product spaces.

Definition 6: A class $\mathcal{M}_0$ of subsets of a space is a **monotone class** if for any increasing or decreasing sequence of sets $\{E_n\}$ of $\mathcal{M}_0$, $\lim E_n \in \mathcal{M}_0$.

Theorem 2: If $\mathcal{Y}$ is any class of subsets of X, there exists a smallest monotone class, denoted by $\mathcal{M}_0(\mathcal{Y})$, containing $\mathcal{Y}$.

Proof: Obviously, $\mathcal{P}(X)$ is a monotone class. Also, the intersection of monotone classes is a monotone class. So the intersection of all the monotone classes which contain $\mathcal{Y}$ provides the required monotone class $\mathcal{M}_0(\mathcal{Y})$. □

Theorem 3: If $\mathcal{A}$ is an algebra, $\mathcal{S}(\mathcal{A}) = \mathcal{M}_0(\mathcal{A})$, that is, the σ-algebra generated by $\mathcal{A}$ is the smallest monotone class containing $\mathcal{A}$.

Proof: For brevity write $\mathcal{M}_0$ in place of $\mathcal{M}_0(\mathcal{A})$. Since every σ-algebra is a monotone class, $\mathcal{S}(\mathcal{A}) \supseteq \mathcal{M}_0$. To prove the opposite inclusion it is sufficient to show that $\mathcal{M}_0$ is an algebra, since each countable union can be written as the limit of finite unions. Let $\mathcal{M}'_0 = [A: \complement A \in \mathcal{M}_0]$; then it easily seen that $\mathcal{M}'_0$ is a monotone class and that $\mathcal{A} \subseteq \mathcal{M}'_0$. So $\mathcal{M}_0 \subseteq \mathcal{M}'_0$, that is, $\mathcal{M}_0$ is closed under the taking of complements. We wish to show that the same is true for finite unions. For each $F \in \mathcal{M}_0$ let

$$\mathcal{K}(F) = [E: E \in \mathcal{M}_0, E \cup F \in \mathcal{M}_0]. \tag{10.1}$$

Then it is sufficient to show that $\mathcal{M}_0 = \mathcal{K}(F)$ for each $F \in \mathcal{M}_0$. Now, $\mathcal{K}(F)$ is a monotone class since, for example, if $E_n \in \mathcal{K}(F)$, $E_n \subseteq E_{n+1}$, $n = 1, 2, \ldots$, we have $\bigcup_{n=1}^{\infty} E_n \in \mathcal{M}_0$ and $\left(\bigcup_{n=1}^{\infty} E_n\right) \cup F = \lim (E_n \cup F) \in \mathcal{M}_0$. So $\lim E_n \in \mathcal{K}(F)$. Also, if $G \in \mathcal{A}$, $\mathcal{K}(G)$ contains $\mathcal{A}$, since $\mathcal{M}_0 \supseteq \mathcal{A}$, so $\mathcal{K}(G) = \mathcal{M}_0$. So, for any $H \in \mathcal{M}_0$, $H \in \mathcal{K}(G)$. But by inspection of (10.1) this implies that $G \in \mathcal{K}(H)$. So $\mathcal{A} \subseteq \mathcal{K}(H)$ for each $H \in \mathcal{M}_0$ and as $\mathcal{K}(H)$ is a monotone class, $\mathcal{M}_0 = \mathcal{K}(H)$ as required. □

Corollary: $\mathcal{S} \times \mathcal{T} = \mathcal{M}_0(\mathcal{E})$.

Definition 7: If $E \subseteq X \times Y$, we define the ***x*-section** of E to be the set $E_x = [y: (x,y) \in E]$, and the ***y*-section** of E to be the set $E^y = [x: (x,y) \in E]$.

Note that $E_x \subseteq Y$ and $E^y \subseteq X$. The next theorem states that measurable sets have measurable sections. However, in Example 4 below, we give an example of a set of which each x-section and y-section is measurable but which is not itself measurable.

Example 2: Show that if $\{A_i\}$ is a monotone sequence of sets, then $\lim A_i^y = (\lim A_i)^y$ and $\lim (A_i)_x = (\lim A_i)_x$.

Solution: This follows from the fact that $\left(\bigcup_{i=1}^{\infty} A_i\right)^y = \bigcup_{i=1}^{\infty} A_i^y$ and $\left(\bigcap_{i=1}^{\infty} A_i\right)^y = \bigcap_{i=1}^{\infty} A_i^y$, and similarly for x-sections.

Theorem 4: If $E \in \mathcal{S} \times \mathcal{T}$, then for each $x \in X$ and $y \in Y$, $E_x \in \mathcal{T}$ and $E^y \in \mathcal{S}$.

Proof: Let $\Omega = [E: E \in \mathcal{S} \times \mathcal{T}, E_x \in \mathcal{T}$ each $x \in X]$. If $A \in \mathcal{S}$ and $B \in \mathcal{T}$, then $(A \times B)_x = B$ or $\emptyset$ according as $x \in A$ or $x \in \complement A$. So Ω contains the measurable rectangles. If $E \in \Omega$ then since

$$(\complement E)_x = [y: (x,y) \in \complement E] = \complement [y: (x,y) \in E] = \complement E_x,$$

we have $\complement E \in \Omega$. Also if $E_n \in \Omega$, $n = 1, 2, \ldots$, we have for each $x \in X$,

$$\left(\bigcup_{n=1}^{\infty} E_n\right)_x = \bigcup_{n=1}^{\infty} (E_n)_x.$$

So Ω is a σ-algebra and hence $\Omega = \mathcal{S} \times \mathcal{T}$. Similarly, $E^y \in \mathcal{S}$ for each $y \in Y$. □

We have corresponding statements for functions on $X \times Y$.

Definition 8: Let f be a function defined on $X \times Y$. Then, given $x \in X$, the **x-section** of f is the function on Y: $f_x(y) = f(x,y)$, and given $y \in Y$, the **y-section** of f is the function on X: $f^y(x) = f(x,y)$.

Theorem 5: Let f be an $\mathcal{S} \times \mathcal{T}$-measurable function on $X \times Y$; then for each $x \in X$ and $y \in Y$, f_x is a $\mathcal{T}$-measurable function and f^y is an $\mathcal{S}$-measurable function.

Proof: Let $E = [(x,y): f(x,y) > \alpha]$. Then for a fixed $x \in X$, $E_x = [y: f_x(y) > \alpha]$ belongs to $\mathcal{T}$ for each α, by Theorem 4, that is f_x is $\mathcal{T}$-measurable. Similarly for f^y. □

Exercises

1. The representation of a rectangle in the form $A \times B$ need not be unique. Find when it is.
2. Let A be a non-measurable set in X. Is $A \times \emptyset$ a measurable rectangle?
3. Show that a rectangle $A \times B$ is non-measurable iff either A is non-measurable and $B \neq \emptyset$, or B is non-measurable and $A \neq \emptyset$.
4. Show that if $V \subseteq X \times Y$, then $(\chi_V)_x = \chi_{V_x}$ and $(\chi_V)^y = \chi_{V^y}$.
5. Show that if f is $\mathcal{S}$-measurable and g is $\mathcal{T}$-measurable then fg is $\mathcal{S} \times \mathcal{T}$-measurable.

10.2 THE PRODUCT MEASURE AND FUBINI'S THEOREM

We obtain in this section, in Theorems 8 and 9 the main results on integration with respect to a product measure. They are due to Fubini and Hobson and usually known as Fubini's theorem. In order to define a measure on $\mathcal{S} \times \mathcal{T}$ we use the following theorem.

Theorem 6: Let $[\![X, \mathcal{S}, \mu]\!]$ and $[\![Y, \mathcal{T}, \nu]\!]$ be σ-finite measure spaces. For $V \in \mathcal{S} \times \mathcal{T}$ write $\phi(x) = \nu(V_x)$, $\psi(y) = \mu(V^y)$, for each $x \in X, y \in Y$. Then ϕ is $\mathcal{S}$-measurable, ψ is $\mathcal{T}$-measurable and

$$\int_X \phi \, d\mu = \int_Y \psi \, d\nu. \tag{10.2}$$

Proof: We suppose first that the result holds if μ and ν are finite measures. We may write $X = \bigcup_{n=1}^{\infty} X_n$, $Y = \bigcup_{m=1}^{\infty} Y_m$, decomposing X and Y into disjoint sequences of sets of finite measure, and the result is therefore assumed true for each rectangle $X_n \times Y_m$ where we are considering μ and ν restricted to the measurable subsets of X_n and Y_m respectively. Let $V \in \mathcal{S} \times \mathcal{T}$ and write

$V_{n,m} = V \cap (X_n \times Y_m)$; then for each x, $V_x = \bigcup_{n,m} (V_{n,m})_x$. By hypothesis $\nu((V_{n,m})_x)$ is a measurable function of x on X_n for each m, so $\sum_{m=1}^{\infty} \nu((V_{n,m})_x)$ is measurable on X_n. Hence, by Example 12, p. 103,

$$\phi(x) = \nu(V_x) = \sum_{m,n=1}^{\infty} \nu((V_{n,m})_x)$$

is measurable with respect to $\mathcal{S}$. Similarly, $\psi(y) = \sum_{m,n=1}^{\infty} \mu(V_{n,m}^y)$ is measurable with respect to $\mathcal{T}$. But using Theorem 17, p. 105, we have

$$\int \phi \, d\mu = \sum_{n=1}^{\infty} \int_{X_n} \phi \, d\mu = \sum_{n=1}^{\infty} \int_{X_n} \sum_{m=1}^{\infty} \nu((V_{n,m})_x) \, d\mu$$

$$= \sum_{n=1}^{\infty} \sum_{m=1}^{\infty} \int_{X_n} \nu((V_{n,m})_x) \, d\mu = \sum_{n=1}^{\infty} \sum_{m=1}^{\infty} \int_{Y_m} \mu(V_{n,m}^y) \, d\nu,$$

by hypothesis, and this last term is similarly equal to $\int \psi \, d\nu$.

So we suppose that μ and ν are finite measures on $\mathcal{S}$ and $\mathcal{T}$ respectively, and write Ω for the class of sets $V \in \mathcal{S} \times \mathcal{T}$ for which (10.2) holds. Then Ω contains every measurable rectangle $A \times B$, since $\nu((A \times B)_x) = \chi_A(x)\,\nu(B)$, $\mu((A \times B)^y) = \chi_B(y)\,\mu(A)$, so (10.2) holds. Then it can be seen that (10.2) holds for any elementary set, that is Ω contains the algebra $\mathcal{E}$. If $V_1 \subseteq V_2 \subseteq \ldots$, where $V_i \in \Omega$ for each i, and if $V = \bigcup_{i=1}^{\infty} V_i$, then $V \in \Omega$. For, write $\phi_i(x) = \nu((V_i)_x)$, $\psi_i(y) = \mu(V_i^y)$. These are measurable functions by hypothesis, and by Example 2, $\phi_i(x) \uparrow \phi(x) = \nu(V_x)$, $\psi_i(y) \uparrow \psi(y) = \mu(V^y)$. So ϕ and ψ are measurable and $\int \phi \, d\mu = \lim \int \phi_i \, d\mu = \lim \int \psi_i \, d\nu = \int \psi \, d\nu$.

If $V_1 \supseteq V_2 \supseteq \ldots$, where $V_i \in \Omega$ for each i, and if $V = \bigcap_{i=1}^{\infty} V_i$, we obtain similarly sequences $\phi_i \downarrow \phi$, $\psi_i \downarrow \psi$. Since μ and ν are finite we may use Theorem 21, p. 107, since $\phi_i \leqslant \nu(Y)$, a function integrable over X and $\psi_i \leqslant \mu(X)$ which is integrable over Y, to get that (10.2) holds for ϕ and ψ. So Ω is monotone class contained in $\mathcal{S} \times \mathcal{T}$ and containing $\mathcal{E}$ and the result follows by Theorem 3. □

Corollary: With the notation and the σ-finiteness condition of the theorem we have $\int_X d\mu(x) \int_Y \chi_V(x,y) \, d\nu(y) = \int_Y d\nu(y) \int_X \chi_V(x,y) \, d\mu(x)$.

Proof: We have $\phi(x) = \nu(V_x) = \int_Y \chi_V(x,y) \, d\nu(y)$, and similarly for ψ, giving this more intuitive version of (10.2). □

Definition 9: Let $[\![X, \mathcal{S}, \mu]\!]$ and $[\![Y, \mathcal{T}, \nu]\!]$ be σ-finite measure spaces. Then the product measure $\mu \times \nu$ on $\mathcal{S} \times \mathcal{T}$ is given by

$$(\mu \times \nu)(V) = \int_X \nu(V_x)\, d\mu = \int_Y \mu(V^y)\, d\nu,$$

for each $V \in \mathcal{S} \times \mathcal{T}$, the last equality holding by Theorem 6. The alternative expression for these integrals given in the last corollary makes it clear that $\mu \times \nu$ is a measure on $\mathcal{S} \times \mathcal{T}$. Clearly $\mu \times \nu$ is σ-finite.

Example 3: Show that if μ and ν are σ-finite measures, then $\mu \times \nu$ as given in Definition 9 is the only measure on $\mathcal{S} \times \mathcal{T}$ giving to each measurable rectangle $A \times B$ the measure $\mu(A)\, \nu(B)$.

Solution: The required measure must have value $\sum_{i=1}^{n} \mu(A_i)\, \nu(B_i)$ on the elementary set which decomposes into disjoint measurable rectangles as $\bigcup_{i=1}^{n} (A_i \times B_i)$. Now $\mu \times \nu$ clearly takes the correct value on measurable rectangles and by Theorem 6, p. 179, it is a measure on $\mathcal{E}$ so it takes the correct value on the sets of $\mathcal{E}$ and indeed is clearly a σ-finite measure on the σ-algebra $\mathcal{E}$. But the extension from $\mathcal{E}$ to $\mathcal{S}(\mathcal{E}) = \mathcal{S} \times \mathcal{T}$ is then unique by Theorem 7, p. 100.

The main results on integration in a product space are given in the next three theorems; their content provides methods of calculating integrals with respect to product measures. For these results we need, as in Theorem 6 on which they depend, the fact that $[\![X, \mathcal{S}, \mu]\!]$ and $[\![Y, \mathcal{T}, \nu]\!]$ are σ-finite measure spaces.

Theorem 7: Let f be a non-negative $\mathcal{S} \times \mathcal{T}$-measurable function and let $\phi(x) = \int_Y f_x\, d\nu$, $\psi(y) = \int_X f^y\, d\mu$ for each $x \in X, y \in Y$; then ϕ is $\mathcal{S}$-measurable, ψ is $\mathcal{T}$-measurable and

$$\int_X \phi\, d\mu = \int_{X \times Y} f\, d(\mu \times \nu) = \int_Y \psi\, d\nu. \tag{10.3}$$

Proof: The last theorem gives the result in the case where f is the characteristic function of an $\mathcal{S} \times \mathcal{T}$ measurable set, and hence for measurable simple functions. In the general case, let $\{f_n\}$ be a sequence of measurable simple functions such that $f_n \uparrow f$. So, by the last theorem, $\phi_n(x) = \int (f_n)_x\, d\nu$ is $\mathcal{S}$-measurable and

$$\int_X \phi_n\, d\mu = \int_{X \times Y} f_n\, d(\mu \times \nu). \tag{10.4}$$

As n tends to infinity, $(f_n)_x \uparrow f_x$, so by Theorem 15, p. 105, $\phi_n \uparrow \phi$ and so ϕ is measurable. Applying Theorem 15 again, to (10.4), we get the first identity of (10.3). The second follows similarly. □

Corollary: Writing the first and last terms in (10.3) as iterated integrals of f, we get

$$\int_X d\mu(x) \int_Y f(x,y)\, d\nu(y) = \int_Y d\nu(y) \int_X f(x,y)\, d\mu(x). \qquad (10.5)$$

Theorem 8: Let f be an $\mathcal{S} \times \mathcal{T}$-measurable function and let $\phi^*(x) = \int_Y |f|_x \, d\nu$, $\psi(y) = \int_X |f|^y \, d\mu$ for each $x \in X$, $y \in Y$; then the conditions $\phi^* \in L^1(\mu)$, $\psi^* \in L^1(\nu), f \in L^1(\mu \times \nu)$ are equivalent.

Proof: We apply the last theorem to $|f|$, and (10.3) gives the result. □

Theorem 9 (Fubini's Theorem): If $f \in L^1(\mu \times \nu)$, then $f_x \in L^1(\nu)$ for almost all $x \in X$, $f^y \in L^1(\mu)$ for almost all $y \in Y$, the functions ϕ and ψ defined as in Theorem 7 are in $L^1(\mu)$ and $L^1(\nu)$ respectively, and (10.3) holds.

Proof: From the measurable functions f^+, f^- we obtain the functions ϕ_1, ϕ_2 as ϕ was obtained from f in Theorem 7. Since $f^+, f^- \in L^1(\mu \times \nu)$, (10.3) implies that $\phi_1, \phi_2 \in L^1(\mu)$. So, for almost all x, both $\phi_1(x)$ and $\phi_2(x)$ are finite, and for such x, since $f_x = f_x^+ - f_x^-$ we have $f_x \in L^1(\nu)$ and $\phi(x) = \phi_1(x) - \phi_2(x)$; hence ϕ is integrable. Also the first equality of (10.3) holds for ϕ_1 and f^+ and for ϕ_2 and f^-, so subtracting we get result for ϕ and f. The second equality and the statements about f^y and ψ are proved similarly. □

Corollary: Theorems 8 and 9 imply that if either of the iterated integrals of $|f|$ is finite, then so is the other and f satisfies (10.5).

It is the form of Fubini's theorem given in this Corollary which is most useful in applications. We now give some examples which show that the conditions in Theorems 6 to 9 may not in general be omitted.

Example 4: The condition $V \in \mathcal{S} \times \mathcal{T}$ is necessary in Theorem 6.

Solution: For X and Y take the set of ordinals $[\alpha: \alpha \leqslant \omega_1]$ where ω_1 is the first uncountable ordinal ([11], p. 79). For $\mathcal{S}$ and $\mathcal{T}$ take the σ-algebra generated by the countable subsets. Let μ and ν be zero for countable sets and 1 for uncountable measurable sets. Let $V = [(x,y): x < y]$. Then if $x = \omega_1$, the x-section $V_x = \emptyset$; otherwise V_x is uncountable but $\mathbf{C}V_x$ is countable and so V_x is measurable. If $y = \omega_1$, $V^y = [x: x < \omega_1]$, a measurable set, and for $y < \omega_1$, V^y is countable and so measurable. But $\int_X d\mu \int_Y \chi_V \, d\nu = \int \nu(V_x)\, d\mu = \mu[x: x < \omega_1] = 1$, whereas $\int_Y d\nu \int \chi_V \, d\mu = \int \mu(V^y)\, d\nu = 0$. So (10.2) does not hold for this V.

The next example shows how Theorem 6, p. 179, breaks down if μ and ν are not both σ-finite. The same example shows that σ-finiteness is essential in Theorems 7, 8 and 9.

Example 5: Let $X = Y = [0,1]$, $\mathcal{S} = \mathcal{T} = \mathcal{B}$. Take $\mu = m$ on the Borel subsets of $[0,1]$, and for ν take the counting measure on $[0,1]$, that is, $\nu(E) = \text{Card}\, E$. Take $V = [(x,y): x = y, (x,y) \in X \times Y]$. Then V is $\mathcal{S} \times \mathcal{T}$-measurable, for if n is any positive integer put $I_j = [(j-1)/n, j/n], j = 1, \ldots, n$ and $V_n = (I_1 \times I_1) \cup \ldots \cup (I_n \times I_n)$. So V_n is measurable, and so therefore is $V = \bigcap_{n=1}^{\infty} V_n$. (A diagram may assist.) However

$$\int_Y d\nu \int_X \chi_V \, d\mu = 0 \quad \text{but} \quad \int_X d\mu \int_Y \chi_V \, d\nu = 1.$$

Example 6: The condition $f \in L^1(\mu \times \nu)$ in Theorem 9, is necessary if the order of integration is to be interchangeable.

Solution: Take X, Y, $\mathcal{S}$, $\mathcal{T}$ as in the last example and let $\mu = \nu = m$, restricted to $[0,1]$. Let $0 < \alpha_1 < \ldots < \alpha_n < \ldots < 1$, $\lim \alpha_n = 1$. For each n choose a continuous function g_n such that $[t: g_n(t) \neq 0] \subseteq (\alpha_n, \alpha_{n+1})$ and also $\int_0^1 g_n \, dt = 1$.

Let $f(x,y) = \sum_{n=1}^{\infty} g_n(y)(g_n(x) - g_{n+1}(x))$. For each (x, y) only one term in this series can be non-zero, so f is well defined. Also f is measurable, indeed f is continuous except at (1,1). But

$$\int_0^1 f(x,y) \, dx = \int_0^1 \sum_{n=1}^{\infty} g_n(y)(g_n(x) - g_{n+1}(x)) \, dx =$$

$$= g_n(y)\left(\int_{\alpha_n}^{\alpha_{n+1}} g_n \, dx - \int_{\alpha_{n+1}}^{\alpha_{n+2}} g_{n+1} \, dx\right) = 0$$

for each y. However

$$\int_0^1 f(x,y) \, dy = \sum_{n=1}^{\infty} (g_n(x) - g_{n+1}(x)) \int_0^1 g_n \, dy = g_1(x),$$

so $\int_0^1 dx \int_0^1 f(x, y) \, dy = 1$ and the iterated integrals are therefore unequal. However, Fubini's theorem is not contradicted since f is not integrable. For, writing $I_i = (\alpha_i, \alpha_{i+1})$, we have

$$\int |f(x,y)| \, dx \, dy = \sum_{i,j=1}^{\infty} \int_{I_i \times I_j} \left|\sum_{n=1}^{\infty} g_n(y)(g_n(x) - g_{n+1}(x))\right| dx \, dy$$

$$= \sum_{i,j=1}^{\infty} \int_{I_i \times I_j} |g_j(y)(g_j(x) - g_{j+1}(x)| \, dx \, dy$$

$$= \sum_{j=1}^{\infty} \int_{I_j \times I_j} + \int_{I_{j+1} \times I_j} |g_j(y)(g_j(x) - g_{j+1}(x))| \, dx \, dy = \infty.$$

In some product spaces, for example in the plane, the 'natural' measure to consider is not $\mu \times \nu$ but its completion, obtained in Theorem 8, p. 101. Results equivalent to Theorem 9 can be obtained in this case but details of the statements are no longer the same; for example, the functions ϕ and ψ of Theorem 6 onwards need not now be measurable but only equal a.e. to measurable functions. For details of an approach using complete measures see, for example, [12], Chapter 12.

Exercises

6. Let μ and ν be complete measures. Show that $\mu \times \nu$ need not be complete.
7. Let f be a measurable function on $\mathsf{R}, f > 0$. Let the ordinate set of f be $0_f = [(x, y): 0 \leqslant y < f(x)]$.
   (i) Show that 0_f is measurable.
   (ii) Show that $(m \times m)(0_f) = \int f \, dx$.
   (iii) The graph of f is the set $G = [(x, y): y = f(x)]$. Show that G is measurable.
   (iv) Show that G has measure zero.
8. Let f be continuous on $A = [a, b] \times [c, d]$ and let $\mathcal{S} = \mathcal{T} = \mathcal{M}$. Show that
$$\int_A f \, dx dy = \int_a^b dx \int_c^d f(x, y) \, dy = \int_c^d dy \int_a^b f(x, y) \, dx.$$
9. Show that if $f \in L^1(\mu)$ and $g \in L^1(\nu)$, then $fg \in L^1(\mu \times \nu)$.
10. Let $f \in L(0, a)$ and $g(x) = \int_x^a f(t)/t \, dt$ $(0 < x \leqslant a)$. Use the results of this section to show that $g \in L(0, a)$ and $\int_0^a g \, dx = \int_0^a f \, dx$ (cf. Exercise 20, Chapter 9).
11. By integrating $e^{-y} \sin 2xy$ with respect to x and y, show that
$$\int_0^{\infty} e^{-y} (\sin^2 y)/y \, dy = \tfrac{1}{4} \log 5.$$
12. By integrating $e^{-xy} \sin 2y$ with respect to x and y, show that
$$\int_0^{\infty} e^{-y} (\sin 2y)/y \, dy = \arctan 2.$$
13. Let $f \in L(0, \infty)$ and for $\alpha > 0$ and $x > 0$ let $g_\alpha(x) = \int_0^x (x - t)^{\alpha - 1} f(t) \, dt$. Show that $\alpha \int_0^y g_\alpha(x) \, dx = g_{\alpha+1}(y), y > 0$.

14. Let $\alpha > 0, f \geqslant 0, f \in L(0, a)$ and $F(x) = \int_0^x f\,dt$; show that

$$\int_0^a f(x) \log \frac{1}{x}\,dx = \int_0^a \frac{F(x)}{x}\,dx + F(a) \log \frac{1}{a}$$

whenever either integral is finite.

15. Let $f(x, y) = \dfrac{xy}{(x^2 + y^2)^2}$, $-1 \leqslant x \leqslant 1, -1 \leqslant y \leqslant 1$, defining $f(0,0) = 0$. Show that the iterated integrals of f over the square are equal but that f is not integrable.

16. Show that if $f(x, y) = \dfrac{x^2 - y^2}{(x^2 + y^2)^2}$, $(x, y) \neq (0,0)$, then

$$\int_0^1 dx \int_0^1 f(x,y)\,dy = \frac{\pi}{4} \text{ and } \int_0^1 dy \int_0^1 f(x,y)\,dx = -\frac{\pi}{4}\,.$$

Show that this does not contradict Theorem 9, p. 182.

17. For each $n \geqslant 1$ let $f(x, y) = 2^{2n}$ when $2^{-n} \leqslant x < 2^{-n+1}, 2^{-n} \leqslant y < 2^{-n+1}$, let $f(x, y) = -2^{2n+1}$ when $2^{-n-1} \leqslant x < 2^{-n}, 2^{-n} \leqslant y < 2^{-n+1}$, and let $f = 0$ on the remainder of the square $[0,1] \times [0,1]$ not contained in one of these rectangles for any n. Show that

$$\int_0^1 dy \int_0^1 f(x,y)\,dx = 0, \text{ but } \int_0^1 dx \int_0^1 f(x,y)\,dy = 1.$$

18. Let $X = Y = \mathsf{N}$, $\mu = \nu =$ counting measure on $\mathcal{P}(\mathsf{N})$. Let $f(x,y) = 2 - 2^{-x}$ for $x = y$, $f(x, y) = -2 + 2^{-x}$ for $x = y + 1$, and let $f = 0$ otherwise. Show that the iterated integrals of f are not equal.

10.3 LEBESGUE MEASURE IN EUCLIDEAN SPACE

In this section we show how, starting with Lebesgue measure on R and forming the completion of the product measure, we may define a measure on the plane and, more generally, in R^n. The resulting measure, like Lebesgue measure on the line, is invariant under translation and rotation, so the original set of axes lose any special significance. We also show how integrals with respect to this measure may be transformed into integrals with respect to polar coordinates, in two or three dimensions.

Definition 10: The completion of the product measure $m \times m$ is Lebesgue measure m_2 on R^2. The class of Lebesgue measurable sets, written $\mathcal{M}_2$ consists of the sets measurable with respect to m_2.

Examples such as $A \times [y]$ where A is a non-measurable set, show that $\mathcal{M}_2 \neq \mathcal{M} \times \mathcal{M}$ so that $m_2 \neq m \times m$. Using induction we may form in R^k the product of k copies of Lebesgue measure.

Definition 11: For $k > 1$ let $m^{(k)} = m^{(k-1)} \times m$ where $m^{(1)} = m$. Similarly $\mathcal{M}^{(k)} = \mathcal{M}^{(k-1)} \times \mathcal{M}$; and m_k on $\mathcal{M}_k$ is the completion of $m^{(k)}$ on $\mathcal{M}^{(k)}$.

The next theorem gives a more convenient description of the Lebesgue measurable sets.

Theorem 10: (i) $\mathcal{M}^{(k)}$ is generated by the sets $A_1 \times \ldots \times A_k$ where $A_i \in \mathcal{M}$ for each i.

(ii) $\mathcal{M}_k$ is the result of completing the measure $m^{(k)}$ on $\mathcal{B} \times \ldots \times \mathcal{B}$ (k terms).

(iii) m_k on $\mathcal{M}_k$ is given by its values on the Borel sets of R^k, the fact that these are measurable and the fact that it is complete.

Proof: (i) This is true, by definition, when $k = 2$. Suppose that it is true for some $k \geqslant 2$. By definition, $\mathcal{M}^{(k+1)}$ is the σ-algebra generated by sets of the form $E \times F$ where $E \in \mathcal{M}^{(k)}$, $F \in \mathcal{M}$. So by hypothesis $\mathcal{M}^{(k+1)}$ contains all the sets $A_1 \times \ldots \times A_{k+1}$, $A_i \in \mathcal{M}$. Now take $A_{k+1} = F$, and in this restricted class of sets consider countable unions and complements in $\mathsf{R}^k \times F$. We generate, by the inductive hypothesis, all sets of the form $E \times F$, $E \in \mathcal{M}^{(k)}$, $F \in \mathcal{M}$, so the σ-algebra generated by the sets $A_1 \times \ldots \times A_{k+1}$ is just $\mathcal{M}^{(k+1)}$, and by induction this holds for all positive integers k.

(ii) Let $\mathcal{S}$ be the σ-algebra obtained by this completion, so $\mathcal{S} \subseteq \mathcal{M}_k$. Also $\mathcal{S} \supseteq \mathcal{B} \times \ldots \times \mathcal{B}$. If $A_1 \in \mathcal{B}$, $m(A_1) = 0$, and $A_2, \ldots, A_k \in \mathcal{B}$, then all the sets $C \times A_2 \times \ldots \times A_k$ with $C \subseteq A_1$ belong to $\mathcal{S}$. So $\mathcal{S}$ contains $\mathcal{M} \times \mathcal{B} \times \ldots \times \mathcal{B}$ and hence by induction $\mathcal{S} \supseteq \mathcal{M}^{(k)}$. So $\mathcal{S} \supseteq \mathcal{M}_k$ as required.

(iii) As on the line the Borel sets are the σ-algebra generated by the open sets, by definition. Now $\mathcal{B} \times \ldots \times \mathcal{B}$ is generated by the open sets $G = G_1 \times \ldots \times G_k$, G_i open, and so consists of Borel sets. But every open set may be written as a countable union of sets G so $\mathcal{B} \times \ldots \times \mathcal{B}$ is just the Borel sets. Now (iii) follows from (ii). □

Corollary 1: The measure m_k is given by its values on the half-open 'intervals' $[x\colon a_i \leqslant x_i < b_i, 1 \leqslant i \leqslant k]$ and the fact that it is complete.

Proof: The details are obvious, but see the solution to Exercise 19 below. □

Corollary 2: $\mathcal{M}^{(k+n)} = \mathcal{M}^{(k)} \times \mathcal{M}^{(n)}$ and $m^{(k+n)} = m^{(k)} \times m^{(n)}$.

Proof: This follows from the fact that $\mathcal{M}^{(k+n)}$ is generated by the sets $A_1 \times \ldots A_{k+n}$, $A_i \in \mathcal{M}$. □

Example 7: Lebesgue measure in R^k is translation invariant, that is if $x \in \mathsf{R}^k$ and $E \in \mathcal{M}_k$ then $E + x \in \mathcal{M}_k$ and their measures are the same.

Solution: The mapping $y \to y + x$, $y \in \mathsf{R}^k$, is continuous and has a continuous inverse, so the open sets on translation give again the open sets. Also, a σ-algebra translates to a σ-algebra, so the Borel sets on translation give just the Borel sets. Also, the half-open 'intervals' translate to half-open intervals of the same measure, so the Borel sets have their measures invariant. But every set in $\mathcal{M}_k$ is the union of a Borel set and a subset of a Borel set of measure zero, just as for the real line, so on translation $\mathcal{M}_k$ is unchanged and so is m_k.

Example 8: Let $E \in \mathcal{M}_k$ and $r > 0$. Show that $rE \in \mathcal{M}_k$ and $m_k(rE) = r^k m_k(E)$.

Solution: As in the last example, the Borel sets are transformed into the Borel sets by $E \to rE$. If I is any half-open 'interval', its measure is multiplied by r^k, by the corresponding result for the real line (Exercise 8, p. 34). So the Borel sets have their measures multiplied by r^k. Since the same sets of zero measure are obtained, the result follows as for Example 7.

Example 9: Show that measurability and measure remain invariant under rotation in k dimensions.

Solution: From the result of Example 7 we may restrict ourselves to rotations about the origin. Let $A^{(r)}$ denote the set obtained from A by the rotation r. Under r, open balls become open balls and a σ-algebra is mapped to a σ-algebra. So open sets and Borel sets remain measurable on rotation. We show that measure is unchanged for half-open 'intervals'. It follows that Borel sets under rotation give Borel sets of the same measure. Also subsets of Borel sets of zero measure rotate into sets of the same kind. So it will follow that the class of measurable sets remains measurable under rotation and that their measures are unchanged.

So let I be the 'interval' $[x\colon a_i \leqslant x_i < b_i,\ 1 \leqslant i \leqslant k]$. Then the interior G of the unit ball is a countable disjoint union of intervals similar to I, say $G = \bigcup I_i$, where $I_i = a + tI, a \in \mathrm{R}^k, t > 0$. Then $I_i^{(r)} = a^{(r)} + tI^{(r)}$ so $m_k(I_i^{(r)}) = t^k m_k(I^{(r)})$. Also $m_k(I_i) = t^k m_k(I)$. So $m_k(I_i^{(r)}) = K\, m_k(I_i)$, where $K = m_k(I^{(r)})/m_k(I)$ depends only on I and r. So $m_k(G^{(r)}) = K\, m_k(G)$, but $G^{(r)} = G$ so $K = 1$ and the result follows.

If $u = (u_1, \ldots, u_k) \in \mathrm{R}^k$ write $|u| = \left(\sum_{i=1}^{k} u_i^2\right)^{1/2}$.

Definition 12: The **unit sphere** in R^k is the set $S_{k-1} = [u\colon |u| = 1]$. For the purpose of considering open sets and hence Borel sets, S_{k-1} is assumed to have the relative topology as a subset of R^k (cf. Chapter 1, p. 18).

We define *angular measure* on the sphere.

Definition 13: Let A be a Borel set on S_{k-1} and write $\tilde{A} = [x\colon x = ru, 0 < r < 1, u \in A]$. Then define $\sigma_{k-1}(A) = k m_k(\tilde{A})$. It is easy to see that $\tilde{A}$ is a Borel set, so σ_{k-1} is well defined and is a measure on the Borel sets of S_{k-1}.

Theorem 11: Let f be a Borel measurable integrable function on R^k, then

$$\int f\, \mathrm{d}m_k = \int_0^\infty r^{k-1}\, \mathrm{d}r \int_{S_{k-1}} f(ru)\, \mathrm{d}\sigma_{k-1}(u). \tag{10.6}$$

Proof: Let $0<r_1<r_2$, let A be an open set, $A \subseteq S_{k-1}$, and let $E=[ru: r_1<r\leqslant r_2, u\in A]$. Then (10.6) holds for χ_E. For,

$$\int_0^\infty r^{k-1}\,\mathrm{d}r \int_{S_{k-1}} \chi_E(ru)\,\mathrm{d}\sigma_{k-1}(u) = \int_{r_1}^{r_2} r^{k-1}\,\sigma_{k-1}(A)\,\mathrm{d}r =$$

$$= k^{-1}(r_2^k - r_1^k)\,\sigma_{k-1}(A).$$

But $E = r_2\tilde{A} - r_1\tilde{A}$, so $m_k(E) = (r_2^k - r_1^k)m_k(\tilde{A})$ by Example 9, and so

$$m_k(E) = k^{-1}(r_2^k - r_1^k)\,\sigma_{k-1}(A)$$

as required. So if $E=[ru: r_1<r<r_2, u\in G]$, for some open set $G\subseteq S_{k-1}$, we have $E=\bigcup_{n=N}^{\infty} E_n$, $E_n = [ru: r_1<r\leqslant r_2 - 1/n]$, where N is such that $r_2 - 1/N > r_1$. But $\chi_{E_n}\uparrow \chi_E$ and so the result holds for χ_E by Theorem 15, p. 105. Let Ω be the class of sets for whose characteristic functions (10.6) holds. Then Ω is closed under disjoint countable unions by Theorem 17, p. 105. Also Ω is closed under the formation of differences of bounded sets since we may write $\chi_{A-B} = (\chi_A - \chi_B)^+$, all the functions being integrable, and use the fact that (10.6) is linear in f. So Ω is closed under countable unions contained in a fixed bounded set, for example $D_N = (-N, N)\times\ldots\times(-N,N) - [\mathbf{0}]$ where $\mathbf{0}$ denotes the origin. Hence Ω contains all open sets in D_N, for each such set is the union of open sets of the same type as E above, and hence by Theorem 8, p. 23, is the union of a countable number of such sets. Taking unions and differences, we find that Ω contains all Borel sets in D_N. So if E is any Borel set in $\mathsf{R}^k - [\mathbf{0}]$, (10.6) holds for the characteristic function of $E\cap D_N$, and letting $N\to\infty$ we find $E\in\Omega$ by Theorem 15, p. 105. Hence the result holds for Borel measurable simple functions and taking increasing sequences of such functions we get (10.6) for f^+ and f^-. Since the left-hand side of (10.6) may be replaced by the integral over $\mathsf{R}^k - [\mathbf{0}]$, the result follows. □

In R^2, if $G\subseteq S_1$ is the intersection of an open disc with S_1, then $\sigma_1(G)=\theta$ where θ is the angle subtended by G at the origin. Then we may write (10.6) in the form

$$\int f\,\mathrm{d}m_2 = \int_0^\infty r\,\mathrm{d}r \int_0^{2\pi} F(r,\theta)\,\mathrm{d}\theta$$

where $F(r,\theta) = f(r\cos\theta, r\sin\theta)$.

Finally we show how the elementary procedure of integrating with respect to three dimensional polar coordinates may be fitted into this context.

Example 10: If S_2^* denotes the sphere S_2 minus the poles (0,0,1) and (0,0,–1); there is a one-to-one mapping of S_2^* onto $(0,\pi)\times[0,2\pi)$ such that $(x,y,z)\to(\theta,\phi)$ where $x=\cos\phi\sin\theta$, $y=\sin\phi\sin\theta$, $z=\cos\theta$. Then if f is a Borel measurable function and is integrable,

$$\int f \, dm_3 = \int_0^\infty r^2 \, dr \int_0^{2\pi} d\phi \int_0^\pi \sin\theta \, F(r, \theta, \phi) \, d\theta$$

where $F(r, \theta, \phi) = f(r \cos\phi \sin\theta, r \sin\phi \sin\theta, r \cos\theta)$.

Solution: By Theorem 11, we need only show that for fixed r,

$$\int_{S_2} f(ru) \, d\sigma_2 = \int_0^{2\pi} d\phi \int_0^\pi \sin\theta \, F(r, \theta, \phi) \, d\theta \tag{10.7}$$

and indeed the integral on the left need be taken over S_2^* only. Let A be the open set in S_2^* given by $\theta_1 < \theta < \theta_2, \phi_1 < \phi < \phi_2$ and take $f(ru) = \chi_A(u)$. Then the right-hand side of (10.7) is

$$(\phi_2 - \phi_1) \int_{\theta_1}^{\theta_2} \sin\theta \, d\theta = (\phi_2 - \phi_1)(\cos\theta_1 - \cos\theta_2),$$

while the left-hand side is $3m_3(A)$, in the notation of Definition 13, which may be checked by a laborious integration with respect to x, y, z to have the same value. Each open set in S_2^* may be written as a countable union of sets such as A, so as in Theorem 11 the result holds if $f(ru) = \chi_E(ru)$ for any Borel set $E \subseteq S_2^*$. Hence we obtain the result for an arbitrary Borel measurable function f on taking a sequence of Borel measurable simple functions with f as limit.

Exercises

19. Let $[\![X, \mathcal{S}]\!]$ and $[\![Y, \mathcal{T}]\!]$ be measurable spaces such that $\mathcal{S}$ is the σ-algebra generated by $\mathcal{E}$, and $\mathcal{T}$ is that generated by $\mathcal{F}$. Show that $\mathcal{S} \times \mathcal{T}$ is the σ-algebra generated by $[E \times F: E \in \mathcal{E}, F \in \mathcal{F}] = \mathcal{E} \times \mathcal{F}$, say.
20. Show that if $f(x)$ is a Borel measurable function on $\mathsf{R}^k - [\mathbf{0}]$ and we write $x = ru, u \in S_{k-1}$, then $g(u) = f(ru)$ is Borel measurable on S_{k-1}.
21. Show that every open set in $\mathsf{R}^2 - [\mathbf{0}]$ is the union of sets $E = [ru: r_1 < r < r_2, u \in G, G \text{ open}, G \subseteq S_1]$.
22. Show that $\int_0^\infty e^{-x^2} \, dx = \frac{1}{2}\sqrt{\pi}$.

10.4 LAPLACE AND FOURIER TRANSFORMS

It is easy to extend the definitions and results on integration obtained for real-valued functions to complex-valued functions. If $f = f_1 + if_2$, f is said to be *measurable* if, and only if, the real functions f_1 and f_2 are measurable and almost everywhere finite-valued. Then $|f|$ is clearly measurable in the previous sense. We define f to be *integrable* provided f is measurable and $|f|$ is integrable, and in this case f_1 and f_2 are clearly integrable. If f is complex we then define the integral of f to be $\int f \, d\mu = \int f_1 \, d\mu + i \int f_2 \, d\mu$. The following familiar result now needs a different proof.

Theorem 12: If f is a complex-valued function integrable with respect to μ, then $|\int f\,d\mu| \leqslant \int |f|\,d\mu$.

Proof: Let $z = \int f\,d\mu$ and choose α, $|\alpha| = 1$, such that $\alpha z = |z|$. Then $|\int f\,d\mu| = \alpha \int f\,d\mu = \int \alpha f\,d\mu = \int Re(\alpha f)\,d\mu \leqslant \int |\alpha f|\,d\mu = \int |f|\,d\mu$, since the linearity of the integral holds as for real functions and since $\int Im(\alpha f)\,d\mu = 0$ by inspection of the first term in the string of equalities. □

Example 11: Show that $|\int f\,d\mu| = \int |f|\,d\mu$ if, and only if, there exists θ, real and constant, such that $e^{i\theta} f \geqslant 0$ a.e. (μ).

Solution: From the proof of Theorem 12 we must have $\int Re(\alpha f)\,d\mu = \int |\alpha f|\,d\mu$. So $Re(\alpha f) = |\alpha f|$ a.e. (μ), and $Im(\alpha f) = 0$ a.e. (μ). Writing $\alpha = e^{i\theta}$ we see that the condition is necessary. It is clearly also sufficient.

Earlier theorems concerning non-negative functions we do not extend explicitly, but where in earlier proofs we split functions into positive and negative parts, we must now split them first into real and imaginary parts and then proceed as before, the condition of finiteness almost everywhere being imposed since we cannot carry over the conventions regarding the extended real numbers. Lebesgue's dominated convergence theorem, p. 107, and Fubini's theorem, p. 182, in particular, hold for complex functions. The methods of Chapter 8 then suggest the consideration of complex-valued measures. These raise no difficulties but we will not examine them.

The next result is a preliminary to considering the convolution of two functions, which will be important in the applications which follow. The functions are real-valued as usual.

Theorem 13: Let f and $g \in L(-\infty, \infty)$, then $f(y-x)g(x)$ is an integrable function of x for almost all y, and if $h(y)$ is defined for these y by $h(y) = \int f(y-x)g(x)\,dx$, then h is integrable and $\|h\|_1 \leqslant \|f\|_1 \cdot \|g\|_1$.

Proof: For each y, $f(y-x)\,g(x)$ is a measurable function of x, so write $h^*(y) = \int |f(y-x)g(x)|\,dx$. If we may apply Theorem 9 (Corollary) to h^* we get

$$\int h^*(y)\,dy = \int dy \int |f(y-x)g(x)|\,dx = \int |g(x)|\,dx \int |f(y-x)|\,dy.$$

But $\int |f(y-x)|\,dy = \int |f(y)|\,dy$ by Example 25, p. 75. So $\int h^*(y)\,dy \leqslant \|g\|_1 \cdot \|f\|_1$ and so h^*, and hence h, is finite valued a.e., h is integrable and $\|h\|_1 \leqslant \|f\|_1 \cdot \|g\|_1$.

However, to apply Theorem 9, we need to know that $f(y-x)\,g(x)$ is measurable with respect to $m \times m$. By Exercise 11, p. 60, we may replace f by a Borel measurable function equalling it a.e. and show the desired measurability for the altered integrand. As the integrals are the same the subsequent argument is unaffected. Write $F(x, y) = f(y-x)$; then F is a Borel measurable function. Indeed if for $E \subseteq \mathsf{R}$, we write $\mathcal{D}E = [(x,y): x-y \in E]$, we have $F^{-1}(\alpha, \infty) = \mathcal{D}f^{-1}(\alpha, \infty)$. As $f^{-1}(\alpha, \infty)$ is a Borel set, we need to show that if $E \in \mathcal{B}$, $\mathcal{D}E$ is a Borel set in R^2. If E is open, $\mathcal{D}E$ is open, since when $x_0 - y_0$ is contained in a neighbourhood of radius ϵ in E, (x_0, y_0) is contained in a neighbourhood of radius $\epsilon/2$ in the strip $\mathcal{D}E$. Also $\Omega = [E\colon \mathcal{D}E$ is a Borel set$]$ is a σ-algebra, since $\mathcal{D}(\bigcup_{i=1}^{\infty} E_i) = \bigcup_{i=1}^{\infty} \mathcal{D}E_i$ and $\mathcal{D}(\mathbf{C}E) = \mathbf{C}(\mathcal{D}E)$. So Ω contains the Borel sets and so $F(x, y)$ is a Borel measurable function. But $G(x, y) = g(x)$ is measurable with respect to $m \times m$, and so $F(x, y)\, G(x, y) = f(y - x)\, g(x)$ is measurable with respect to $m \times m$ and the theorem is proved. □

Definition 14: If $f, g \in L(-\infty, \infty)$, the **convolution** $f * g$ of f and g is given by

$$(f * g)(y) = \int f(y - x)\, g(x)\, dx.$$

By Theorem 13, $f * g$ is well defined and is an integrable function.

Theorem 14: If $f, g, h \in L(-\infty, \infty)$, then (i) $f * g = g * f$ a.e., (ii) $(f * g) * h = f * (g * h)$ a.e.

Proof. (i) Let y be such that $g(y - x)\, f(x)$ is integrable with respect to x, and substitute $t = y - x$ in $\int g(y - x) f(x)\, dx$ to get, by Example 12, p. 171,

$$\int g(y - x) f(x)\, dx = \int g(t) f(y - t)\, dt = (f * g)(y).$$

So $f * g = g * f$ a.e.

(ii) $(f * (g * h))(y) = \int f(y - x)(\int g(x - z)\, h(z)\, dz)\, dx$. $((f * g) * h)(y) = \int h(z)(\int f(y - z - t)\, g(t)\, dt)\, dz = \int h(z)(\int f(y - x)\, g(x - z)\, dx)\, dz$ using the substitution $t = x - z$ and Example 25, p. 75. To establish the result we need only change the order of integration, so we need to show that Theorem 9, p. 182, can be applied to $F(x, z) = f(y - x)\, g(x - z)\, h(z)$. Replacing f and g by Borel measurable functions equal to them almost everywhere, we obtain as in Theorem 13, an integrand measurable with respect to $m \times m$. Also, for almost all y, $F(x, z)$ is integrable; for

$$\int dy \int |h(z) f(y - x) g(x - z)|\, dx\, dz$$
$$= \int |h(z) g(x - z)| (\int |f(y - x)|\, dy)\, dx\, dz \leqslant \|f\|_1 \cdot \|g\|_1 \cdot \|h\|_1$$

as in Theorem 13. So for these y we may change the order of integration to get the result. □

An important aspect of the convolution of two functions is its relation to the Fourier transform (Definition 16, below) and to the Laplace transform, which we now define.

Definition 15: If $f \in L(0, \infty)$ the function

$$(\mathcal{L}f)(x) = \int_0^{\infty} e^{-xt} f(t)\, dt$$

is the **Laplace transform** of f. Clearly it is well defined on $(0, \infty)$.

Theorem 15: If $f, g \in L(0, \infty)$ then $\mathcal{L}(f * g) = \mathcal{L}f \cdot \mathcal{L}g$.

Proof: For such functions, in defining $f * g$ we take f and g to be zero on $(-\infty, 0)$, so that

$$(f * g)(x) = \int_0^x f(x-t)\, g(t)\, \mathrm{d}t,$$

an integrable function on $(0, \infty)$. So for $y \geqslant 0$, writing $f * g = h$,

$$(\mathcal{L}h)(y) = \int_0^\infty \mathrm{d}x \int_0^x e^{-yx} f(x-t)\, g(t)\, \mathrm{d}t.$$

From Theorem 13, p. 190, we may interchange the order of integration to get

$$(\mathcal{L}h)(y) = \int_0^\infty g(t)\left(\int_t^\infty e^{-yx} f(x-t)\, \mathrm{d}x\right) \mathrm{d}t$$

$$= \int_0^\infty g(t)\left(\int_0^\infty e^{-y(x+t)} f(x)\, \mathrm{d}x\right) \mathrm{d}t,$$

by Example 25, p. 75. So

$$(\mathcal{L}h)(y) = \int_0^\infty e^{-yx} f(x)\, \mathrm{d}x \int_0^\infty e^{-yt} g(t)\, \mathrm{d}t,$$

as required. □

One important way in which integrals of complex functions arise naturally is in connection with Fourier transforms. In their definition a factor $(2\pi)^{-1/2}$ is sometimes introduced, the object being to remove the factor $(2\pi)^{-1/2}$ appearing in Theorem 18.

Definition 16: If $f \in L(-\infty, \infty)$ the function

$$\hat{f}(x) = \int e^{ixt} f(t)\, \mathrm{d}t$$

is the **Fourier transform** of f. Clearly it is well defined.

Theorem 16: If $f \in L(-\infty, \infty)$, $\hat{f}$ is a continuous function on R and $|\hat{f}| \leqslant \int |f|\, \mathrm{d}x$.

Proof: Clearly $|\hat{f}| \leqslant \int |f|\, \mathrm{d}x$. To show that $\hat{f}$ is continuous consider

$$\hat{f}(x+h) - \hat{f}(x) = \int e^{ixt}(e^{iht} - 1) f(t)\, \mathrm{d}t. \tag{10.8}$$

Since $|e^{ixt}(e^{iht} - 1) f(t)| \leqslant 2|f(t)| \in L(-\infty, \infty)$, we may let $h \to 0$ in (10.8) to obtain the result by Example 15, p. 64. □

Theorem 17: Let $f, g \in L(-\infty, \infty)$; then $\widehat{f * g} = \hat{f} \cdot \hat{g}$.

Proof: We have

$$(\widehat{f * g})(s) = \int \mathrm{d}t \int e^{ist} f(t-x)\, g(x)\, \mathrm{d}x. \tag{10.9}$$

As in Theorem 13, p. 190, $|f(t-x)\, g(x)|$ is integrable with respect to (x, t), so by Theorem 9, p. 182, we may interchange the order of integration in (10.9) to get

$$(\widehat{f * g})(s) = \int g(x)(\int e^{ist} f(t-x)\,\mathrm{d}t)\,\mathrm{d}x = \int g(x)(\int e^{is(t+x)} f(t)\,\mathrm{d}t)\,\mathrm{d}x,$$

by Example 25, p. 75. So

$$(\widehat{f * g})(s) = \int e^{ist} f(t)\,\mathrm{d}t \int e^{isx} g(x)\,\mathrm{d}x$$

as required. □

Example 12: Show that if $f(x) = e^{-x^2/2}$, then $\hat{f} = (2\pi)^{1/2} f$.

Solution: $\hat{f}(y) = \int e^{-x^2/2} e^{iyx}\,\mathrm{d}x$. By Exercise 51, p. 76, we may differentiate under the integral sign. So using Theorem 9, p. 163, and letting the limits go to infinity, we have

$$\frac{\mathrm{d}}{\mathrm{d}y}\hat{f}(y) = i \int x e^{-x^2/2} e^{iyx}\,\mathrm{d}x$$

$$= i[-e^{-x^2/2} e^{iyx}]_{-\infty}^{\infty} - \int e^{-x^2/2} y e^{iyx}\,\mathrm{d}x = -y\hat{f}(y).$$

So $\hat{f}$ satisfies the equation $\mathrm{d}u/\mathrm{d}y = -yu$. So $\hat{f}(y) = Ce^{-y^2/2}$. Then $\hat{f}(0) = C = \int e^{-x^2/2}\,\mathrm{d}x = (2\pi)^{1/2}$ by Exercise 22, p. 189.

The following results are easily obtained by a change of variable.

Example 13: (i) $\hat{f}(\beta y) = (\beta^{-1} f(\beta^{-1} x))\hat{}$. (ii) If $k(x) = \overline{f(-x)}$, then $\hat{k} = \overline{\hat{f}}$.

Theorem 18 (Parseval's Theorem): Let f be a complex-valued function defined on R and such that $f \in L^1 \cap L^2$, then $\hat{f} \in L^2$ and $\|\hat{f}\|_2 = (2\pi)^{1/2} \|f\|_2$.

Proof: Write $h(t) = (2\pi)^{-1/2} e^{-t^2/2}$ and $h_n(t) = nh(nt)$. Put $g = f * k$ where $k(x) = \overline{f(-x)}$. Then $g \in L^1$ by Theorem 13, p. 190, so using Example 12 and 13(i) with f replaced by h and β by n,

$$(g * h_n)(x) = \int g(x-y)\, h_n(y)\,\mathrm{d}y = (2\pi)^{-1/2} \int\int g(x-y)\, h(t/n)\, e^{ity}\,\mathrm{d}t\,\mathrm{d}y.$$

If we put $s = x - y$, this becomes

$$\begin{aligned}(g * h_n)(x) &= (2\pi)^{-1/2} \int h(t/n)\,\mathrm{d}t \int g(s) e^{it(x-s)}\,\mathrm{d}s \\ &= (2\pi)^{-1/2} \int h(t/n)\,\hat{g}(-t) e^{itx}\,\mathrm{d}t \\ &= (2\pi)^{-1/2} \int h(t/n)\,\hat{g}(t)\, e^{-itx}\,\mathrm{d}t, \qquad (10.10)\end{aligned}$$

as h is an even function. So

$$(g * h_n)(0) = (2\pi)^{-1/2} \int h(t/n)\,\hat{g}(t)\,\mathrm{d}t. \qquad (10.11)$$

Now $g(y) = \int f(y-x)\,\overline{f(-x)}\,\mathrm{d}x = \int f(y+x)\,\overline{f(x)}\,\mathrm{d}x$ is a continuous function of y, since Exercise 35, p. 120, extends to complex functions. Also it is bounded; indeed by Hölder's inequality using now, as in Example 25, p. 75, the notation $f_x(y) = f(y+x)$,

$$|g(y)| \leqslant \|f_x\|_2 \cdot \|f\|_2 = \|f\|_2^2.$$

Since $\int h_n\,\mathrm{d}y = 1$ we get

$$(g * h_n)(x) - g(x) = \int (g(x-y) - g(x))\, h_n(y)\, dy$$
$$= \int g(x - s/n) - g(x))\, h(s)\, ds.$$

Letting $n \to \infty$ and using the continuity and boundedness of g we get

$$\lim\, (g * h_n)(0) = g(0) = \|f\|_2^2. \tag{10.12}$$

From Example 13(ii) we have $\hat{g} = |\hat{f}|^2 \geqslant 0$, so letting $n \to \infty$ in (10.11) we may use Theorem 4, p. 000, and (10.12) to get

$$\|f\|_2^2 = (2\pi)^{-1} \int \hat{g}(t)\, dt = (2\pi)^{-1}\, \|\hat{f}\|_2^2$$

as required. □

Theorem 19 (Fourier Inversion Theorem): If $f \in L^1$ and $\hat{f} \in L^1$, then

$$f(x) = 1/2\pi \int \hat{f}(t) e^{-ixt}\, dt \text{ a.e.} \tag{10.13}$$

Proof: By the same argument which gave (10.10) we get

$$(f * h_n)(x) = (2\pi)^{-1/2} \int h(t/n)\, \hat{f}(t) e^{-ixt}\, dt. \tag{10.14}$$

Now $|h(t/n)\, \hat{f}(t) e^{-ixt}| \leqslant |\hat{f}| \in L^1$, so Theorem 10, p. 63, gives that as $n \to \infty$ the right-hand side of (10.14) tends to the right-hand side of (10.13). Also $|(f * h_n)(x) - f(x)| \leqslant \int |f(x-y) - f(x)|\, h_n(y)\, dy$. Integrate with respect to x and use Fubini's theorem to get

$$\|f * h_n - f\|_1 \leqslant \int \|f_{-y} - f\|_1\, h_n(y)\, dy. \tag{10.15}$$

Now $\forall\, \epsilon > 0$, $\exists\, \delta > 0$ such that $\|f_{-y} - f\|_1 < \epsilon$ whenever $|y| \leqslant \delta$, as in Exercise 43, p. 75. So the right-hand side of (10.15) equals $I_1 + I_2$ where

$$I_1 = \int_{|y| \leqslant \delta} \|f_{-y} - f\|_1\, h_n(y)\, dy \leqslant \epsilon \int h_n(y)\, dy = \epsilon,$$

while $$I_2 = \int_{|y| > \delta} \|f_{-y} - f\|_1\, h_n(y)\, dy \leqslant 2\|f\|_1 \int_{|y| > \delta} h_n(y)\, dy$$

which has limit zero by Theorem 10, p. 63. So $\lim \|f * h_n - f\|_1 = 0$. So by Theorem 10, p. 118, for some subsequence $\{n_i\}$

$$\lim\, (f * h_{n_i})(x) = f(x) \text{ a.e.}$$

So (10.14) gives the result. □

Corollary: If $f \in L^1$ and $\hat{f} = 0$ a.e., then $f = 0$ a.e.

So the Fourier transform of an integrable function determines it a.e. From Exercises 32, 33 of Chapter 6 we see that $L^1 \cap L^2$ is dense in L^2 and Theorem 18 then shows that T: $f \to \hat{f}$ is a continuous mapping of a dense subset of L^2 into L^2. It has, therefore, a well defined extension T^* with domain L^2 giving the Fourier transform on L^2 (sometimes called the Plancherel transform). It is customary to use the same notation, and write $T^*f = \hat{f}$ for $f \in L^2$, although $\hat{f}$ is not defined pointwise but only as an element of L^2. The next theorem collects the important properties of T^*. In (ii) we extend Parseval's theorem to L^2; (iii) constitutes an L^2-inversion theorem.

Theorem 20 (Plancherel's Theorem): Let $f, g \in L^2$, then

(i) T^* maps L^2 onto L^2,

(ii) $\int \hat{f}\bar{\hat{g}}\, dx = 2\pi \int f\bar{g}\, dx$.

(iii) If $\phi_n(t) = \int_{-n}^{n} f(x)\, e^{itx}\, dx$ and $\psi_n(x) = 1/2\pi \int_{-n}^{n} \hat{f}(t)\, e^{-itx}\, dt$, then $l.i.m.\ \phi_n = \hat{f}$, and $l.i.m.\ \psi_n = f$ where $l.i.m.$ stands for L^2 limit.

Proof: Write $f_n = f\chi_{(-n,n)}$. Then $f_n \in L^1 \cap L^2$ and $\phi_n = (f_n)\hat{}$. Since $\{f_n\}$ is convergent in L^2, Theorem 18 gives that $\{\phi_n\}$ is a fundamental sequence in L^2, with a unique limit, namely $T^*f = \hat{f}$, proving the first part of (iii). Note that by Theorem 10, p. 118, if $l.i.m.\ F_n$ and $\lim F_n$ both exist, they are equal a.e., so this definition of $\hat{f}$ and that of Definition 17 give rise to the same element of L^2, for $f \in L^1 \cap L^2$.

We wish to construct an inverse to T^* so we define S^* on L^2 by

$$(S^*g)(x) = l.i.m.\ 1/2\pi \int_{-n}^{n} g(t)\, e^{-ixt}\, dt.$$

Since $(S^*g)(x) = (2\pi)^{-1}\ (T^*g)(-x)$, S^*g is well defined. So S^*, T^* are continuous mappings to L^2 into itself.

For $f \in L^1 \cap L^2$, (10.14) gives

$$\begin{aligned}(f * h_n)(x) &= (2\pi)^{-1/2} \int h(t/n)\, \hat{f}(t)\, e^{-ixt}\, dt \\ &= (2\pi)^{-1} \int \hat{h}_n(t)\, \hat{f}(t)\, e^{ixt}\, dt \\ &= (2\pi)^{-1} \int (\widehat{f * h_n})\, e^{-ixt}\, dt.\end{aligned}$$

So
$$f * h_n = S^*T^*(f * h_n). \tag{10.16}$$

Also, $l.i.m.\ f * h_n = f$ if $f \in L^1 \cap L^2$, for

$$|(f * h_n)(x) - f(x)| \leqslant \int |f(x-y) - f(x)|\, h_n(y)\, dy.$$

So, by Jensen's inequality, p. 113, using the measure μ where $\mu(E) = \int_E h_n\, dy$,

$$|(f * h_n)(x) - f(x)|^2 \leqslant \int |f(x-y) - f(x)|^2\, h_n(y)\, dy.$$

Integrate with respect to x and use Fubini's theorem to get

$$\|f * h_n - f\|_2^2 \leqslant \int \|f_{-y} - f\|_2^2\, h_n(y)\, dy.$$

So, as in Theorem 19, we get the result.

So by continuity, taking $l.i.m.$ as $n \to \infty$ in (10.16), $S^*T^*f = f$ for $f \in L^1 \cap L^2$, that is, (iii) has been proved for $f \in L^1 \cap L^2$. But $L^1 \cap L^2$ is dense in L^2 and S^*, T^* are continuous, so $S^*T^*f = f$ for $f \in L^2$. This gives the second part of (iii) and shows that the range of S^* is L^2. But $(S^*f)(x) = (2\pi)^{-1}\,(T^*g)(-x)$ shows that the range of T^* is the same as that of S^*, proving (i).

The result of Parseval's theorem extends to L^2 since $\|\phi_n\|_2 = (2\pi)^{1/2}\,\|f_n\|_2$, so $\|\hat{f}\|_2 = \lim \|\phi_n\|_2 = (2\pi)^{1/2} \lim \|f_n\|_2 = (2\pi)^{1/2}\,\|f\|_2$. The result (ii) now follows, since

$$4f\bar{g} = |f+g|^2 - |f-g|^2 + i|f+ig|^2 - i|f-ig|^2;$$

so if the L^2-norm is multiplied by a factor $(2\pi)^{1/2}$ when f, g go to $\hat{f}, \hat{g}$, the inner product $\int f\bar{g}\,dx$ is multiplied by 2π. □

Exercises

23. Let z, w be complex numbers such that $Re\, z > 0$, $Re\, w > 0$, then

$$\Gamma(z)\,\Gamma(w) = \Gamma(z+w)\int_0^1 u^{w-1}\,(1-u)^{z-1}\,du.$$

24. Following Definition 16, p. 192, we define the Fourier transform of a finite measure μ on R to be the function $\hat{\mu}(x) = \int e^{ixt}\,d\mu(t)$. Show that if L is the Lebesgue function corresponding to the Cantor set P, p. 26, then the Fourier transform of μ_L is

$$\hat{\mu}_L(x) = e^{ix/2} \prod_{k=1}^{\infty} \cos x/3^k.$$

25. If $f \in L(-\infty, \infty)$ then $\hat{f}(x) \to 0$ as $x \to \infty$ or $x \to -\infty$.
26. Find a function f for which the upper bound given for $|\hat{f}|$ in Theorem 16, p. 192, is attained.
27. Let $K(x, t) \in L^2(m \times m)$ and write $(\mathcal{K}f)(x) = \int K(x, t)\, f(t)\, dt$ for $f \in L^2(-\infty, \infty)$. Show that $f \to \mathcal{K}f$ transforms $L^2(-\infty, \infty)$ linearly into itself.

Hints and Answers to Exercises

HINTS AND ANSWERS TO EXERCISES: CHAPTER 1

1. (i) $x \sim y$ clearly defines an equivalence relation. The equivalence classes are sets of points such that, in each set, any two points are at zero distance apart (as measured by ρ) and points in distinct classes are a positive distance apart.

   (ii) ρ^* is well defined. For let $z \in [x]$, $w \in [y]$; then we have $[z] = [x]$ and $[w] = [y]$, but $\rho(z, w) \leqslant \rho(z,x) + \rho(x,y) + \rho(y,w) \leqslant \rho(x,y)$. Similarly $\rho(x, y) \leqslant \rho(z, w)$. So ρ^* is independent of the points chosen in the equivalence classes. To show that ρ^* is a metric we need only show $\rho^*(a, b) \geqslant 0$ for $a \neq b$. Let $\rho^*(a, b) = 0$ where $a = [x]$, $b = [y]$. Then $\rho(x, y) = 0$, so $y \in [x]$, i.e. $a = b$, as required. (The result of this exercise is worth keeping in mind in considering the conventions of Section 6.1.)
2. F dense means that $\forall x \in \complement F$, every ϵ-neighbourhood of x meets F, so $\complement\bar{A}$ dense means that $\forall\, x \in \bar{A}$ every ϵ-neighbourhood of x meets $\complement\bar{A}$. So $\bar{A}$ contains no neighbourhood of any of its points. So A is nowhere dense.
3. Every interval is easily seen to be both an F_σ-set and a G_σ-set, so these provide examples.
4. $[x] = \bigcap_{n=1}^{\infty} [y\colon \rho(x, y) < 1/n]$. If $y \neq x$, then $y \in [z\colon \rho(z, y) < \rho(x, y)]$, an open ball containing y but not x; so $[x]$ is closed.
5. Let $A = [1/n\colon n \in \mathsf{N}]$ and let G_n be the open interval, centre $1/n$, of length $1/(n+1)^2$. Then $A \subset \bigcup_{n=1}^{\infty} G_n$ but, clearly, A is contained in no finite subcollection; A is not closed as $0 \in \bar{A}$ but $0 \notin A$. For boundedness, take $A = \mathsf{N}$ and let G_n be the open interval, centre n, of length 1. Then $A \subset \bigcup_{n=1}^{\infty} G_n$ but we may not go to a finite subcollection.

6. Take the case of ϕ monotone increasing and consider $\phi(a+)$. We have

$$\limsup_{x\to a+} \phi(x) = \inf[\sup_{0<t<h} \phi(a+t): h>0]$$
$$= \inf[\phi(a+\delta): \delta>0] = \rho,$$

say, while

$$\liminf_{x\to a+} \phi(x) = \sup[\inf_{0<t<h} \phi(a+t): h>0] = \sup \rho = \rho.$$

The other cases are similar.

7. Adding the lengths of the removed intervals $I^{(\alpha)}_{n,k}$ gives

$$l(J^{(\alpha)}_{n,k}) = \frac{1}{2^n}\,(1-\alpha(1-\frac{2^n}{3^n})).$$

8. It is easily seen from a diagram that

$$|L_1 - L_2| \leqslant \frac{1}{2^2} - \frac{1}{3}\cdot\frac{1}{2} = \frac{1}{2.6}, \ldots,$$

$$|L_n - L_{n+1}| \leqslant \frac{1}{2^{n+1}} - \frac{1}{3.2^n} = \frac{1}{2^n.6}, \ldots$$

So $$|L_m - L| \leqslant \frac{1}{6}\sum_{n=m}^{\infty} \frac{1}{2^m} = \frac{1}{6.2^{m-1}},$$ as required.

HINTS AND ANSWERS TO EXERCISES: CHAPTER 2

1. $m^*(B) \leqslant m^*(A\cup B) \leqslant m^*(A) + m^*(B) = m^*(B)$.
2. If the set is $x_1, x_2, \ldots$, then $m^*(\bigcup_{i=1}^{\infty}[x_i]) \leqslant \sum_{i=1}^{\infty} m^*([x_i]) = 0$.
3. From Theorems 2 and 3 it is sufficient to show $[0,1] \subseteq \cup I_n$. If $x \in [0,1] - \cup I_n$ and its nearest left-hand end-point of an interval I_n is a, where $a > x$, then $[x, a)$ contains rationals not covered by any I_n. (The case $x = 1$ does not arise.)
4. Let $m_A^*(E)$ be the outer measure of E obtained when the end-points lie in the dense set A. From Definition 1, $m_A^*(E) \geqslant m^*(E)$. As in Example 3, for each I_n consider an interval I_n' with end-points in A such that $I_n \subseteq I_n'$ but $l(I_n') \leqslant (1+\epsilon)\, l(I_n)$ and obtain $m^*(E) \geqslant m_A^*(E)$.
5. We may include $\cap$ as one of the operations since $A \cap B = \mathbf{C}(\mathbf{C}A \cup \mathbf{C}B)$. But $(c,d) = (\mathbf{C}[d,\infty)) \cap \bigcup_{n=1}^{\infty} [c + 1/n, \infty)$.
6. If $m(E_j) < \infty$, we may apply the theorem to $E_j, E_{j+1}, \ldots$ In the opposite direction, let $E_i = (i, \infty)$ for each i. So $\lim E_i = \emptyset$ and $m(\lim E_i) = 0$. But $m(E_i) = \infty$ for all i.

7. Write $F_i = \bigcup_{j \geq i} E_j$ and $G_i = \bigcap_{j \geq i} E_j$. So $F_i \supseteq E_i \supseteq G_i$ and $\lim F_i = \lim E_i = \lim G_i$ since $\lim E_i$ exists. Since $m(F_1) < \infty$ we have from Theorem 9 that $m(\lim F_i) = \lim m(F_i) \geq \lim\sup m(E_i) \geq \lim\inf m(E_i) \geq \lim m(G_i) = m(\lim G_i)$. So we have equality throughout and $\lim m(E_i)$ exists and equals $m(\lim G_i) = m(\lim E_i)$.

8. (i) $\forall \epsilon > 0$, $\exists \{I_i\}$ such that $kA \subseteq \bigcup_{i=1}^{\infty} I_i$ and $m^*(kA) \geq \sum_{i=1}^{\infty} l(I_i) - \epsilon$. But then $A \subseteq \bigcup_{i=1}^{\infty} k^{-1} I_i$. So $m^*(A) \leq \sum_{i=1}^{\infty} l(k^{-1} I_i) = k^{-1} \sum_{i=1}^{\infty} l(I_i) \leq k^{-1}(m^*(kA) + \epsilon)$. So $km^*(A) \leq m^*(kA)$. Replace A by kA and k by k^{-1} for the opposite inequality.

   (ii) To show that $m^*(B) = m^*(B \cap kA) + m^*(B \cap \mathbf{C}(kA))$, for A measurable, write $B = kC$ and note that $\mathbf{C}(kA) = k(\mathbf{C}A)$. Then (i) gives kA measurable and the converse follows similarly.

9. Argue as in the last Exercise, using now $l(-I_i) = l(I_i)$ and the result of Example 3.

10. If E is measurable the identity is obvious. In the opposite direction: by Example 6 we have $E \subseteq E'$ and $M - E \subseteq F'$ where E', F' are measurable, $m^*(E) = m(E')$, $m^*(M - E) = m(F')$. Replacing E' by $E' \cap M$, F' by $F' \cap M$ we may suppose that $E' \subseteq M, F' \subseteq M$. Since E', F' are measurable and $E' \cup F' = M$ we have $m(E' \,\Delta\, F') + m(E' \cap F') = m(E' \cup F') = m(M)$. Also $m(E' - F') + m(E' \cap F') = m(E'), m(F' - E') + m(F' \cap E') = m(F')$. So $m(E' \,\Delta\, F') + 2m(E' \cap F') = m(E') + m(F') = m^*(E) + m^*(M - E) = m(M)$. The two equations for $m(M)$ give $m(E' \cap F') = 0$ as all the measures are finite. But $E' - E = E' \cap (M - E) \subseteq E' \cap F'$. So $E' - E$ is measurable, and so is E.

11. Since the sequence of sets is monotone $\lim m^*(E_n)$ exists (it may be infinite). Since $m^*(E_n) \leq m^*(\cup E_i)$ we have $\lim m^*(E_n) \leq m^*(\cup E_i)$. In the opposite direction, choose, as in Example 6, $F_n \supseteq E_n$, F_n measurable, $m(F_n) = m^*(E_n)$. Writing $B_n = \bigcap_{j \geq n} F_j$ we have $B_1 \subseteq B_2 \subseteq \ldots$, B_n measurable, and $E_n \subseteq B_n \subseteq F_n$. So $m^*(E_n) = m(B_n)$. Then $m^*(\cup E_i) \leq m^*(\cup B_i) = \lim m(B_i) = \lim m^*(E_i)$.

12. By Example 2, $\exists\ 0$ open with $0 \supseteq E$, $m(0) < \alpha^{-1}\ m(E)$. But $0 = \cup I_n$, disjoint open intervals. So $\alpha\, \Sigma\, l(I_n) < m(E) \leq \Sigma\, m(E \cap I_n)$. But then we must have $\alpha\, l(I_n) < m(E \cap I_n)$, for some n.

13. Since $\cos x^{-1} = \cos(-x^{-1})$ we wish to find $\lim_{\delta > 0} 1/\delta\ m(E \cap [0,\delta])$, where $E = [x : x > 0, \cos x^{-1} > 1/2]$. Then

$$E = \bigcup_{n=1}^{\infty} \left(\frac{3}{(6n+1)\pi}, \frac{3}{(6n-1)\pi} \right) \cup \left(\frac{3}{\pi}, \infty \right)$$

so $$m(E \cap [0,\delta]) = \sum_{n=k}^{\infty} \frac{3}{(6n-1)\pi} - \frac{3}{(6n+1)\pi} + O(k^{-2})$$

where $k \leqslant \dfrac{1}{2\delta\pi} - \dfrac{1}{6} \leqslant k+1$. So

$$\begin{aligned} m(E \cap [0,\delta]) &= 6 \sum_{n>k} \frac{1}{(36n^2-1)\pi} + O(k^{-2}) \\ &= (6/\pi) \int_k^{\infty} \frac{dx}{36x^2-1} + O(k^{-2}) \\ &= (2\pi)^{-1} \log \left(\frac{6k-1}{6k+1} \right) + O(k^{-2}) \\ &= \frac{1}{(6k+1)\pi} + O(k^{-2}). \end{aligned}$$

Since $\delta^{-1} = 2k\pi + O(1)$, the result follows.

14. The residual set at the nth stage has measure $2^n \xi^n$, and since $\xi < 1/2$ the result follows.
15. Measurability follows as for P in Example 8. Also, in the notation of Chapter 1, p. 26, $l(I_{n,r}^{(\alpha)}) = \alpha/3^n$, so

$$m\left(\bigcup_{n=1}^{\infty} \bigcup_{r=1}^{2^{n-1}} I_{n,r}^{(\alpha)}\right) = \sum_{n=1}^{\infty} \frac{2^{n-1}\alpha}{3^n} = \alpha \text{ as required.}$$

16. G is a Cantor-like set with $I_{1,1} = (1/5, 4/5)$, $I_{2,2} = (21/25, 24/25)$, etc. So

$$m(G) = 1 - \left(\frac{3}{5} + 2 \cdot \frac{3}{5^2} + 4\frac{3}{5^3} + \ldots \right) = 0.$$

17. To obtain the set in question we remove from $[0,1]$ first the interval $(5/10, 6/10)$, then the intervals $(5/100, 6/100)$, $(15/100, 16/100)$, . . . , $(95/100, 96/100)$, and so on, to get a residual set of measure

$$1 - \left(\frac{1}{10} + \frac{9}{10^2} + \frac{9^2}{10^3} + \ldots \right) = 0.$$

18. If one integer h, $0 \leqslant h \leqslant k-1$ is not to appear in the expansion we get a set of zero measure as in the last exercise; indeed the set for which h need not occur in the first n places has measure $(k-1)^n/k^n$ giving 0 as limit when $n \to \infty$.

    Let the finite sequence be $x_r, x_{r+1}, \ldots x_{r+n}$ and write $h = k^n x_r + \ldots + x_{r+n}$. Let A be the set such that h occurs in the expansion of x to the base k^{n+1}, and B the set in which the sequence $x_r, \ldots, x_{r+n}$ occurs in the expansion to the base k. Then A is a (proper) subset of B. But from the first part $m(A) = 1$ so $m(B) = 1$, giving the result.
19. Let $\alpha \in (0,1)$ and let $P^{(\alpha)}$ be the Cantor-like set of Chapter 1, p. 26; so that

$m(P^{(\alpha)}) = 1 - \alpha$ (Exercise 15). Now $l(I) = m(I \cap P_\alpha) + m(I - P_\alpha)$, so we wish to show $m(I - P_\alpha) > 0$ for each I. But P_α is closed and nowhere dense so $I - P_\alpha$ is a non-empty open set and Example 7(i) gives the result.

20. We take again the sets of type $P^{(\alpha)}$ and note that if $E_n = P^{(n^{-1})}$ then for $n > m, E_n \supset E_m$. But $m(\bigcup_{n=1}^{\infty} E_n) = \lim m(E_n) = 1$. Note: it follows that there is a set of zero measure which is of the second category, that is: not a union of a sequence of nowhere dense sets ([16], p. 74).

21. Using n sets $A_1, A_2, \ldots, A_n$ and the σ-algebra operations we may write X as the union of x disjoint sets such as $A_1 \cap A_2 \cap \mathbf{C}A_3 \cap \ldots$ There are 2^n such sets though some may be empty. Now form unions of these disjoint sets. The number of possible distinct unions is 2^{2^n} and comprises all sets of the σ-algebra.

22. Let $\mathcal{F}$ contain the sequence of distinct sets $\{E_i\}$, so that $E_i \,\Delta\, E_j \neq \emptyset$ for $i \neq j$. From this sequence form a family of disjoint sets $[F_\alpha]$ as in the last solution. This family must contain an infinite sequence of non-empty disjoint sets $\{F_i\}$, for otherwise it could generate only a finite σ-algebra by the last exercise. Now consider all sets of the form $F_{n_1} \cup F_{n_2} \cup \ldots$ for $n_1 < n_2 < \ldots$ These are distinct sets of $\mathcal{F}$. But clearly the collection of sequences $\{n_1, n_2, \ldots\}$ has the same cardinality as the set of numbers $0 \cdot \epsilon_1 \epsilon_2 \ldots (\epsilon_i = 0$ or $1)$, that is, cardinality c.

23. If $m^*(S) = 0$ the result is obvious for any interval. Suppose $m^*(S) > 0$ and let x be the given number. Since $m^*(A + x) = m^*(A)$ we may assume $x = 0$. Write $f(\alpha) = m^*(S \cap (-\alpha, \alpha))$, $g(\alpha) = m^*(S \cap \mathbf{C}(-\alpha, \alpha))$. Then $f(\alpha) < \frac{1}{2}m^*(S)$ for small α. Since $(-\alpha, \alpha)$ is measurable, $f(\alpha) + g(\alpha) = m^*(S)$. Suppose $S \subseteq (-b, b)$, so $f(b) = m^*(S)$. Since m^* is subadditive, for $h > 0$, $m^*(S \cap (-\alpha - h, \alpha + h)) - m^*(S \cap (-\alpha, \alpha)) \leqslant 2h$. So f is continuous and so there exists α with $f(\alpha) = g(\alpha) = \frac{1}{2}m^*(S)$.

24. $E \cup \bigcup_{i=1}^{n} I_i = (E \,\Delta \bigcup_{i=1}^{n} I_i) \cup \bigcup_{i=1}^{n} (E \cap I_i)$. If any I_i is infinite, the left-hand side has infinite outer measure. But the right-hand side has finite outer measure.

25. Let $E = \bigcup_{n=0}^{\infty} (n, n + \frac{1}{2})$ and let $\epsilon < \frac{1}{2}$. Then clearly, $l(I_i) < 1$ for each i. But then $m(E - \bigcup_{i=1}^{n} I_i) = \infty$.

26. Let $E = \bigcup_{k=1}^{\infty} (k, k + 1/2^k)$; so E is measurable and $m(E) < \infty$. Then

$$m^*(E \,\Delta \bigcup_{i=1}^{n} I_i) \geqslant \sum_{n+1}^{\infty} 2^{-k} = 2^{-n}$$

for the best possible approximation to E by n intervals. So $n \to \infty$ as $\epsilon \to 0$.

27. The intervals $J_{n,k}$, $k = 1, \ldots, 2^n$ cover P, and at the nth stage

$$m(P \,\Delta \bigcup_{k=1}^{2^n} J_{n,k}) = 2^n/3^n < \epsilon$$

provided $n > (\log \epsilon)/\log 2/3$. This approximation uses $N = 2^n$ intervals with $N > \exp(\log 2 \log \epsilon/\log 2/3)$.

28. $f \geqslant 0$ on the set $[\pi^{-1}] \cup [0] \cup \bigcup_{n=1}^{\infty} (1/(2n+1)\pi,\ 1/(2n\pi))$, disjoint intervals with measure $\pi^{-1} \sum_{n=1}^{\infty} ((2n)^{-1} - (2n+1)^{-1}) = \pi^{-1}(1 - \log 2)$.

29. $f^{-1}(\alpha, \infty)$ is an interval.

30. By Theorem 15, g is measurable. Also $f - g = 0$ a.e. so ess sup $(f-g) = 0$. So, by Example 17, ess sup $f \leqslant$ ess sup g + ess sup $(f-g) \leqslant$ ess sup g, and similarly for the converse inequality. Since f is continuous, $[x: f(x) > \alpha]$ is open, so if $m[x: f(x) > \alpha] = 0$ we have $f \leqslant \alpha$ by Example 7(i); that is: $f \leqslant \alpha$ a.e. $\Rightarrow f \leqslant \alpha$ so ess sup $f = \inf[\alpha: f(x) \leqslant \alpha] = \sup f$.

31. $\inf[\alpha: f \leqslant \alpha \text{ a.e.}] \leqslant \inf[\alpha: f \leqslant \alpha]$.

32. Use Example 16.

33. Let $E_n = [x: f(x) < n]$. Then $\lim m(E_n) = m(\bigcup_{n=1}^{\infty} E_n) > 0$. So $m(E_n) > 0$ for all large n.

34. Let $g(x) = f'(x)$ where f' exists and define g arbitrarily otherwise. We may suppose that f is defined and constant on $[b, b+1]$. Then

$$g(x) = \lim n(f(x + 1/n) - f(x)) = \lim g_n(x), \text{ say a.e.}$$

But each g_n is measurable, so g is.

35. $[x: f(g(x)) > \alpha] = [x: g(x) \in f^{-1}(\alpha, \infty)]$. But $f^{-1}(\alpha, \infty) = \bigcup_{n=1}^{\infty} I_n$, where the I_n are open intervals. So $[x: f(g(x)) > \alpha] = \bigcup_{n=1}^{\infty} g^{-1}(I_n)$, a measurable set.

36. $f_n(0) = 0, f_n(x) = \sum_{r=0}^{l-1} \sum_{k=0}^{l^{n-1}-1} r\, \chi_{I_{r,k}}$, where $I_{r,k} = (rl^{-n} + kl^{-n+1}, (r+1)l^{-n} + kl^{-n+1}]$, for $x \neq 0$, so f_n is measurable.

37. Let E be a non-measurable set and let $f_\alpha = \chi_{E \cap [\alpha]}$. Then each f_α is measurable but $\sup f_\alpha = \chi_E$ is non-measurable.

38. Let E be a non-measurable set. Then $\chi_E - 1/2$ is not measurable, but $|\chi_E(x) - 1/2| = 1/2$ for all x so $|\chi_E - 1/2|$ is measurable.

39. $2/3 \in P$ and has ternary expansions $0.200\ldots$ and $0.122\ldots$ neither of which is in the range of f.

40. As in the solution to Exercise 20 we may write $[0,1]$ as the union of a set of zero measure and of a sequence of Cantor-like sets E_n. Then if V is the

non-measurable set of Theorem 17, $V = \cup(E_n \cap V) \cup A$ where $m(A) = 0$. So for some $n, E_n \cap V$ is non-measurable.

41. As in the last exercise we can find a Cantor-like set E_n and a non-measurable set $B \subset E_n$. Let F be the continuous function defined in Chapter 1, p. 26, mapping E_n into P. Then $F(B) = B_1$, say, where $B_1 \subseteq P$ and so is measurable. So χ_{B_1} is measurable and χ_B is non-measurable. But $\chi_{B_1} \circ F = \chi_B$, giving the result.
42. Suppose not, then by Example 4 all measurable sets are Borel sets, contradicting Theorem 18.
43. As in the last exercise, let E be a set of zero measure which is not a Borel set. Then $g = \chi_E$ and f identically zero give a counter-example.
44. Let f be Cantor's function, defined in Theorem 18, which maps [0,1] into P and is one-to-one. So c = Card [0,1] $\leqslant$ Card P. But since $P \subset [0,1]$, $c \geqslant$ Card P. Every subset of P is measurable so if α = Card $\mathcal{M}$ we have 2^c = Card $\mathcal{P}(P) \leqslant \alpha$. Since $\alpha \leqslant$ Card $\mathcal{P}(\mathsf{R}) = 2^c$ we have $\alpha = 2^c$.
[Note: The proof may be shortened by using the continuum hypothesis: every countable set has cardinality $\geqslant c$. Then [5] p. 26, Exercise 9, and the fact that the cardinality of the class of intervals is c gives Card $\mathcal{B} = c$, which with the result of this exercise proves a considerably strengthened version of Theorem 8.].
45. Each measurable set has a measurable characteristic function so, by the last exercise, Card [measurable functions] $\geqslant 2^c$. But 2^c is the cardinality of the set of all real functions on R. So Card [measurable functions] $= 2^c$.
46. Let $p > 0$ and $A = \cup [x_i]$. Then by Theorem 19(iv) and Theorem 21, $H_p(A) = 0$, giving the result.
[Note: There exist uncountable sets in R with Hausdorff dimension zero.]
47. As $h(t) < t$ we have $H(A) \leqslant m(A)$. Also $\forall\, \epsilon > 0$, $\exists\, \delta > 0$ such that $t < (1 + \epsilon)\, h(t)$ for $0 < t < \delta$. So $H_{1,\delta}(A) \leqslant (1 + \epsilon) H_\delta(A)$. So $m(A) \leqslant (1 + \epsilon)$ $H(A)$, giving the result.
[A similar result will hold for any $h(t)$ with $h'(0) \neq 0$.]
48. We have $\alpha_n \leqslant 1$ for all n. So let $\alpha = \sup \alpha_n$ and let $p > \alpha$. Then $H_p(A_n) = 0$ implies $H_p(A) = 0$. Let $q < \alpha$ so $q < \alpha_n$ for some n. Then $H_q(A_n) = \infty$ implies $H_q(A) = \infty$. So A has Hausdorff dimension α.
49. As in Theorem 20, $H^*_{q,\delta}(A) \leqslant \delta^{q-p} H^*_{p,\delta}(A)$ for $q > p$, giving the result.
50. In the original construction of P_ξ the removed intervals have total Lebesgue measure 1. So continue the removal until the measure of the intervals removed is at least $1 - 2^{-n}$. Then translating the residual intervals transforms P_ξ into a set of the same Hausdorff measure but contained in $[0,2^{-n}]$. So, taking a sequence of disjoint intervals $\{I_n\}$ with $l(I_n) = 2^{-n}$ and $\cup I_n$ compact, we may construct a set A_n with $H_p(A_n) = \alpha, 0 < \alpha < \infty, A_n \subseteq I_n$, and $A = \cup A_n$ has the desired properties as in the example.
51. From a diagram it can be seen that $\omega_f(t) = |\cos(\pi/2 + t/2) - \cos(\pi/2 - t/2)| = 2 \sin t/2$ for $0 \leqslant t \leqslant \pi$, $\omega_f(t) = 2$ for $t \geqslant \pi$.

52. A routine check shows that these properties were the only ones required.
53. Only the statement about the measures is not obvious. By its definition $g \leqslant h$ so, for each A, $H_g(A) \leqslant H_h(A)$. Now let $\delta > 0$. Choose $\delta' \in (0, \delta]$ such that $t \in [0, \delta']$ implies $0 \leqslant h(t) \leqslant g(\delta)$, which is possible by the continuity of h at 0. Let $\eta \in (0, \delta']$ and if $h(\eta) = g(\eta)$, let $\eta_0 = \eta$. Suppose that $h(\eta) > g(\eta)$. If $h(t) > (1+\delta) g(\eta)$ on $[\eta, \delta]$, then by its definition g is constant on $[\eta, \delta]$ and so $h(\delta') > (1+\delta) g(\delta)$, a contradiction (a diagram may assist). So in each case $\exists\, \eta_0 \in [\eta, \delta]$ with $h(\eta_0) < (1+\delta) g(\eta)$. So $\forall\, \epsilon > 0$ and $\delta > 0$ choose a sum $\Sigma\, g(l(I_i)) < H_{g,\delta'}(A) + \epsilon$, $A \subseteq \cup I_i$, $l(I_i) < \delta'$, using an obvious notation for the approximating measure. Replace each I_i of length η by J_i of length η_0 where $J_i \supseteq I_i$, η and η_0, being related as above. Then $H_{g,\delta'}(A) + \epsilon > (1+\delta)^{-1}\, \Sigma\, h(l(J_i))$ and $l(J_i) \leqslant \delta$. So $H_{g,\delta'}(A) + \epsilon \geqslant (1+\delta)^{-1} H_{h,\delta}(A)$. Let $\delta \to 0$, so $\delta' \to 0$ and we get $H_g(A) + \epsilon \geqslant H_h(A)$, giving the result.
[More results of this type may be found in [4].]

HINTS AND ANSWERS TO EXERCISES: CHAPTER 3

1. $\int \phi \,\mathrm{d}x + \int \psi \,\mathrm{d}x = \sum_i \sum_j (a_i + b_j)\, m(A_i \cap B_j)$. Collect terms so that the r.h.s. is of the form required in Definition 1.
2. $f = (f-g) + g$. Apply Theorem 6.
3. Clearly $\limsup \int f_n \,\mathrm{d}x \leqslant \int f \,\mathrm{d}x$. But $\int f \,\mathrm{d}x \leqslant \liminf \int f_n \,\mathrm{d}x$ from Fatou's Lemma.
4. Apply Theorem 4 to the sequence $\{f_n\}$.
5. Replace the sequence $\{\phi_n\}$ of Theorem 5 by $\{\psi_n\}$ where $\psi_n = \varphi_n \chi_{(-n,n)}$, and show that $\psi_n \uparrow f$.
6. We have only to show that Fatou's Lemma may be obtained from Theorem 4. Write $g_n = \inf_{k \geqslant n} f_k$. Then $g_n \uparrow g = \liminf f_n$. So $\int g \,\mathrm{d}x = \int \liminf f_n \,\mathrm{d}x$. But $\int g_n \,\mathrm{d}x \leqslant \int f_n \,\mathrm{d}x$, so taking lim inf, $\int g \,\mathrm{d}x \leqslant \liminf \int f_n \,\mathrm{d}x$.
7. In one direction the result is immediate. Conversely, to obtain Fatou's Lemma as given in the text, use an argument analogous to that of Exercise 6.
8. Apply Theorem 4 to the sequences $\{f_k - f_n, n = k+1, k+2, \ldots\}$. For the last part take the sequence $\{\chi_{(n,\infty)}\}$.
9. f is the sum of non-negative step functions and so is non-negative and measurable. $\int_0^1 f \,\mathrm{d}x = \sum_{p=1}^{\infty} p\, 2^{p-1}\, 3^{-p} = 3$.
10. Write $g_n = f_n \chi_E$ and apply Fatou's Lemma to the sequences $\{f_n - g_n\}$ and $\{g_n\}$ to get inequalities which give the result.
11. Since we may approximate f^+ and f^- separately, suppose $f \geqslant 0$. Then f is the limit of a sequence of measurable simple functions $\{\varphi_n\}$ and by Exercise 5 we may suppose that

$$\varphi_n = \sum_{k=1}^{N} \alpha_k \chi_{E_k}, \text{ where } m(\bigcup_{k=1}^{N} E_k) < \infty.$$

Since each E_k is measurable we have by Theorem 10, p. 36, an F_σ-set $F_k \subseteq E_k$ with $m(E_k - F_k) = 0$. So the Borel measurable function $g_n = \sum_{k=1}^{N} \alpha_k \chi_{F_k} = \varphi_n$ a.e., so $g_n = \varphi_n$ for all n except on a set E with $m(E) = 0$. Choose a G_δ-set $E^* \supseteq E$ with $m(E^*) = 0$ and write χ_0 for the characteristic function of $\mathbf{C}E^*$. Then, for each n, the Borel measurable function $\varphi_n \chi_0 = g_n \chi_0$, so $\chi_0 f = \lim \varphi_n \chi_0$ is Borel measurable. But $\chi_0 f = f$ a.e.

12. Let g be the step function $[1/x]^{-1}$ on (0,1). Then $\int_0^1 g\, dx > \int_{1/N+1}^{1} g\, dx = \sum_{n=1}^{N} 1/(n+1)$. So $\int_0^1 g\, dx = \infty$. Also $g - f = 0$ a.e., so by Example 3 $\int (g - f)\, dx = 0$. But $\int g\, dx = \int (g-f)\, dx + \int f\, dx$, giving the result.
13. Put $f_n = -n\, \chi_{[0,1]} + n\, \chi_{[1,2]}$. Then $\int \liminf f_n\, dx$ is not defined.
14. By Example 11 we may suppose that f_1 is finite-valued. If $g_n = f_n - f_1$, then $g_n \uparrow g = f - f_1$, so Theorem 4 gives $\lim \int g_n\, dx = \int g\, dx$. Add $\int f_1\, dx$ to both sides to get the result.
15. Putting $f'_n = \max(f_n, g)$ we have $f'_n \geqslant g$ for each n, and $f_n = f'_n$ a.e. Also $\liminf f'_n = \liminf f_n$ a.e., since the union of a countable number of sets of measure zero has measure zero. So apply Fatou's Lemma to $\{f'_n - g\}$.
16. In each case $\lim_{n\to\infty} f_n(x) = 0$ a.e. We have to check that Theorem 10 applies; (i) $\log(x + n) < nx + n$ $(x > 0)$, so $|f_n(x)| \leqslant (1 + x)\, e^{-x}$, an integrable function, (ii) $|f_n(x)| < \frac{1}{2} \log 1/x$ which is integrable, (iii) $|f_n(x)| \leqslant 2^{-1}\, x^{-1/2}$, (iv) $|f_n(x)| \leqslant 1/2$, (v) $|f_n(x)| \leqslant x^{-1/2}$ (consider here $h(x) = 1 + n^2x^2 - n^{3/2}\, x^{3/2}$ and proceed as in Example 17), (vi) $f_n(x) = \dfrac{n^p\, x^\epsilon}{1 + n^2x^2}\ \dfrac{\log x}{x^{-r+\epsilon}}$ where $p < \epsilon < \min(2, 1 + r)$. The second factor is integrable and the first can be shown < 1 as in Example 17.
17. Take $g(x)$ as in Example 18 to get limit $= 0$.
18. Substitute $nx = t$. The integral becomes

$$\frac{1}{n} \int_0^\infty \chi_{[0,n]}\, \frac{1+t}{(1 + t/n)^n}\, dt.$$

Using $(1 + t/n)^n \geqslant t^3/27$ $(n \geqslant 3)$ on $t > 1$, we obtain a dominating function, so limit $= 0$ by Theorem 10.
19. Substitute $nx = u^\beta$ and use $1 - x \leqslant e^{-x}$ $(0 \leqslant x \leqslant 1)$ to get an integrand dominated by βe^{-u^β}. Then Theorem 10 gives the result.
20. Show that $d/dt \log f_t(x) < 0$. Also $\lim_{n\to\infty} f_n(x) = e^{-x}$. So in (i) $\int \lim = 2$,

but $f_n(x) > 1/2^n$ for all x. So $\int_0^\infty f_n(x)\, e^{x/2}\, dx = \infty$ for all n. (ii): Theorem 10 applies as $f_n \leqslant f_1 \leqslant 1$ so $\lim \int = \int \lim = 2/3$.

21. $\chi_{[0,n]}\, (1 + (x/n))^n\, e^{-2x} < e^{-x}$ for all $x > 0$. So $\lim \int = \int \lim = 1$.
22. $(1 - (x/n))^n \leqslant e^{-x}$. Proceed as in Exercise 21.
23. R.h.s. $= 0$ for all α. For $\alpha > 0$ substitute $t = x\sqrt{n}$ to get l.h.s. limit $= 0$. For $\alpha = 0$, $\int_0^\infty \sqrt{n}\, e^{-nx^2}\, dx > \int_0^{1/\sqrt{n}} \sqrt{n}\, e^{-nx^2}\, dx > e^{-1}$.
24. (i) l.h.s. $= 0$; r.h.s., on calculation, is $0, 1, \infty$ for $\alpha < 2, \alpha = 2, \alpha > 2$ respectively. (ii) For $0 < \alpha < 2$ show that $n^\alpha(1 - x)\, x^n < \alpha^\alpha/((1 - x)^{\alpha-1})$ (an integrable function), by considering the function $n^\alpha\, (1 - x)^\alpha\, x^n$ and proceeding as in Example 17. For $\alpha \leqslant 0$ the integrand is clearly bounded. For $\alpha \geqslant 2$ the conditions of Theorem 10 cannot hold by (i).
25. Integrand $= \sum_{n=1}^\infty x^{a-1}\, e^{-nx}$. Apply Theorem 7; a change of variable gives the result.
26. Integrand $= (\log x)^2 \sum_{n=1}^\infty nx^{n-1}$; apply Theorem 7.
27. Integrand $= \sum_{n=1}^\infty \frac{x^n}{n!} \log x + \sum_{n=1}^\infty \frac{x^{n-1}}{n!}$. Apply Theorem 7 to each sum.
28. Integrand $= e^{-(t^2+1)x}\, (e^{2tx} - e^{-2tx})(1 - e^{-2x})^{-1}$

$$= (e^{-(t^2-2t+1)x} - e^{-(t^2+2t+1)x}) \sum_{n=0}^\infty e^{-2nx},$$

to which Theorem 7 applies, giving the first result. Second part: above integrand $= 2e^{-(t^2+1)x} \sinh 2tx + 2e^{-(t^2+1)x}\, (e^{2x} - 1)^{-1} \sinh 2tx$. The first term integrates to give $4t(t^2 - 1)^{-2}$. Since $\sinh x \leqslant xe^x$ for $x \geqslant 0$, the second is dominated by $8x(e^{2x} - 1)^{-1}$ for $0 \leqslant t \leqslant 2$. So Example 15 applies and gives the result.
29. Expand $(1 - x)^{-1}$ and apply Theorem 7.
30. Integrand $= 2e^{-ax} \sinh bx \sum_{n=0}^\infty e^{-2anx}$. Theorem 7 applies and the value of the integral is $\sum_{n=0}^\infty \frac{2b}{(2n+1)^2\, a^2 - b^2}$.
31. Integrand $= \sum_{n=0}^\infty 2(-1)^n\, e^{-(2n+1)x^2}$ $(x > 0)$, with partial sums dominated by the first term, $2e^{-x^2}$. So Theorem 10 applies and gives the result. ($\int_0^\infty e^{-x^2}\, dx = \sqrt{\pi}/2$ is assumed, cf. Exercise 22, p. 189.)

32, 33. On expansion an alternating series is obtained, as in Exercise 31, to which Theorem 10 applies.

34. For $0<b<1$, $\chi_{[0,b]}\sum_{n=1}^{\infty} x^{n-1}/\sqrt{n}\leqslant\chi_{[0,1]}\sum_{n=1}^{\infty} x^{n-1}/\sqrt{n}$. R.h.s., by Theorem 7, has integral $\sum_{n=1}^{\infty} n^{-3/2}$. So we may apply Exercise 15 to l.h.s. to obtain the result.

35. Integrand $=\cos^2 t\sum_{n=1}^{\infty}(-1)^{n-1}e^{-nt}x^{-n}$ $(t>0)$, and Theorem 10 applies to the partial sums to give the result.

36. Expand $\cos\sqrt{x}$ and apply Theorem 11.

37. Substitute for J_m and J_0 in (i) and (ii) respectively, and apply Theorem 11.

38. Expand $(1+x^2)^{-1}$ and apply Theorem 11.

39. Expand r.h.s. in powers of a^{-1}. Then apply Theorem 11 to l.h.s. to get the same series.

40. Integrand $=\frac{1}{2}f(x)((1-re^{ix})^{-1}+(1-re^{-ix})^{-1})$. Expand and apply Theorem 11.

41. For $S=(a,b)$ integrate explicitly and use periodicity to get the result in the limit. Since $m(S)<\infty$, $\forall\ \epsilon>0$, $\exists$ disjoint intervals $I_1,\ldots,I_k$ such that if $E=\bigcup_{j=1}^{k} I_j$ then $m(S\ \Delta\ E)<\epsilon$. But then $\left|\int_{S\Delta E}\right|<\epsilon$ for each n, and the result follows.

42. f integrable implies that $\forall\ \epsilon>0$, $\exists\ n$ such that $\int_{|x|>n}|f|\,dx<\epsilon/3$. Then by Theorem 16 (Corollary) $\exists\ g$ continuous such that $\int_{|x|\leqslant n}|f-g|\,dx<\epsilon/3$, with $g=0$ outside a finite interval (possibly extending beyond $[-n,n]$). But in Theorem 16 we may suppose without loss of generality that

$$\int_{|x|\geqslant n}|g|\,dx<\epsilon/3,$$

and the result follows.

43. $\forall\ \epsilon>0$, $\exists$ a continuous function K (by Exercise 42) such that $\int|K(x)-f(x)|\,dx<\epsilon$ and $K\equiv 0$ outside $[a,b]$, say. Then $K_h(x)$ $(\equiv K(x+h))=0$ outside $[a-1,b+1]$ for all h with $|h|<1$, and then $\int|f(x+h)-f(x)|\,dx$

$$\leqslant\int|f(x+h)-K_h(x)|\,dx+\int|f(x)-K(x)|\,dx+\int_{a-1}^{b+1}|K(x+h)-K(x)|\,dx$$

$<3\epsilon$ for $|h|<\delta$, say, as K is continuous.

44. Write $I_k=\int f(x)\,\varphi(kx)\,dx=\int f(x+\beta k^{-1})\,\varphi(kx+\beta)\,dx$

$$=-\int f(x+\beta k^{-1})\,\varphi(kx)\,dx.$$

So $2|I_k|\leqslant\int|f(x)-f(x+\beta k^{-1})|\,dx.$ (ess sup $|\varphi|$) and the result follows by Exercise 43.

45. Defined to be 1 at $x=0$, the integrand is continuous. If

$$a_n = \int_{n\pi}^{(n+1)\pi} x^{-1} \sin x \, dx$$

then the integral is $2 \sum_{n=0}^{\infty} a_n$, an alternating series, so the Riemann integral exists. But if $b_n = \int_{n\pi}^{(n+1)\pi} |x^{-1} \sin x| \, dx$, then $b_n > 2/n\pi$, $\Sigma\, b_n = \infty$ and the (Lebesgue) integral does not exist.

46. Write $E_n = [x: n \leqslant f(x) < n+1]$. So $\sum_{k=n}^{\infty} m(E_k) > 0$ for all n. So there exists a subsequence $\{n_r\}$ with $m(E_{n_r}) > 0$. Then define g to equal $r^{-2}(m(E_{n_r}))^{-1}$ on E_{n_r} $(r = 1, 2, \ldots)$ and define $g = 0$ on $\complement(\bigcup_{r=1}^{\infty} E_{n_r})$. Then g is integrable but $\int fg \, dx > \sum_{r=1}^{\infty} n_r/r^2 = \infty$, as $n_r \geqslant r$.
47. For each n, f_n has a 'saw-tooth' graph with zeros at $k \cdot 10^{-n}$ and $\int_0^1 f_n \, dx = 4^{-1} \cdot 10^{-n}$. By Theorem 7 the result follows.
48. Define $g(0) = 0, g(x) = a^{-1}$ if a is the first finite non-zero integer in the expansion of x. So $\int g \, dx = \int f \, dx$. Then $g = n^{-1}$ on $\bigcup_{k=1}^{\infty} [n \cdot 10^{-k}, (n+1)10^{-k})$ for $n = 1, \ldots, 9$, giving $\int_0^1 g \, dx = (\sum_{n=1}^{9} n^{-1}) \cdot 9^{-1}$.
49. Elementary integration; Theorem 10 is not contradicted.
50. $\forall\, \epsilon > 0$, $\exists\, \delta > 0$ such that $|f(t) - f(x)| < \epsilon$ for $|t - x| < \delta$. Then $F_n(x) - f(x)$ may be written as an integral over $(x - \delta, x + \delta)$, on which $|f(t) - f(x)| < \epsilon$ and this integral is easily calculated to be less than ϵ, and two integrals over finite intervals on which the integrands are monotone decreasing as n increases, and integrable for $n = 1$. So Theorem 10 applies to give the result.
51. Write $g_h(x, t) = h^{-1}(f(x, t+h) - f(x, t))$. Then $\lim_{h \to 0} g_h = \partial f/\partial t \, (x, t)$. But $|g_h(x, t)| \leqslant \varphi(x)$, so Example 15 applied to $\int g_h \, dx$ gives the result.
52. Considering separately $x \leqslant 1$ and $x > 1$, we get $|x^\gamma f(x)| \leqslant (x^\alpha + x^\beta)|f(x)|$ an integrable function, so the integral exists. For continuity use the fact that for small h, $\gamma + h \in (\alpha, \beta)$ so $|(x^{\gamma+h} - x^\gamma) f(x)| \leqslant 2(x^\alpha + x^\beta)|f(x)|$, and apply Example 15.
53. Fatou's Lemma applied to $\{g + f_n\}$ and $\{g - f_n\}$ gives the first and third inequalities respectively, and the second is trivial. Define the functions f_n on $[0,1]$ by $f_1 = \chi_{(1/2,1]}$, $f_2 = \chi_{(1/4,1]}$, $f_3 = \chi_{[0,3/4]}$, $f_4 = 1 - f_1$, $f_5 = f_1$, $f_6 = f_2$, etc. Then the inequalities read $0 < 1/2 < 3/4 < 1$.

HINTS AND ANSWERS TO EXERCISES: CHAPTER 4

1. $D^+f = \inf_{\delta} \sup_{0<h<\delta} \sin h^{-1} = 1$. Similarly $D_+f = -1, D^-f = 1, D_-f = -1$.
2. If $x \in Q$, $D^+f = \infty$, $D_+f = 0$, $D^-f = 0$, $D_-f = -\infty$. If $x \notin Q$, $D^+f = 0$, $D_+f = -\infty$, $D^-f = \infty$, $D_-f = 0$.
3. Consider D^+f: $\sup_{0<h<\delta} (f(x+h) - f(x))/h$ is continuous with respect to δ, so we may take its infimum over $\delta = n^{-1}$, i.e. $D^+f = \inf_n \sup_{m>n} m(f(x+m^{-1}) - f(x)) = \lim\sup n(f(x+n^{-1}) - f(x))$, a measurable function. Similarly for the other derivates.
4. Since $f(x+h) - f(x) = hf'(x) + o(h)$,
$$D^+(f+g)(x) = f'(x) + \inf_{\delta} \sup_{0<h<\delta} [o(1) + (g(x+h) - g(x))/h]$$
$$= f'(x) + \inf_{\delta} [o(1) + \sup_{0<h<\delta} (g(x+h) - g(x))/h] = f'(x) + D^+g(x).$$
Similarly for the other derivates.
5. Let $f(0) = 0$, $f(x) = x$ for $x \in Q$, $f(x) = -x$ for $x \notin Q$, $g(x) = -f(x)$. Then $D^+f = 1$, $D^+g = 1$, $D^+(f+g) = 0$ (at $x = 0$).
6. From (4.1) it is sufficient to consider right-hand end-points and we may suppose in fact that $x = 0$. Then on $[0, 2^{-1}\,3^{-m}]$ we have $L_{m+1} \geqslant L_m$ and also $L_{m+2} \geqslant L_m$. Similarly $L_{m+3} \geqslant L_{m+1} \geqslant L_m$ on $[0, 2^{-1}\,3^{-m-1}]$ and a comparison of the graphs shows that then $L_{m+3} \geqslant L_m$ on $[0, 2^{-1}\,3^{-m}]$. Similarly for L_{m+4} and we deduce that, for $n > m$, L_n, and hence $L \geqslant L_m$ on $[0, 2^{-1}\,3^{-m}]$. So the graph of L does not lie below a line of slope $(3/2)^m$ in this interval. But $D_+L < \infty$ at $x = 0$ implies that $\exists\, l < \infty$, and a sequence $\{x_n\}$, $x_n \to 0+$, such that $L(x_n)$ lies below $(l+1)x$, giving a contradiction.
7. Suppose that $g(1) + f(1) - g(0) - f(0) = -p$ $(p > 0)$. Now if h is any function such that $h(1) - h(0) = -p$, we have $(h(1) - h(1/2)) + (h(1/2) - h(0)) = -p$, so at least one bracket $\leqslant -p/2$. Call the corresponding end-points (x_1, y_1). Bisect (x_1, y_1) to obtain similarly (x_2, y_2) such that $y_2 - x_2 = 2^{-2}$, $h(y_2) - h(x_2) \leqslant -p \cdot 2^{-2}$. So by induction we obtain sequences $x_n \uparrow$ and $y_n \downarrow$, $y_n - x_n = 2^{-n}$ and $h(y_n) - h(x_n) \leqslant -p(y_n - x_n)$. Take $h = g + f$ to get
$$\frac{f(y_n) - f(x_n)}{y_n - x_n} - \frac{g(y_n) - g(x_n)}{y_n - x_n} \leqslant -p$$
where $\lim x_n = \lim y_n = \alpha$, say. If $\alpha \notin E$, the first ratio is non-negative as f is increasing, the second has limit $g'(\alpha) > 0$, as $n \to \infty$, giving a contradiction. If $\alpha \in E$ the first ratio $\to \infty$ as $n \to \infty$, and the lower limit of the second is $> -\infty$, again giving a contradiction.
8. In each interval $[2/2n + 1, 2/2n - 1]$, f is monotone and $T_f = 2$. So by Example 7, $T_f[0,1] = \infty$.

9. g is clearly continuous on $(0,1]$ but $g(0+) = g(0)$, so g is continuous on $[0,1]$. Consider the partition

$$0 < \frac{2}{2n-1} < \frac{2}{2n-3} < \ldots < \frac{2}{3} < 1 \text{ of } [0,1];$$

then
$$t_g = \frac{2}{2n-1} + \left(\frac{2}{2n-1} + \frac{2}{2n-3}\right) + \ldots + \left(\frac{2}{5} + \frac{2}{3}\right) + \frac{2}{3} =$$

$$= 2\sum_{k=2}^{n} \frac{2}{2k-1} \to \infty \text{ as } n \to \infty. \text{ So } T_g[0,1] = \infty.$$

10. Suppose that $|f'| \leqslant M$ on $[a,b]$. Then for any $x, y \in [a,b]$, $|f(y) - f(x)| \leqslant M|y - x|$; so for any partition, $t \leqslant M(b-a)$.
11. By Theorem 1, p. 81, $f(x) - f(a) = P_f[a,x] - N_f[a,x)$. So $|f(x)| \leqslant |f(a)| + T_f[a,b] < \infty$.
12. By the last exercise, $|f| \leqslant M_1 < \infty$, $|g| \leqslant M_2 < \infty$. Let $a = x_0 < x_1 < \ldots < x_n = b$ be any partition. We have

$$f(x_i)g(x_i) - f(x_{i-1})g(x_{i-1}) =$$
$$= (f(x_i) - f(x_{i-1}))g(x_i) + (g(x_i) - g(x_{i-1}))f(x_{i-1}).$$

So, taking moduli and adding,

$$t_{fg} \leqslant M_2 t_f + M_1 t_g \leqslant M_2 T_f + M_1 T_g,$$

giving the result.
13. By the decomposition of Theorem 2, p. 82, it is sufficient to consider monotone functions. But then the result is trivial.
14. Show that f' is bounded and apply Exercise 10.
15. f is non-differentiable everywhere, so by Theorem 8, p. 85, is not of bounded variation.
16. Let $\{r_i\}$ be an enumeration of the rationals in $[0,1]$ and define f on $[0,1]$ by $f(x) = \sum_{r_n \leqslant x} 2^{-n}$. Clearly f is monotone increasing and for each rational r_i, $f(r_i) - f(r_i-) = 2^{-i} > 0$.
17. In the intervals on which f is continuous, $F' = f$ by Theorem 12, p. 65.
18. Let $f(x) = 1$, $x \in \mathbf{Q}$, $f(x) = 0$, $x \notin \mathbf{Q}$. Then $F = 0$ so $F' = f$ on $\mathbf{CQ}$, $F' \neq f$ on $\mathbf{Q}$.
19. Take $f = 0$ on $[0,1)$, $f(1) = -1$. Then $\int_0^1 f'\,\mathrm{d}x = 0$ but $f(1) - f(0) = -1$.
20. (i) Consider $-f$ to get $\int_a^b f'\,\mathrm{d}x \geqslant f(b) - f(a)$. (ii) Write $f = g - h$, as in the proof of Theorem 2, to get $\left|\int_a^b f'\,\mathrm{d}x\right| \leqslant g(b) - g(a) + h(b) - h(a) = T_f[a,b]$.

21. Consider Lebesgue's function L on $[0,1]$. Then $L \in BV[0,1]$, L is continuous and $L' = 0$ a.e. But if $L(x) = \int_0^x l(t)\,dt$ then $l = 0$ a.e. by Theorem 12, p. 89, so $L(1) = L(0)$. But $L(1) = 1, L(0) = 0$, contradiction.

22. Write $F(x) = \int_{-\infty}^x \chi_X\,dt$, so F is finite and $F' = \chi_X$ a.e. by Theorem 12. Also the required limit is easily seen to be that of $1/2h\,(F(x+h) - F(x-h))$. But a.e., for $h > 0$,

$$\frac{F(x+h)-F(x)}{h} \to F'(x) \text{ and } \frac{F(x-h)-F(x)}{h} \to -F'(x).$$

So subtracting gives the result.

HINTS AND ANSWERS TO EXERCISES: CHAPTER 5

1. Let $\mathcal{R}$ be the class of all finite unions of finite intervals, which may be open, closed or half-open and where we note that points are special cases of closed intervals. All intervals of each type may be obtained from the open intervals by the ring operations. Also $\mathcal{R}$ is a ring, so it is the required class.
2. This follows directly from Definitions 1 and 2, p. 93.
3. $\mu(A) + \mu(B) = \mu(A \cap B) + \mu(A \cup B)$. But $\mu(A) = \mu(A \cup B) = \mu(C) < \infty$, giving the result.
4. If $\mu(E) < \infty$, then as $\mu(E) = \mu(E) + \mu(\emptyset)$ we have $\mu(\emptyset) = 0$, so μ is a measure.
5. The conditions for a pseudometric are easily checked: (i) $\rho(A, B) \geqslant 0$, (ii) $\rho(A, A) = 0$, (iii) $\rho(A, B) = \rho(B, A)$, (iv) $\rho(A, C) \leqslant \rho(A, B) + \rho(B, C)$, the latter holding since $A \,\Delta\, C \subseteq (A \,\Delta\, B) \cup (B \,\Delta\, C)$.
6. This follows as in the proof of Theorem 4, p. 97, since the definition of measurability is the same.
7. Let $\mathcal{R}$ be the ring of finite unions of intervals $[a, b)$ and define μ on $\mathcal{R}$ by $\mu(E) = \infty$ if $E \neq \emptyset$, $\mu(\emptyset) = 0$. Let ξ be any real number and extend the measure μ to μ_ξ on the ring generated by $\mathcal{R}$ and $[\xi]$ by putting $\mu_\xi([\xi]) = 0$, $\mu_\xi(E) = \infty$ if $E - [\xi] \neq \emptyset$, $\mu_\xi(\emptyset) = 0$. Then μ_ξ can be extended as in Theorem 3, p. 96, to $\mathcal{S}(\mathcal{R})$ (the Borel sets) to get $\bar{\mu}_\xi$, say, which extends μ on $\mathcal{R}$ and is finite only on the sets $\emptyset$, $[\xi]$. Clearly, varying ξ gives different extensions of μ.
8. We have $D = B \,\Delta\, N$ where $B \in \mathcal{S}$ and $\bar{\mu}(N) = 0$. But $D \,\Delta\, B = B \,\Delta\, (B \,\Delta\, N) = N$, giving the result.

9, 10 Direct verification gives the results.

11. $[x\colon \chi_E(x) \neq 0]$ is not Borel measurable, so we cannot say that $\chi_E = 0$ a.e.
12. Take μ to be Borel measure and let E be a measurable set of measure zero which is not a Borel set. Then $E \subset G$, a Borel set of measure zero. Let $g = \chi_E + \chi_G + f$ so that $[x\colon f(x) \neq g(x)] = G$ and so $f = g$ a.e. (μ). But g is not

Borel measurable for then $\chi_E = g - f - \chi_G$ would be Borel measurable, which is false.

13. The relations follow directly from Definition 6, p. 33.
14. These follow from the characterizations of lim sup and lim inf given after Definition 6, p. 33.
15. (i) As in Theorem 9, p. 33,

$$\mu(\liminf E_n) = \mu\Big(\bigcup_{n=1}^{\infty} \bigcap_{m \geqslant n} E_m\Big) = \lim \mu\Big(\bigcap_{m \geqslant n} E_m\Big).$$

So $\forall\, \epsilon > 0$, $\exists\, N$ such that $\mu(\liminf E_n) \leqslant \mu\Big(\bigcap_{m=N}^{\infty} E_m\Big) + \epsilon \leqslant \mu(E_m) + \epsilon$ for all $m \geqslant N$. So $\mu(\liminf E_n) \leqslant \liminf \mu(E_n)$.

(ii) $\mu(X) - \mu(\limsup E_n) = \mu(X - \limsup E_n) = \mu(\liminf (X - E_n)) \leqslant \mu(X) - \limsup \mu(E_n)$, using Exercise 13. To see that the result may fail when $\mu(X) = \infty$, let $E_n = (n, \infty)$, so $\limsup E_n = \emptyset$, but $\mu(E_n) = \infty$ for all n.

16. $F_j = [x\colon \sum_{i=1}^{k} \chi_{E_i}(x) = j]$, a measurable set. Also $\sum_{i=1}^{k} \mu(E_i) = \int \sum_{i=1}^{k} \chi_{E_i}\, d\mu$. But $\sum_{i=1}^{k} \chi_{E_i}$ is a measurable simple function taking the value j on F_j, so its integral equals $\sum_{j=1}^{k} j\, \mu(F_j)$.
17. $g \geqslant h$ implies $g^- \leqslant h^-$ so $\int g^-\, d\mu < \infty$.
18. $\int |g|\, d\mu \leqslant k \int |f|\, d\mu$.
19. $f \chi_E \in L(X, \mu)$. But $\chi_E = \chi_F$ a.e. since $[x\colon \chi_E(x) \neq \chi_F(x)] = E \,\Delta\, F$. So $f \chi_F \in L(X, \mu)$ and $f \chi_F = f \chi_E$ a.e., giving the result.
20. Since $f \chi_A \geqslant 0$ the integral is well defined. Let $C = [x\colon f(x) > c]$. Then $\int_A f\, d\mu \geqslant \int_C f\, d\mu \geqslant c\, \mu(C)$, as required.
21. Let $G_n = [x\colon f(x) > n^{-1}]$, for each n. Then $n^{-1} \mu(G_n) \leqslant \int f\, d\mu < \infty$, so $\mu(G_n) < \infty$. But $[x\colon f(x) \neq 0] = \bigcup_{n=1}^{\infty} G_n$.
22. Write $F_n = [x\colon f_n(x) \neq 0]$. If $g(x) \neq 0$, then $f_n(x) \neq 0$ for some n, so $x \in \bigcup_{n=1}^{\infty} F_n$, a set of σ-finite measure by the last exercise. So $G \subseteq \bigcup_{n=1}^{\infty} F_n$ gives the result for G; similarly for H.
23. By Exercise 21, $[x\colon f(x) \neq 0]$ has σ-finite measure, that is, it can be written as the union of a countable number of sets each containing a finite number of points, and so consists of a countable set of points which may be enumerated $x_1, x_2, \ldots$ But if the sets E_j are disjoint

$$\int_{\cup E_i} f\, d\mu = \Sigma \int_{E_i} f\, d\mu$$

so $\int_X f \, d\mu = \sum_{i=1}^{\infty} f(x_i)$. Also $|f|$ is integrable and is non-zero on the same sequence, so $\sum_{i=1}^{\infty} |f(x_i)| = \int |f| \, d\mu < \infty$, so $\Sigma f(x_i)$ is absolutely convergent and so its value is independent of the ordering of the sequence.

24. If $\int f \, d\mu < \infty$, Theorem 21, p. 107, gives the result. If $\int f \, d\mu = \infty$, note that $\int f \, d\mu \leqslant \liminf \int f_n \, d\mu$ by Fatou's Lemma, p. 105, so $\lim \int f_n \, d\mu = \infty$.
25. Clearly each f_n is integrable and as $|f| \leqslant |g|$ a.e., f is integrable. Then Fatou's Lemma applied to the sequence $\{g_n - f_n\}$ gives $\int f \, d\mu \geqslant \limsup \int f_n \, d\mu$ and applied to the sequence $\{g_n + f_n\}$ gives $\int f \, d\mu \leqslant \liminf \int f_n \, d\mu$, giving the result.
26. Using the result and notation of Exercise 14 we have $\mu(\liminf E_n) = \int \chi_* \, d\mu = \int \liminf \chi_{E_n} \, d\mu \leqslant \liminf \int \chi_{E_n} \, d\mu = \liminf \mu(E_n)$ as required.
27. $\int \Sigma \chi_{E_n} \, d\mu = \Sigma \int \chi_{E_n} \, d\mu = \Sigma \mu(E_n) < \infty$. So $\Sigma \chi_{E_n} < \infty$ except possibly on a set of zero measure, but this is the set, $\limsup E_n$, such that x is in infinitely many E_n.
28. Write $\varphi(E) = \int_E f^+ \, d\mu$ so that φ is a measure on $\mathcal{S}$ (Theorem 18, p. 106). Also $E_n \supseteq E_{n+1}$ for each n, $\bigcap_{n=1}^{\infty} E_n$ has μ-measure and therefore φ-measure zero, and $\varphi(E_1) < \infty$. So $\lim \varphi(E_n) = 0$ by Theorem 10, p. 103. But $\int_{E_n} f \, d\mu = \int_{E_n} f^+ \, d\mu = \varphi(E_n)$, giving (i). We obtain (ii) from $n\lambda\mu(E_n) \leqslant \varphi(E_n)$.
29. The set of points x such that for some sequence $\{n_i\}$, $x \in \cup F_{n_i}$ is just $\liminf F_n$. So by Exercise 15(i) it is sufficient to show that $\liminf m(F_n) = 0$. Now if $E_n = [y: f(y) \geqslant n]$ we have $F_n = nE_n$ so $m(F_n) = nm(E_n)$ by Exercise 8, Chapter 2. But $\lim nm(E_n) = 0$ by the last exercise.
30. Write $\varphi(E) = \int_E f \, dx$. Then φ vanishes on intervals and also on open sets, so φ vanishes on G_δ-sets. But if $E = [x: f(x) > 0]$, $E \subseteq G$, a G_δ-set, such that $m(G - E) = 0$. So $\varphi(E) = \varphi(G) = 0$, and so $m(E) = 0$, i.e. $f \leqslant 0$ a.e. Similarly $f \geqslant 0$ a.e.
31. Without loss of generality we may suppose $0 \leqslant f < \infty$. Let $E_n = [x: f(x) \geqslant n]$, $F_n = E_n - E_{n+1}$, and $F_0 = [x: 0 \leqslant f(x) < 1]$. Then for $1 \leqslant k \leqslant n$, $\mu(E_{n+1}) + \sum_{i=k}^{n} \mu(F_i) = \mu(E_k)$. So, adding over k,

$$n\mu(E_{n+1}) + \sum_{i=1}^{n} i\mu(F_i) = \sum_{i=1}^{n} \mu(E_i). \qquad (*)$$

Let f be integrable, then by Exercise 28 and this identity we have

$$\sum_{i=1}^{\infty} i\mu(F_i) = \sum_{i=1}^{\infty} \mu(E_i).$$

But on $F_n, f \geqslant n$ and the sets F_n are disjoint, so

$$\sum_{n=1}^{\infty} \mu(E_n) = \sum_{n=1}^{\infty} n\mu(F_n) \leqslant \int_{\cup F_i} f \, \mathrm{d}\mu < \infty.$$

If $\sum_{n=1}^{\infty} \mu(E_n) < \infty$, we have

$$\int f \, \mathrm{d}\mu = \int_{F_0} f \, \mathrm{d}\mu + \sum_{i=1}^{\infty} \int_{F_i} f \, \mathrm{d}\mu$$

$$\leqslant \mu(F_0) + \sum_{n=1}^{\infty} (n+1)\mu(F_n) \leqslant \mu(X) + \sum_{n=1}^{\infty} n\mu(F_n).$$

But from the identity (*) above

$$\sum_{n=1}^{\infty} n\mu(F_n) \leqslant \sum_{n=1}^{\infty} \mu(E_n), \text{ so } \int f \, \mathrm{d}\mu < \infty.$$

Suppose $\mu(X) = \infty$. Then if e.g. $f = 1/2$, $\sum_{n=1}^{\infty} \mu(E_n) = 0$ but $f \notin L(X, \mu)$.

But if $f \in L(X, \mu)$ we still have $\sum_{n=1}^{\infty} \mu(E_n) < \infty$, as above. Summation from $n = 0$: $\sum_{n=0}^{\infty} \mu(E_n) = \mu(X) + \sum_{n=1}^{\infty} \mu(E_n)$ so the l.h.s. converges iff f is integrable and $\mu(X) < \infty$.

32. Write $E_1 = [x: f(x) = 1]$, $E_2 = [x: f(x) > 1]$, $E_3 = [x: f(x) = -1]$, $E_4 = [x: f(x) < -1]$, $E_5 = [x: |f(x)| < 1]$. Then

$$\int f^n \, \mathrm{d}\mu = \mu(E_1) + \sum_{i=2}^{5} \int_{E_i} f^n \, \mathrm{d}\mu.$$

Now $\int_{E_2} f^n \, \mathrm{d}\mu$ has limit ∞ if $\mu(E_2) > 0$ since, for some $\delta > 0$, E_2 has a subset F of positive measure on which $f > 1 + \delta$; so $\mu(E_2) = 0$. If $\mu(E_3) > 0$ and $\mu(E_4) > 0$ we have that $\int_{E_3} f^n \, \mathrm{d}\mu = (-1)^n \mu(E_3)$ oscillates finitely, while $\int_{E_4} f^n \, \mathrm{d}\mu = (-1)^n \int_{E_4} |f|^n \, \mathrm{d}\mu$ oscillates infinitely. So for the limit of the sum to exist, $\mu(E_3) = \mu(E_4) = 0$, since by Theorem 21, $\lim \int_{E_5} f^n \, \mathrm{d}\mu = 0$.

33. Write $E_1 = [x: f(x) \geqslant 1]$ and $E_2 = [x: f(x) < 1]$. Applying Theorem 21 to the integral over $E \cap E_1$ we get the limit $= \mu(E \cap E_1)$, and Theorem 15 applied to that over $E \cap E_2$ gives $\mu(E \cap E_2)$.

HINTS AND ANSWERS TO EXERCISES: CHAPTER 6

1. $|f^2 + g^2|^{1/2} \leqslant |f| + |g|$, an integrable function.
2. (i) is obvious. (ii) For $\sigma = 0$, integrate over (n^{-1}, a) to get $(\log a^{-1})^{-1} - (\log n)^{-1}$ with a finite limit as $n \to \infty$. For $\sigma > 0$, that the integral g is infinite follows from the fact that if $\alpha > 0, x^\alpha \log x \to 0$ as $x \to 0$.
3. $(\log x^{-1})^p \leqslant x^{-1/2}$ for all small x.
4. Since, for $\alpha > 0$, $t^\alpha e^{1/t} \to \infty$ as $t \to 0$, we have $(e^{1/x})^p > x^{-1}$ for all small x.
5. (i) For $p = 2$ integrate explicitly over (0,1) and $(1,\infty)$ to get a finite integral.

   (ii) If $p > 2$, $\int_0^1 x^{-p/2}(1 - \log x)^{-p}\, dx$ is seen to be infinite as in Exercise 2(ii).

   (iii) If $p < 2$, $\int_1^\infty x^{-p/2}(1 + \log x)^{-p}\, dx$ is again infinite, by comparison with $\int_1^\infty x^{-1}\, dx$.
6. Let $M = \text{ess sup}\, f$, so $f \leqslant M$ a.e. Hence $\forall\, \epsilon > 0$, $I_n \leqslant M(\mu(X))^{1/n} < M + \epsilon$ for $n > n_1$. Write $X(\epsilon) = [x: f(x) > M - \epsilon]$. Since $\mu(X(\epsilon)) > 0$,

   $$\int f^n\, d\mu \geqslant \int_{X(\epsilon)} f^n\, d\mu \geqslant (M - \epsilon)^n\, \mu(X(\epsilon)).$$

   So $I_n \geqslant (M - \epsilon)(\mu(X(\epsilon)))^{1/n} \geqslant M - 2\epsilon$ for $n > n_2$, which with the previous inequality gives the result.
7. If the x_i are equal, equality is obvious. The converse follows from Example 6, p. 113, since 'a.e.' is now equivalent to 'except for those numbers i for which $\alpha_i = 0$'.
8. This is a special case of the next result.
9. In Example 7 put $x_i = \log y_i$ and $\psi(x) = e^x$ to get

   $$\exp\left(\sum_{i=1}^{n} \alpha_i \log y_i\right) \leqslant \sum_{i=1}^{n} \alpha_i y_i,$$

   and, by Exercise 7, equality occurs iff the y_i are equal.
10. Suppose that ψ is not convex. Then there exist numbers a, b, λ, μ such that $a > b$, $\lambda + \mu = 1$, $\lambda > 0$, $\mu > 0$ and $\psi(\lambda a + \mu b) > \lambda\psi(a) + \mu\psi(b)$. Let l be any number $0 < l < \lambda$, let $m = 1 - l/\lambda$ and let $n = l(1/\lambda - 1)$. Then l, m, n are positive and $l + m + n = 1$. Define f on [0,1] by $f = a$ on $[0, l)$, $f = \lambda a + \mu b$ on $[l, l + m)$, $f = b$ on $[l + m, 1]$. Then $\int_0^1 f\, dx = \lambda a + \mu b$, but

$$\int_0^1 \psi \circ f \, dx = l\psi(a) + m\psi(\lambda a + \mu b) + n\psi(b)$$
$$= \psi(\lambda a + \mu b) - l/\lambda\,(\psi(\lambda a + \mu b) - \lambda\psi(a) - \mu\psi(b))$$
$$< \psi(\lambda a + \mu b).$$

So $\int_0^1 \psi \circ f \, dx < \psi\left(\int_0^1 f \, dx\right)$, a contradiction.

11. (i) If $a_i \geqslant 0, b_i \geqslant 0, i = 1, 2, \ldots, n$ and $p > 1$, $1/p + 1/q = 1$, then
$$\sum_{i=1}^n a_i b_i \leqslant \left(\sum_{i=1}^n a_i^p\right)^{1/p} \left(\sum_{i=1}^n b_i^q\right)^{1/q}.$$
(ii) If $p \geqslant 1$, $\left(\sum_{i=1}^n |a_i + b_i|^p\right)^{1/p} \leqslant \left(\sum_{i=1}^n |a_i|^p\right)^{1/p} + \left(\sum_{i=1}^n |b_i|^p\right)^{1/p}$.
(iii) If $a_i \geqslant 0, b_i \geqslant 0, i = 1, \ldots, n$, $\sum_{i=1}^n a_i b_i \leqslant \left(\sum_{i=1}^n a_i\right) \max_{1 \leqslant i \leqslant n} b_i$. The proof of (i), for example, is obtained by taking $X = [1, \ldots, n]$, $a(i) = a_i$, $\mu([i]) = 1$ so that $\sum_{i=1}^n a_i = \int a \, d\mu$, and applying Hölder's inequality.
12. They imply $\|\sin x - \cos x\|_2 = \|(f - \sin x) - (f - \cos x)\|_2 \leqslant \|f - \sin x\|_2 + \|f - \cos x\|_2 \leqslant 1$. But the first term is $\sqrt{\pi}$.
13. Apply the Schwarz inequality.
14. (i) is a special case of (ii). Write $|f|^p = F$, $|g|^q = G$, $\alpha = 1/p$, $\beta = 1/q$, then $F \in L^\alpha(\mu)$, $G \in L^\beta(\mu)$, so by Theorem 7, $FG \in L^1(\mu)$.
15. (i) Minkowski's inequality gives $|\,\|f\|_2 - \|f_n\|_2| \leqslant \|f - f_n\|_2$.
(ii) $\left|\int_a^t f \, dx - \int_a^t f_n \, dx\right| = \left|\int_a^b \chi_{(a,t)}\,(f - f_n) \, dx\right| \leqslant \sqrt{(t-a)}\,\|f - f_n\|_2$, by Hölders inequality.
(iii) To verify (i), integrate explicitly and use $\sum_{n=1}^\infty 1/n^2 = \pi^2/6$. To verify (ii), integrate and use the standard Fourier Series for t^2.
16. By Minkowski's inequality $|\,\|f_n\|_p - \|f\|_p| \leqslant \|f_n - f\|_p \to 0$.
17. By Example 20, p. 67, we can find ψ such that $\psi(t)t^{p-1} f^p \in L^1$, $\psi \geqslant 1$ on $[0,1]$, $\psi(0+) = \infty$. Then

$$F(x) = \int_x^1 f \, dt = \int_x^1 \frac{1}{\psi^{1/p}\, t^{(p-1)/p}} \psi^{1/p}\, t^{(p-1)/p} f \, dt$$
$$\leqslant \left(\int_x^1 \psi^{-q/p}\, t^{-1} \, dt\right)^{1/q} \left(\int_x^1 t^{p-1}\, \psi f^p \, dt\right)^{1/p},$$

by Hölder's inequality, p and q being conjugate indices. So

$$F(x) \leqslant M\left(\int_x^1 \psi^{-q/p}\, t^{-1}\, dt\right)^{1/q}$$

where M is a constant. But $\forall\, \epsilon > 0$, $\exists\, x_0$ such that $\psi^{-q/p} < \epsilon$ for $x < x_0$ and then

$$\int_x^1 \psi^{-q/p}\, t^{-1}\, dt = \int_x^{x_0} + \int_{x_0}^1 \leqslant \epsilon(\log 1/x - \log 1/x_0) + \log 1/x_0$$
$$\leqslant 2\epsilon \log 1/x \text{ for small } x.$$

18. At $x = 0$, Hölder's inequality gives $F(x) \leqslant x^{1/q}(\int_0^x f^p\, dt)^{1/p}$. But $\int_0^x f^p\, dt \to 0$ as $x \to 0$ by Theorem 18, p. 106. At $x = \infty$, $\forall\, \epsilon > 0$, $\exists\, y$ such that $\int_y^\infty f^p\, dt < \epsilon^p$; so let $x > y$. Then $F(y) - F(x) \leqslant (x - y)^{1/q}\left(\int_y^x f^p\, dt\right)^{1/p} \leqslant \epsilon x^{1/q}$. So $F(x) < 2\epsilon x^{1/q}$ for all large x.
19. For $n = 2$ we get Hölder's inequality. So suppose that the result holds for $n = m - 1$, $m > 2$. Then if $\alpha = (1 - k_m^{-1})^{-1}$ we have $\sum_{i=1}^{m-1} \alpha k_i^{-1} = 1$, so $\int |f_1|^\alpha \ldots |f_{m-1}|^\alpha\, d\mu \leqslant \prod_{i=1}^{m-1} (\int |f_i|^{k_i}\, d\mu)^{\alpha/k_i}$.
   But from Hölder's inequality
   $$\int |f_1 \ldots f_m|\, d\mu \leqslant (\int |f_1|^\alpha \ldots |f_{m-1}|^\alpha\, d\mu)^{1/\alpha} (\int |f_m|^{k_m}\, d\mu)^{1/k_m}.$$
   So the result holds for m, and by induction the result follows.
20. If any $f_i = 0$ a.e. or if (6.13) holds, equality is trivial. Conversely, the case $n = 2$ is given in Example 10, p. 115. We suppose that the result is true for $n - 1$ functions and consider the case of n functions, supposing that no $f_i = 0$ a.e. Then the fact that (6.13) holds for $n - 1$ functions, together with the chain of inequalities which gives the result for n functions, gives us (6.13) in this case.
21. In Exercise 19, take $n = 3$, $f_1(x) = x^{-p\alpha}$, $f_2(x) = |x - 1|^{-p\beta}$, $f_3(x) = |x - 2|^{-p\gamma}$ and if $\delta = p(\alpha + \beta + \gamma) < 1$ take $k_1 = \delta/(p\alpha)$, $k_2 = \delta/(p\beta)$, $k_3 = \delta/(p\gamma)$ to get the result.
22. Let $p \geqslant 1$ and let $f_i \in L^p(\mu)$, $i = 1, \ldots, n$. Then $\|\sum_{i=1}^n f_i\|_p \leqslant \sum_{i=1}^n \|f_i\|_p$ follows immediately, by induction, from the case $n = 2$.
23. As we may consider f/M, we may suppose that ess sup $f = 1$. Then $f^{n+1} \leqslant f^n$ a.e., so the required limit exists and is not greater than 1. But $\int f^n\, d\mu = \int f^n \cdot 1\, d\mu \leqslant (\int f^{n+1}\, d\mu)^{n/n+1} (\mu(X))^{1/n+1}$. So
   $$\frac{\int f^n\, d\mu}{\mu(X)} \leqslant \frac{(\int f^{n+1}\, d\mu)^n}{(\int f^n\, d\mu)^n}.$$

Take nth roots, let $n \to \infty$ and use the result of Exercise 6, p. 110, to get the result.

24. Write $I_n = \int f^n \, d\mu$. Then, for $n \geq 2$, $I_n^2 = (\int f^{(n-1)/2} f^{(n+1)/2} \, d\mu)^2 \leq I_{n-1} \cdot I_{n+1}$ by Schwarz's inequality.
25. Write $p_1' = p_1\lambda$, $p_2' = p_2\lambda$ where $\lambda = 1/p_1 + 1/p_2$, so that p_1' and p_2' are conjugate indices. As $f \in L^{p_1}(\mu)$, we have $|f|^{1/\lambda} \in L^{p_1'}(\mu)$, and similarly $|g|^{1/\lambda} \in L^{p_2'}(\mu)$. So Theorem 7, p. 115, gives $|fg|^{1/\lambda} \in L^1(\mu)$, and so the required index p is $(p_1 p_2)/(p_1 + p_2)$.
26. By Hölder's inequality $P_f^{1/2} \geq \int_E 1 \, d\mu = \mu(E)$, giving the first part. By Example 10, p. 000, equality holds when $s\sqrt{f} + t/\sqrt{f} = 0$ a.e., i.e. when f is constant a.e. For the last part, take $E = (a, b)$, $f = (x-a)^2$ to get $P_f = \infty$.
27. We have $\frac{P}{q} g \geq f \geq \frac{p}{Q} g$ a.e. So $\left(f - \frac{P}{q}g\right)\left(f - \frac{p}{Q}g\right) \leq 0$ a.e. Write $F_y = f^2 + yfg(\sqrt{(PQ)}/\sqrt{(pq)} + \sqrt{(pq)}/\sqrt{(PQ)}) + y^2g^2 = (f + yg\sqrt{(PQ)}/\sqrt{(pq)})(f + yg\sqrt{(pq)}/\sqrt{(PQ)})$. So if $\alpha = \sqrt{(Pp)}/\sqrt{(Qq)}$, then $F_{-\alpha} \leq 0$ a.e., but $F_\alpha > 0$ on a set of positive measure. So $\int F_y \, d\mu$ is not greater than zero for $y = -\alpha$, and is positive for $y = \alpha$. But $\int F_y \, d\mu = \int f^2 \, d\mu + y \int fg \, d\mu(\sqrt{(PQ)}/\sqrt{(pq)} + \sqrt{(pq)}/\sqrt{(PQ)}) + y^2 \int g^2 \, d\mu$ and this quadratic in y must have a real root, giving the result.
28. If $0 < k < 1$ we have $m < 0$, and conversely, so we may suppose that $0 < k < 1$. Suppose that $\int g^m \, d\mu < \infty$, so that $g > 0$ a.e. So writing $p = k^{-1}$, the equations $f = (uv)^p$, $g = v^{-p}$ define non-negative measurable functions u and v a.e., and hence $\int uv \, d\mu \leq \|u\|_p \, \|v\|_q$ where p and q are conjugate indices. So
$$\int f^k \, d\mu \leq (\int fg \, d\mu)^k \, (\int g^m \, d\mu)^{1-k},$$
giving the result.
29. Let $p + q = pq$; then by the last exercise we have
$$\begin{aligned}\int (f+g)^p \, d\mu &= \int f(f+g)^{p-1} \, d\mu + \int g(f+g)^{p-1} \, d\mu \\ &\geq (\int f^p \, d\mu)^{1/p} \, (\int (f+g)^{(p-1)q} \, d\mu)^{1/q} \\ &\quad + (\int g^p \, d\mu)^{1/p} \, (\int (f+g)^{(p-1)q} \, d\mu)^{1/q}\end{aligned}$$
which gives the result, as in Theorem 8, p. 115.
30. $0 \leq f_n^p \leq f^p$, so $f_n \in L^p(\mu)$. Also $0 \leq (f - f_n)^p \leq f^p$, an integrable function, so by Theorem 21, p. 107, as $\lim f_n = f$ we have $\lim \int (f_n - f)^p \, d\mu = 0$.
31. Given $f \in L^p(\mu)$ and $\epsilon > 0$ it is required to find $g \in L^p(\mu)$, where g is bounded and measurable and $\|f - g\|_p < \epsilon$. Minkowski's inequality allows us to consider f^+ and f^- separately. So suppose that $f \geq 0$. But then the last exercise gives the result.
32. By the last exercise we may suppose that f is bounded and measurable. Then the proof of Theorem 16, p. 73, may be used with appropriate modifications. For example, we have by Minkowski's inequality that if $f = g + h$,

$\int_a^b |g-g_1|^p \, dx < \epsilon$ and $\int_a^b |h-h_1|^p \, dx < \epsilon$, then

$$\int_a^b |f-(g_1+h_1)|^p \, dx < 2^p \epsilon.$$

33. We wish to show that if $f \in L^p(a, b)$, then ∃ sequences $\{f_n\}$ of the desired type such that $\|f_n - f\|_p = 0$. By Minkowski's inequality it is sufficient to consider $f \geqslant 0$. Then, for simple functions the result follows from Theorem 5, p. 58, and Theorem 10, p. 63. For step functions and continuous functions the result is that of the last exercise.
34. Let $f_n = f\chi_{(-n,n)}$. Then Theorem 10, p. 63, gives $\lim \|f-f_n\|_p = 0$ since $|f-f_n|^p \leqslant |f|^p$. So Minkowski's inequality and the last exercise give the result.
35. $|F(t+h)-F(t)| \leqslant (\int |f(x+t+h)-f(x+t)|^p \, dx)^{1/p} . (\int |g|^q \, dx)^{1/q}$. So it is sufficient to show that $\int |f(x+t+h)-f(x+t)|^p \, dx$ tends to zero with h, and hence by Example 25, p. 75, to show the same for $\int |f(x+h) - f(x)|^p \, dx$. By Exercise 31, there is, for any $\epsilon > 0$, a continuous function k vanishing outside a bounded set, such that $\|f-k\| < \epsilon$. So if $f_h(x) = f(x+h)$, $k_h(x) = k(x+h)$, we have
$$\|f-f_h\|_p \leqslant \|f-k\|_p + \|f_h - k_h\|_p + \|k-k_h\|_p \leqslant 2\epsilon + \|k-k_h\|_p$$
by Example 25, p. 75, again. But since k is uniformly continuous and k and k_h vanish outside a bounded set, $\|k-k_h\|_p < \epsilon$ for all small h, giving the result.
36. $\|f_n g_n - fg\|_1 \leqslant \|fg - fg_n\|_1 + \|fg_n - f_n g_n\|_1 \leqslant \|f\|_p \, \|g-g_n\|_q + \|f - f_n\|_p \, \|g_n\|_q$. But $| \, \|g_n\|_q - \|g\|_q | \leqslant \|g_n - g\|_q$ by Minkowski's inequality, so for all large n, $\|g_n\|_q \leqslant \|g\|_q + 1$ and the result follows.
37. (i) $|\int (f_n - f)g \, d\mu| \leqslant \|f_n - f\|_2 \cdot \|g\|_2 \to 0$ as $n \to \infty$. For a counterexample take $X = (-\pi, \pi)$, $f_n(x) = \sin nx$ so that $\|f_n - f_m\|_2 = (2\pi)^{1/2}$ for $n \neq m$, and $\{f_n\}$ does not converge in L^2. But if $g \in L^2(-\pi, \pi)$, $\lim \int_{-\pi}^{\pi} g(x) \sin nx \, dx = 0$ by, for example, Exercise 44, p. 75.
(ii) If for any sequence $\{n_i\}$ we have $\|f_{n_i}\|_2 \leqslant C$, then $\|f\|_2 \leqslant C$, for then $C \, \|f\|_2 \geqslant |\int f_{n_i} f \, d\mu| \to (\|f\|_2)^2$. We may suppose that $\|f\|_2 \neq 0$ and divide to get the result.
(iii) If $f_n \to f$ weakly, then $\int_a^b \chi_{[a,x]} f_n \, dt \to \int_a^b \chi_{[a,x]} f \, dt$. In the opposite direction: if this holds, we get $\int_a^b g f_n \, dx \to \int_a^b gf \, dx$ for any step function g, and so $\|gf_n - gf\|_1 < \epsilon$ for all large n. Now if $h \in L^2(a, b)$ and $\epsilon > 0$ there exists a step function g, by Exercise 33, such that $\|g-h\|_2 < \epsilon$, so $\|hf - hf_n\|_1 \leqslant \|hf - gf\|_1 + \|gf - gf_n\|_1 + \|gf_n - hf_n\|_1$. Let $M =$

$\sup \|f_n\|_2$. Then $\|hf - hf_n\|_1 \leqslant \epsilon \|f\|_2 + \epsilon + M\epsilon$ for all large n. So $f_n \to f$ weakly.

(iv) $0 \leqslant (\|f_n - f\|)^2 = \int f^2 \, d\mu + \int f_n^2 \, d\mu - 2\int f_n f \, d\mu \to \alpha^2 - \|f\|^2 \leqslant 0$. So $\alpha = \|f\|_2$ and $f_n \to f$ in L^2.

38. Let $p_1 = p/p'$ and q_1 be conjugate indices. Then

$$\int |f_n - f|^{p'} \, d\mu \leqslant \|(f_n - f)^{p'}\|_{p_1} (\mu(X))^{1/q_1} = (\|f_n - f\|_p)^{p'} \mu(X)^{1/q_1} \to 0$$

as required.

39. As in Theorem 10, p. 118, $\exists$ a subsequence $\{n_i\}$ such that $f_{n_i} \to f$ a.e. So $|f| \leqslant K$ and $|f_n - f| \leqslant 2K$ a.e. for all n. Suppose $p'' > p$ (otherwise apply the last exercise). Then $(\|f_n - f\|_{p''})^{p''} \leqslant (2K)^{p''-p} (\|f_n - f\|_p)^p \to 0$ as $n \to \infty$.

HINTS AND ANSWERS TO EXERCISES: CHAPTER 7

1. This follows from $[x: |f_n + g_n - (f+g)| > \epsilon] \subseteq [x: |f_n - f| > \epsilon/2] \cup [x: |g_n - g| > \epsilon/2]$.

2. If $\alpha = 0$, the result is obvious; otherwise

$$[x: |\alpha f_n - \alpha f| > \epsilon] = [x: |f_n - f| > \epsilon/|\alpha|],$$

the measure of which tends to zero as $n \to \infty$.

3. From the definition of convergence in measure it can be seen that the limit function f is finite valued a.e. So $\forall\, \epsilon > 0$, $\exists$ set G, and $K > 0$ such that $\mu(G) < \epsilon/2$ and $|f| < K$ on $\mathbf{C}G$. Write $E_\gamma = [x: |f_n - f| > \gamma]$, then on $\mathbf{C}(G \cup E_\gamma)$, $|f_n^2 - f^2| = |f_n + f| \cdot |f_n - f| \leqslant (\gamma + 2K)\gamma < \epsilon$ for an appropriate $\gamma > 0$. But $\mu(E_\gamma) < \epsilon/2$ for all large n. So for large n, $\mu[x: |f_n^2 - f^2| > \epsilon] < \epsilon$, giving the result.

4. Use $f_n g_n = \frac{1}{4}(f_n + g_n)^2 - \frac{1}{4}(f_n - g_n)^2$, and the last three exercises.

5. Take $X = (0, \infty)$, $f_n = f = x$ and let g_n be positive constants α_n where $\lim \alpha_n = 0$. Then α_n, as a function of x, tends to zero in measure, but $m[x: |f_n g_n - fg| > \epsilon] = m[x: \alpha_n x > \epsilon] = \infty$ for all n.

6. Use $|f_n - f| \geqslant ||f_n| - |f||$.

7. A set K is convex if when $k_i \in K, i, \ldots, n$, we have $\sum_{i=1}^{n} \alpha_i k_i \in K$ whenever $\alpha_i \geqslant 0$, $\sum_{i=1}^{n} \alpha_i = 1$. So S is clearly convex. Write J_i for the interval $[(i-1)/(n+1), i/(n+1))$, $i = 1, \ldots, n+1$. Then for $f \in S$, $(n+1)\chi_{J_i} f \in U_n$. But then

$$f = 1/(n+1) \sum_{i=1}^{n+1} (n+1)\chi_{J_i} f$$

lies in any convex set containing U_n, giving the result.

8. In Example 1, p. 124, let X be the set of positive integers, and let $\mu([n]) = a_n$ where $a_n > 0$ and $\sum_{n=1}^{\infty} a_n < \infty$. Then if $\xi = \{\xi_n\}$ and $\eta = \{\eta_n\}$ are any two real sequences, define $\rho(\xi, \eta) = \sum_{n=1}^{\infty} \frac{a_n |\xi_n - \eta_n|}{1 + |\xi_n - \eta_n|}$.

9. Clearly $\rho(f, g) = \rho(g, f)$, $\rho(f, g) \geqslant 0$, $\rho(f, f) = n(0) = 0$. If $n(f) = 0$ then, as tan is continuous at the origin, if $\epsilon > 0$ there exists $\alpha < \epsilon$ such that $\alpha + \mu[x: |f(x)| > \alpha] < \epsilon$, so $\mu[x: |f(x)| > \epsilon] < \epsilon$, i.e., $f = 0$ a.e. So if $\rho(f, g) = 0$, $f = g$ a.e. Finally, for $\xi, \eta \geqslant 0$ we obtain from the addition formula for tan that

$$\arctan \xi + \arctan \eta \geqslant \arctan (\xi + \eta),$$

and it easily follows that $n(f + g) \leqslant n(f) + n(g)$, so that $\rho(f, g) + \rho(g, h) \geqslant \rho(f, h)$ and ρ is a metric. Convergence in measure implies convergence with respect to ρ since if $n(f_n - f) > \delta > 0$ and $0 < \epsilon < \frac{1}{2}\tan \delta$, we easily obtain $\mu[x: |f_n(x) - f(x)| > \epsilon] > \frac{1}{2}\tan \delta$. Conversely, let $\epsilon, \delta > 0$ and let $\mu[x: |f_n(x) - f(x)| > \epsilon] > \delta$. Then for $\alpha \geqslant \epsilon$ we have $\alpha + \mu[x: |f_n(x) - f(x)| > \epsilon] \geqslant \epsilon$, and for $\alpha < \epsilon$ we have $\alpha + \mu[x: |f_n(x) - f(x)| > \alpha] > \delta$. So $\rho(f_n, f) \geqslant \min(\arctan \delta, \arctan \epsilon)$.

10. For each n we have, a.e., $f_n g \leqslant F$. ess sup $|g|$, an integrable function. But $\lim f_n g = fg$ a.e., so by Theorem 21, p. 107, $f_n g \to fg$ in the mean.

11. For each k, there exists E_k such that on $\complement E_k$, $\{f_n\}$ is uniformly fundamental and $\mu(E_k) < k^{-1}$. We may suppose $E_1 \supseteq E_2 \supseteq \ldots$ Then $E = \bigcap_{k=1}^{\infty} E_k$ has measure zero and if $x \in \complement E$, $f_n(x) \to f(x)$, defining a measurable function. Also $f_n \to f$ uniformly on each set $\complement E_k$, giving the result.

12. (i) μ is well defined on $\mathcal{S}$, $\mu \geqslant 0$, $\mu(\emptyset) = 0$. Let $\{B_i\}$ be a sequence of disjoint sets of $\mathcal{S}$; then

$$\mu\left(\bigcup_{i=1}^{\infty} B_i\right) = \lim \mu_n\left(\bigcup_{i=1}^{\infty} B_i\right) = \lim \sum_{i=1}^{\infty} \mu_n(B_i) = \sum_{i=1}^{\infty} \lim \mu_n(B_i) = \sum_{i=1}^{\infty} \mu(B_i).$$

The interchange of summation and lim is allowable by Theorem 15, p. 105, so μ is a measure.

(ii) It is sufficient to prove the result for $f \geqslant 0$. As $\mu \geqslant \mu_n$, $\int f \, d\mu_n < \infty$ implies $f \in L(X, \mu_n)$ for each n. There exists a sequence of measurable simple functions $\phi_k \uparrow f$, and clearly $\lim_{n\to\infty} \int \phi_k \, d\mu_n = \int \phi_k \, d\mu$. So

$$\int f \, d\mu = \lim_{k\to\infty} \int \phi_k \, d\mu = \lim_{k\to\infty} \lim_{n\to\infty} \int \phi_k \, d\mu_n$$
$$= \lim_{n\to\infty} \lim_{k\to\infty} \int \phi_k \, d\mu_n = \lim_{n\to\infty} \int f \, d\mu_n,$$

the interchange of limits being justified, since the double sequence increases with respect to both k and n.

13. $\mu_{n+1} \geqslant \mu_n$ easily gives $\int |f| \, d\mu_n \leqslant \int |f| \, d\mu_{n+1}$. To prove the last assertion, take X to be a single point and $\mu_n(X) = n$. Then every finite function belongs to each $L(\mu_n)$, but only $f \equiv 0 \in L(\mu)$.
14. Routine verification.
15. By the last exercise $\mu_1 - \mu_n$ is a measure, for each n. Applying Exercise 12 to this sequence of measures we get that $\mu_1 - \mu$ is a measure. So the last exercise, again, gives that $\mu = \mu_1 - (\mu_1 - \mu)$ is a measure.
16. Take $X = [1, \infty)$ and define $\mu_n(E) = \int_E (nx)^{-1} \, dx$. Let $B_i = [i, i+1)$, $i = 1, 2, \ldots$ Then $\mu_n(B_i) \downarrow 0$ as $n \to \infty$, so $\mu(B_i) = 0$, $\sum_{i=1}^{\infty} \mu(B_i) = 0$. But $\mu_n\left(\sum_{i=1}^{\infty} B_i\right) = \infty$, each n, so $\mu\left(\bigcup_{i=1}^{\infty} B_i\right) = \infty$ and so μ is not countably additive.
17. The first part is obvious. For the second, considering simple functions and taking limits, we get $\int f \, d\mu_1 - \int f \, d\mu_n = \int f \, d(\mu_1 - \mu_n)$, since $\mu_1 - \mu_n$ is a measure by Exercise 14. Applying Exercise 12(ii) to $\{\mu_1 - \mu_n\}$ we get the result.
18. Finite additivity and the other properties being obvious, we wish to check that μ is countably additive. Let $\{B_i\}$ be disjoint measurable sets. Then

$$\mu\left(\bigcup_{i=1}^{\infty} B_i\right) - \sum_{i=1}^{N} \mu(B_i) \leqslant \left|\mu\left(\bigcup_{i=1}^{\infty} B_i\right) - \mu_n\left(\bigcup_{i=1}^{\infty} B_i\right)\right| + \mu_n\left(\bigcup_{i=N+1}^{\infty} B_i\right)$$
$$+ \sum_{i=1}^{N} |\mu(B_i) - \mu_n(B_i)|.$$

We have $\mu\left(\bigcup_{i=k}^{\infty} B_i\right) \to 0$ as $k \to \infty$, so since $\{\mu_n\}$ converges uniformly, $\forall \, \epsilon > 0$, $\exists \, N_1$ such that $\mu_n\left(\bigcup_{i=N+1}^{\infty} B_i\right) < \epsilon/3$ for all n and for $N \geqslant N_1$. Now choose N_2 such that for $n \geqslant N_2$ the first and third terms are $< \epsilon/3$, so that the right-hand side $< \epsilon$.

19. We may suppose $f \geqslant 0$, so there exists a sequence of measurable simple functions $\phi_m \uparrow f$. By Egorov's theorem, $\phi_m \to f$ a.u. with respect to each μ_n and with respect to μ, so $\forall \, \epsilon > 0$, $\phi_m \to f$ uniformly on $\mathbf{C}E_n$ where $\mu_n(E_n) < \epsilon/2^{n+1}$, and we may suppose also when choosing E_n that $\int_{E_n} f \, d\mu_n < \epsilon/2^n$. By uniformity, for $n, k \geqslant n_0$, $\mu_k(E_n) < \epsilon/2^n$, so we may suppose, when taking limits with respect to n, that $\mu_k(E_n) < \epsilon/2^n$ for all k and n. Then if $F = \bigcup_{n=1}^{\infty} E_n$, $\mu_k(F) < \epsilon$ for all k, and $\phi_m \uparrow f$ uniformly on $\mathbf{C}F$. Now

$\int f\,d\mu_n - \int \phi_m\,d\mu_n = \int_F (f-\phi_m)\,d\mu_n + \int_{CF} (f-\phi_m)\,d\mu_n < \int_F f\,d\mu_n +$ $\epsilon\mu(X) < \epsilon\,(1+\mu(X))$ for $m \geqslant m_0(\epsilon)$. We have ϕ_{m_0} bounded a.e., so $\int \phi_{m_0}\,d\mu_n < M$, say, for all n. But then $\int f\,d\mu_n < M'$, say, for all n. This implies that $\int f\,d\mu < \infty$, for otherwise $\forall\, i$, $\exists\, m(i)$ such that $\int \phi_m\,d\mu > i$ and then $\exists\, n(i)$ such that $\int \phi_m\,d\mu_n > i-1$. But this gives $\int f\,d\mu_n > i-1$ and a contradiction, so $f \in L^1(\mu)$. Hence $\int f\,d\mu - \int \phi_m\,d\mu < \epsilon$ for all $m \geqslant m_1(\epsilon)$. Let $m_2 = m_0 + m_1$. Then $|\int \phi_{m_2}\,d\mu_n - \int \phi_{m_2}\,d\mu| < \epsilon$ for $n \geqslant n_0$, as ϕ_{m_2} is finite-valued a.e. So $|\int f\,d\mu - \int f\,d\mu_{n_0}| < 2\epsilon + \epsilon\mu(X)$, giving the result.

20. For example, let $f_n \to f$ in the mean of order $p, p>0$. Then $\int |f_n - f_m|^p\,d\mu \leqslant 2^p \int |f_n - f|^p\,d\mu + 2^p \int |f_m - f|^p\,d\mu$, and letting $n, m \to \infty$ gives the result.

21, 22. Obvious.

23. $\forall\ \epsilon > 0$, $\sup f_n$ in $[\epsilon, 1]$ is $ne^{-n\epsilon} \to 0$ as $n \to \infty$. But $\displaystyle\int_{-0}^{1} f_n\,dx = 1 - e^{-n} \not\to 0$ as $n \to \infty$.

24. Similar to Exercise 23.

25. By Hölder's inequality f_n, and the limit function f, are in $L^1(\mu)$, and
$$\|f_n - f\|_1 \leqslant \|f_n - f\|_p\,(\mu(X))^{1/q},$$
where p, q are conjugate indices.

26. $f_n(x) \leqslant x^{-2} \in L^1(1,\infty)$ but, if $n > m$, $\|f_n - f_m\|_{1/2} = \left(\int_m^n x^{-1}\,dx\right)^2 \not\to 0$ as $n, m \to \infty$.

27. If $t > 0$, $e^{-t} < 1/t$; so $n^{3/2}\,xe^{-n^2x^2} < n^{-1/2}\,x^{-1} \to 0$ as $n \to \infty$, for $x > 0$. But, on computation, $\displaystyle\int_0^1 f_n^2\,dx \not\to 0$.

28. As for Exercise 27, using $e^{-t} < t^{-k}\,k!$ where $k > 3/2p$.

29. $f_n < \epsilon^n$ on $[\epsilon, 1]$, $0 < \epsilon < 1$. So $f_n \to f$ a.u. where $f = 0$ on $[0,1)$, $f(1) = 1$. But $\{f_n\}$ does not converge uniformly as its pointwise limit, f, is discontinuous.

30. $\int |f_n - f|^p\,d\mu \leqslant \mu(X)\,\text{ess sup}\,|f_n - f|^p \to 0$ as $n \to \infty$.

31. Let $X = (1, \infty)$, and let $f_n(x) = x^{-2/p} + n^{-1/p}$ on $(1, n)$, $f_n(x) = x^{-2/p}$ on $[n, \infty)$, $f(x) = x^{-2/p}$ on $(1, \infty)$. Then f and each $f_n \in L^p(1,\infty)$ and $f_n \to f$ uniformly. But $\|f_n - f\|_p = 1$ for each n.

HINTS AND ANSWERS TO EXERCISES: CHAPTER 8

1. A is positive iff $m(A \cap (-\infty, 0)) = 0$, B is negative iff $m(B \cap (0,\infty)) = 0$, C is ν-null iff $m(C) = 0$, and $(-\infty, 0)$, $[0, \infty)$ forms a Hahn decomposition.

2. We have $\nu(E) = \nu(A) + \nu\left(\bigcup_{k=1}^{\infty} E_k\right)$ from (8.1), and from the proof the theorem

$$0 > \nu\left(\bigcup_{k=1}^{\infty} E_k\right) > -\infty.$$

So $\nu(A) = \infty$ would imply $\nu(E) = \infty$.

3. We may take $X = \mathsf{R}$, $\mathcal{S} = \mathcal{M}$, $\nu(E) = \int_E f\,dx$, where $f(x) = xe^{-x}$ ($|x| > 1$), $f(x) = 0$ ($|x| \leqslant 1$). Then $(-\infty, a)$, $[a, \infty)$ is a Hahn decomposition of R with respect to ν, for any a such that $|a| \leqslant 1$.
4. Write $a_n^+ = \max(a_n, 0)$, $a_n^- = \max(-a_n, 0)$. Then condition (i) of Definition 1, p. 133, implies that either $\sum_{n=1}^{\infty} a_n^+ < \infty$ or $\sum_{n=1}^{\infty} a_n^- < \infty$. That ν is then always well defined and is a signed measure, is obvious, as is the last part.
5. As in the last exercise, and using the finiteness conditions given, we get that the sum of the positive terms in $\{\nu(E_i)\}$ is finite, and that of the negative terms is finite, as required.
6. $\nu(E) = \nu(F) + \nu(E - F)$, and Definition 1(i), p. 133, implies that neither of the terms on the right is infinite.
7. $\bigcup_{i=1}^{\infty} E_i = E_1 \cup \bigcup_{i=2}^{\infty} (E_i - E_{i-1})$. So

$$\nu\left(\bigcup_{i=1}^{\infty} E_i\right) = \nu(E_1) + \sum_{i=2}^{\infty} \nu(E_i - E_{i-1})$$

$$= \lim\left(\nu(E_1) + \sum_{i=2}^{n} \nu(E_i - E_{i-1})\right) = \lim \nu(E_n),$$

and this limit exists, though it may be infinite.

8. From Exercise 6, $\nu(E_1) - \nu(E_i) = \nu(E_1 - E_i)$ for each i. But $E_1 = \bigcap_{i=1}^{\infty} E_i \cup \bigcup_{i=1}^{\infty} (E_1 - E_i)$, so by Exercise 7,

$$\nu(E_1) = \nu\left(\bigcap_{i=1}^{\infty} E_i\right) + \lim \nu(E_1 - E_i) = \nu\left(\bigcap_{i=1}^{\infty} E_i\right) + \nu(E_1) - \lim \nu(E_i),$$

So, since $|\nu(E_1)| < \infty$, we get the result.

9. There exist sets A, B such that $\mu(A) = \nu_1(\complement A) = \mu(B) = \nu_2(\complement B) = 0$. So $\mu(A \cup B) = (\nu_1 + \nu_2)(\complement A \cap \complement B) = 0$.
10. From Example 5, p. 138, $|\nu|(E) = \int_E |f|\,d\mu$.
11. Obvious from Definition 6, p. 138.
12. Let A, B be a Hahn decomposition with respect to ν. Then $|\nu|(E) = |\nu(E \cap A)| + |\nu(E \cap B)| \leqslant \sup \sum_{i=1}^{n} |\nu(E_i)|$ as A, B are disjoint. But

$$\sup \sum_{i=1}^{n} |\nu(E_i)| \leqslant \sup \sum_{i=1}^{n} |\nu|(E_i) = |\nu|(E),$$

as $|\nu|$ is a measure.

13. Suppose that $\nu = \nu^+ - \nu^-$ is the Jordan decomposition with a corresponding Hahn decomposition A, B. Let $E \in \mathcal{S}$ and write $E_1 = E \cap A, E_2 = E \cap B$. Then $\nu^-(E_1) = 0$ so $\nu_2(E_1) \geqslant \nu^-(E_1)$. But $\nu_1 - \nu^+ = \nu_2 - \nu^-$ and hence $\nu_1(E_1) \geqslant \nu^+(E_1)$. Similarly for E_2, so $\nu^+(E) \leqslant \nu_1(E)$ and $\nu^-(E) \leqslant \nu_2(E)$ for all $E \in \mathcal{S}$, and $|\nu| \leqslant \nu_1 + \nu_2$ follows. For equality we need $\nu^+(E) + \nu^-(E) = \nu_1(E) + \nu_2(E)$ which implies $\nu_1 = \nu^+, \nu_2 = \nu^-$ since $\nu^+ \leqslant \nu_1, \nu^- \leqslant \nu_2$.
14. Obvious.
15. Let $X = [0,1]$ and let μ be given by $\mu(E) = \int_E f\, dx$ where $f = 1$ on $[0, 1/2)$ and $f = -1$ on $[1/2, 1]$. Then $|\mu| = m$ so $|\mu|(E) = 0 \Rightarrow m(E) = 0$ but $\mu(E) = 0 \not\Rightarrow m(E) = 0$ if $E = (1/2 - \alpha, 1/2 + \alpha), 0 < \alpha \leqslant 1/2$.
16. Take, for example, $X = \mathsf{R}$, $\mathcal{S} = \mathcal{M}$, $\mu(E) = \int_{[0,\infty)\cap E} e^{-x^2}\, dx$, $\nu(E) = \int_{(-\infty,1)\cap E} e^{-x^2}\, dx$.
17. ν is a measure by direct verification. Since $\nu([x_0]) = 1$ and $m([x_0]) = 0$ we have not got $\nu \ll m$.
18. Suppose that $\nu(X) = \int f\, d\mu$ where f is non-negative and finite-valued and μ is σ-finite. Then $X = \bigcup_{n=1}^{\infty} X_n$, $\mu(X_n) < \infty$ and also $X = \bigcup_{m=1}^{\infty} Y_m$, $Y_m = [x: f(x) \leqslant m]$, so that $X = \bigcup_{m,n} (X_n \cap Y_m)$. But $\nu(X_n \cap Y_m) \leqslant m\mu(X_n) < \infty$, so ν is σ-finite. So the result of the theorem for ν implies that ν is σ-finite. Note that the next exercise shows that $\nu \ll \mu$ and μ σ-finite do not imply that ν is σ-finite.
19. If X can be written as the union of a sequence of sets of finite ν-measure, so can any subset. But if F is the set on which f is infinite, so that $\mu(F) > 0$, then every subset E of F has ν-measure 0 or ∞ according as $\mu(E) = 0$ or $\mu(E) > 0$; so F provides the required contradiction.
20. Take X such that Card $X = \aleph_0$, and let $\mathcal{S} = [E$: Card $E \leqslant \aleph_0$, or Card $\mathbf{C}E \leqslant \aleph_0]$. Then $[\![X, \mathcal{S}]\!]$ is a measurable space. Let $\mu(E) =$ Card E; then μ is a measure and is not σ-finite. Let $\nu(E) = 0$ if Card $E \leqslant \aleph_0$, $\nu(E) = 1$ if Card $E > \aleph_0$. Then ν is a measure, as in any sequence of disjoint sets of $\mathcal{S}$ at most one is uncountable; also $\nu \ll \mu$. Suppose that f exists such that $\nu(E) = \int_E f\, d\mu$ for each $E \in \mathcal{S}$. Then $\int_{[x]} f\, d\mu = f(x)$ but $\nu([x]) = 0$ so $f \equiv 0$. But $\nu(X) = 1$, so no such f exists.
21. Follows immediately from Definition 5, p. 137, and Definition 8, p. 139.

22. As in the proof of Theorem 6 we may suppose that μ and ν are measures. Then by Theorem 5, for $E \in \mathcal{S}$, $\nu(E) = \int_E f\, d\mu$ where $f \geqslant 0$ and measurable. But then Theorem 18, p. 106, gives the result.
23. Choose a sequence $\{n_i\}$ such that $\lim a_{n_i} = 0$. Then $\mu([n_i]) \to 0$, but $\nu([n_i]) \geqslant \inf b_n > 0$.
24. $\exists\ n_0$ such that for $n, m \geqslant n_0$, $\int |f_n - f_m| < \epsilon/2$. The result of Theorem 18, p. 106, extends immediately to give, for fixed n_0, that $\exists\ \delta$ such that if $\mu(E) < \delta$, then $\int_E |f_n|\, d\mu < \epsilon/2, n = 1, \ldots, n_0$. If $n > n_0$, $\int_E |f_n|\, d\mu \leqslant \int_E |f_{n_0}|\, d\mu + \int_E |f_n - f_{n_0}|\, d\mu < \epsilon$.
25. By decomposing X into four sets as in Theorem 10, p. 145, and adding the results, we may suppose μ and ν to be measures. Let f be such that $\mu(E) = \int_E f\, d\nu$ for each $E \in \mathcal{S}$. Since $\nu \ll \mu$, $[x\colon f(x) = 0]$ has zero measure with respect to μ and ν, so $1/f = (d\mu/d\nu)^{-1}$ is defined a.e. But Theorem 9, p. 144, gives $(d\nu/d\mu)f = d\nu/d\nu = 1$, so $d\nu/d\mu = (d\mu/d\nu)^{-1}\ [\mu]$.
26. $\nu = \nu_0 + \nu_1$ where $\nu_0(E) = \nu((-\infty, 0] \cap E)$, $\nu_1(E) = \nu((0, \infty) \cap E)$.
27. As ν_0 is σ-finite, $D = \bigcup_{i=1}^{\infty} D_i$ where $\nu_0(D_i) < \infty$. But $D_i = \bigcup_{n=1}^{\infty} [x\colon x \in D_i, \nu_0([x]) > n^{-1}]$ and each set of this union is finite, so D is countable.
28. [2]. By a standard argument the proof may be reduced to the case where ν is finite. Let $\mathcal{V} = [E\colon E \in \mathcal{S}, \mu(E) = 0]$ and $\alpha = \sup\, [\nu(E)\colon E \in \mathcal{V}]$. Let $\{E_n\}$ be a sequence of sets in $\mathcal{V}$ such that $\lim \nu(E_n) = \alpha$. If $B = \cup E_n$, then $B \in \mathcal{V}$ and $\nu(B) = \alpha$. If $E \in \mathcal{V}$ we have $\alpha = \nu(E \cup B) = \nu(E - B) + \nu(B)$ so $\nu(E - B) = 0$. Write $\nu_0(E) = \nu(E \cap B)$, $\nu_1(E) = \nu(E - B)$, for each $E \in \mathcal{S}$. So $\nu = \nu_0 + \nu_1$, $\nu_1 \ll \mu$. But $\mu(B) = \nu_0(\complement B) = 0$ so $\nu_0 \perp \mu$. Uniqueness then follows as before.
29. By Minkowski's inequality, $|\,\|g_n\|_q - \|g\|_q| \leqslant \|g_n - g\|_q$, so $\|g_n\|_q \to \|g\|_q$. Also $\|f_n g_n - fg\|_1 \leqslant \|(f_n - f)g_n + f(g_n - g)\|_1 \leqslant \|f_n - f\|_p \|g_n\|_q + \|f\|_p \|g_n - g\|_q$. So the right-hand side has limit zero.
30. Obviously $G, F \in V^*$ implies $aG + bF \in V^*$. $\|G\| = 0$ iff $G = 0$, $\|aG\| = \sup\, [|aG(x)|\colon \|x\| \leqslant 1] = |a|\, \|G\|$, $\|G + F\| = \sup\, [|(G + F)(x)|\colon \|x\| \leqslant 1] \leqslant \sup\, [|G(x)| + |F(x)|\colon \|x\| \leqslant 1] \leqslant \|G\| + \|F\|$.
31. $f^{**}(aG + bF) = (aG + bF)(f) = aG(f) + bF(f) = af^{**}(G) + bf^{**}(F)$. So f^{**} is a linear functional. $\|f^{**}\| = \sup\, [|f^{**}(G)|\colon \|G\| \leqslant 1] = \sup\, [|G(f)|\colon \|G\| \leqslant 1] \leqslant \|f\|$. In fact, by the Hahn-Banach theorem [15] there exists G of the norm 1 such that $|G(f)| = \|f\|$, so that $\|f^{**}\| = \|f\|$. For a particular case see Exercise 32.
32. $f \to f^{**}$ is linear, for $(af + bg)^{**}(G) = G(af + bg) = aG(f) + bG(g) =$

$af^{**}(G) + bg^{**}(G)$ for each G in $(L^p)^*$. To show that the mapping is norm-preserving (and so one-to-one) we need, for each f, to construct G of norm 1 with $|G(f)| = \|f\|_p$. If $f = 0$, any G of norm 1 will do; otherwise take $G(f) = \int fg\, d\mu$ with $g = \alpha|f|^{p-1}/(\|f\|_p)^{p-1}$, where $\alpha = \operatorname{sgn} f$. Then $\int fg\, d\mu = \|f\|_p$ while $\|g\|_q = (\int |f|^p\, d\mu)^{1/q}/(\|f\|_p)^{p-1} = 1$ since $p/q = p - 1$. In general the mapping $V \to V^{**}$ of a normed space into second dual is norm preserving and into. In cases like the above where every linear function on L^q is provided by an element of L^p, we say that L^p is reflexive.

33. This follows from Example 10, p. 115.
34. Theorem 13, p. 151, identifies the dual space. However, L^1 is not in general reflexive, as defined in the solution to Exercise 32.

HINTS AND ANSWERS TO EXERCISES: CHAPTER 9

1. $\bar{\mu}_g([x]) = \lim_{n\to\infty} (g(x + n^{-1}) - g(x)) = 0.$
2. Write $A_n = [x: \bar{\mu}_g([x]) \geqslant n^{-1}, k \leqslant x < k + 1]$. Then Card $A_n \leqslant n(g(k + 1) - g(k))$, so A_n is finite. But $\bigcup_{n=1}^{\infty} A_n = [x: \bar{\mu}_g([x]) > 0, k \leqslant x < k + 1]$ giving the result, since the required set is a countable union of such sets.
3. Theorem 1, p. 153, and Theorem 2, p. 154, hold as before. The proof of Theorem 3(i) becomes: Let $I = (a, b]$, $E_i = (a_i, b_i]$, $i = 1, \ldots, n$. Choose $c > 0$ such that $a < a + c < b$ and $h(a + c) - h(a) < \epsilon$. For each i choose β_i such that $\beta_i > b_i$ and $h(\beta_i) - h(b_i) < \epsilon/2^i$. Write $F = [a + c, b]$ and $U_i = (a_i, \beta_i)$. As $F \subset \bigcup_{i=1}^{\infty} U_i$, we get $F \subset \bigcup_{i=1}^{n} U_i$ for some n. Then Theorem 2 gives $h(b) - h(a + c) \leqslant \sum_{i=1}^{n} (h(\beta_i) - h(a_i)) < \sum_{i=1}^{n} (h(b_i) - h(a_i) + \epsilon/2^i)$. So $h(b) - h(a) - \epsilon < \sum_{i=1}^{n} (h(b_i) - h(a_i)) + \epsilon$, giving the result. Theorem 3(ii) and Theorem 4, p. 155, then hold as for g.
4. $\bar{\mu}_g(-\infty, x) = \lim_{n\to\infty} \bar{\mu}_g([-n, x)) = \lim_{n\to\infty} (g(x) - g(-n)) = g(x).$
5. On each of the removed 'middle thirds' $I_{n,k}$, L is constant, so $\nu(I_{n,k}) = 0$. Hence $\nu([0,1] - P) = 0$ where P is the Cantor ternary set. As $m(P) = 0$ we have $\nu \perp m$. Also $\nu([0,1]) = L(1) - L(0)$, and by Exercise 1, $\nu([x]) = 0$ for each x, giving the result.
6. We are given that there exist sequences $\{x_i\}$, $\{\lambda_i\}$ in R, with $\lambda_i > 0$ for each i, such that if $\delta_i = \chi_{(x_i, \infty)}$ then $g = \sum_{i=1}^{\infty} \lambda_i \delta_i$. Then $\bar{\mu}_g([x_i]) = \lambda_i$. Write $\{x_i'\}$, $\{\lambda_i\}$ for the subsequences obtained when we restrict to $[-N, N)$.

Then $\bar{\mu}_g([-N, N)) = g(N) - g(-N) = \sum_{i=1}^{\infty} \lambda_i' < \infty$. So if $A \subseteq [-N, N) - \bigcup_{i=1}^{\infty} [x_i']$, $\bar{\mu}_g(A) = 0$, so A is $\bar{\mu}_g$-measurable. But each $[x_i]$ is a Borel set and so measurable with respect to $\bar{\mu}_g$, and the result follows on letting $N \to \infty$.

7. (i) is obvious. (ii): As f_n, f are monotone increasing, it is sufficient to show that $\lim_{n\to\infty} \lim_{m\to\infty} f_n(x - 1/m) = \lim_{m\to\infty} \lim_{n\to\infty} f_n(x - 1/m)$. As the double sequence involved is monotone increasing in n and m, this is immediate. (iii): Let $f_n = \chi_{(n^{-1},\infty)}$, so $f = \chi_{(0,\infty)}$. Then $f_n(0+) = 0$ but $f(0+) = 1$. (iv): $\forall\ \epsilon > 0$, $\exists\ N_0$ such that $|f_n - f| < \epsilon$ for $n \geqslant N_0$. So $|f_n(x + 1/m) - f(x + 1/m)| < \epsilon$, for $n \geqslant N_0$, and letting $m \to \infty$ we get $|f_n(x+) - f(x+)| \leqslant \epsilon$ for $n \geqslant N_0$. So $\lim f_n(x+)$ exists and equals $f(x+)$. (v): From (i) we now have $f(x-) = \lim f_n(x-) = \lim f_n(x) = f(x)$.
8. Consider $f(x) \equiv x$.
9. $\sum_{i=1}^{n} |f(b_i) - f(a_i)| \leqslant K \sum_{i=1}^{n} |b_i - a_i| < \epsilon$ whenever $\sum_{i=1}^{n} |b_i - a_i| < \epsilon/K$.
10. Let c', d' be such that $a < c' < c, d < d' < b$. Then if $x, y \in [c, d]$ we have, as in Theorem 3, p. 111, $\phi(c', c) \leqslant \phi(x, y) \leqslant \phi(d', d)$. So $|\phi(y) - \phi(x)| \leqslant M|y - x|$ for some fixed M and all $x, y \in [c, d]$, i.e. ϕ satisfies a Lipschitz condition of order 1, and the previous exercise gives the result.
11. If $\eta > \xi \geqslant 0$, we have $\eta^p - \xi^p = p(\eta - \xi)(\xi + \theta(\eta - \xi))^{p-1}$ where $0 < \theta < 1$. As f is continuous, $|f| < M$, say, on $[a, b]$. So for $x, y \in [a, b]$, $|f^p(y) - f^p(x)| \leqslant p(2M)^{p-1} |f(y) - f(x)|$ and the result follows.
12. Take $[a, b] = [0,1]$ and define $f(0) = 0$, $f(x) = x^2 \sin^2 1/x$ for $0 < x \leqslant 1$. Then it is easily seen that $|f'| \leqslant 3$, so f satisfies a Lipschitz condition of order 1 and is absolutely continuous by Exercise 9. But $f^{1/2}$ is not absolutely continuous as it is not of bounded variation (Exercise 9, p. 83).
13. L is not an indefinite integral, by the solution to Exercise 21, p. 90, and so it is not absolutely continuous by Theorem 8, Corollary 1.
14. Since g is continuous, it is Borel measurable, so clearly the integral exists. By Theorem 8, p. 161, and Example 11, p. 145, we have

$$\int_a^b g(x)\, \mathrm{d}\bar{\mu}_F(x) = \int_a^b g(x) f(x)\, \mathrm{d}x = 0$$

as whenever $g(x) \neq 0, f(x) = 0$.
15. Let $f(x) = \sqrt{x}$, $0 \leqslant x \leqslant 1$ and let $x_i = 0$, $y_i = \delta/2i^2$, $i = 1, 2, \ldots$; then $\sum_{i=1}^{n} |x_i - y_i| < \delta$ for each n but $\sum_{i=1}^{n} |f(x_i) - f(y_i)| = (\delta/2)^{1/2} \sum_{i=1}^{n} i^{-1} \to \infty$ as $n \to \infty$. So f does not satisfy the more restrictive definition obtained by omitting disjointness. But f is absolutely continuous as it is the indefinite integral of an integrable function.

16. No. Let $g(x) = x, x \in (-\pi, \pi], g(x + 2\pi) = g(x)$, and let $f(x) = \sqrt{|x|}$. Take $x_i = 2(i-1)\pi, y_i = x_i + \delta/2i^2, i = 1, \ldots, n\ (0 < \delta < 2\pi)$. Then the intervals (x_i, y_i) are disjoint and $\sum_{i=1}^{n} |x_i - y_i| < \delta, g(x_i) = 0, g(y_i) = \delta/2i^2$. Then $f \circ g$ is not absolutely continuous since $\sum_{i=1}^{n} |(f \circ g)(x_i) - (f \circ g)(y_i)| = (\delta/2)^{1/2} \sum_{i=1}^{n} i^{-1} \to \infty$ as $n \to \infty$.

17. $\int_{-\infty}^{\infty} x\, dg = \lim_{a, b \to \infty} \int_{-a}^{b} x\, dg = \lim_{a, b \to \infty} - \int_{-a}^{b} e^{-x^2}\, dx = -\sqrt{\pi}$, using Theorem 9, p. 163, and a standard integral. Similarly $\int_{-\infty}^{\infty} x^2\, dg = -\int_{-\infty}^{\infty} e^{-x^2}\, d(x^2) = -2\int_{-\infty}^{\infty} e^{-x^2}\, x\, dx = 0$.

18. Write $f(t) = \int_0^t h\, du$, and $g(t) = t$; then (9.9) gives

$$\int_0^x f\, dg = f(x)\, g(x) - f(0)\, g(0) - \int_0^x th(t)\, dt = \int_0^x xh(t)\, dt - \int_0^x th(t)\, dt,$$

as required.

19. For f and g continuous the result is true by Theorem 12, p. 65. For integrable functions f and g choose continuous functions f_n and g_n to approximate f and g in the mean as in Theorem 16, p. 73. Their indefinite integrals F_n and G_n then approximate F and G uniformly, and letting $n \to \infty$ gives the result.

20. (i): For $x \in (0, d]$ and $t \in [x, d]$, $t^{-1}\ |h(t)| \leqslant x^{-1}\ |h(t)| \in L(x, d)$. (ii): $\forall\ \epsilon > 0$, $\exists\ \delta > 0$ such that $\int_0^\delta |h|\, dt < \epsilon$. So for $x \in (0, \delta)$, $\int_x^\delta t^{-1}\, xh(t)\, dt < \epsilon$. Also $\int_\delta^d t^{-1}\, x|h(t)|\, dt \leqslant \delta^{-1} x \int_0^d |h|\, dt < \epsilon$ for $x < x_0$, say. So if $0 < x < \min(\delta, x_0)$, $|xk(x)| \leqslant \int_x^d t^{-1} x\, |h(t)|\, dt < 2\epsilon$. (iii) and (iv): We may suppose $h \geqslant 0$ and hence $k \geqslant 0$. Let $a \in (0, d)$ and apply Theorem 9, p. 163, to (a, d), taking $G(t) = t, g = 1, F(t) = k(t), f(t) = -t^{-1}\ h(t), x = d$. As $k(d) = 0$ we get $-k(a)a = \int_a^d (-h + k)\, dt$. Letting $a \to 0$ gives the result.

21. (i): We are considering the limits of the measures of decreasing sequences of sets with void intersection. The necessity for finiteness is seen from the case $f(x) \equiv x$, $\mu = m$. (ii) As increasing sequences of sets are involved, no finiteness condition is necessary. (iii): $F(x-) = \lim \mu f^{-1}(-\infty, x - 1/n] =$

$\mu f^{-1}(-\infty, x) = G(x)$. Similarly for $F(x) = G(x+)$. (iv): The required continuity follows from Example 2, p. 156. As a counter-example when $\mu(X) = \infty$, take $X = [0,1]$, $\mathcal{S} = \mathcal{P}[0,1]$, $\mu(E) = \text{Card } E, f(x) \equiv x$. Then $F(0) = 0$, $F(0+) = \infty$. (v). Let $\mu(A) > 0, f = \chi_A$; then $F(1) = \mu(A), F(1-) = 0, G(1) = 0, G(1+) = \mu(A)$. (vi): The first part is obvious, the second follows from Example 9, p. 167.

22. $\mathsf{R} = \bigcup_{n=1}^{\infty} E_n$ where $\mu f^{-1}(E_n) < \infty$ for each n, so $X = \bigcup_{n=1}^{\infty} X_n$ where $X_n = f^{-1}(E_n)$ decomposes X as required.
23. Take $[\![X, \mathcal{S}, \mu]\!] = [\![\mathsf{R}, \mathcal{M}, m]\!]$ and define f by $f(x) = x, 0 \leqslant x < 1, f(x+k) = f(x), k = \pm 1, \pm 2, \ldots$ Then if $m(E) = 0$, we have $mf^{-1}(E) = 0$; so $mf^{-1} \ll m$. Also $mf^{-1}(E) = 0$ if $E \subseteq \mathsf{C}[0,1)$. But $mf^{-1}(E) = \infty$ if $m(E \cap [0,1)) > 0$. So, for every measurable set, mf^{-1} is either 0 or ∞, and so mf^{-1} is not σ-finite.
24. We have $X = \bigcup_{n=1}^{\infty} X_n$ where $\mu(X_n) < \infty$. If $E_n = f(X_n)$, E_n is μf^{-1}-measurable and $\mu f^{-1}(E_n) = \mu(X_n) < \infty$. But $\bigcup_{n=1}^{\infty} E_n = f(X)$ whose complement in R has zero μf^{-1}-measure.
25. Regarding f as a function from the space B to $[0, M]$ we may apply Example 10, p. 168, to get $\int_B f^n \, dx = \int_0^M y^n \, de(y)$. But by Theorem 11, p. 165, this latter integral equals $[y^n e(y)]_{y=M^+} - [y^n e(y)]_{y=0} - \int_0^M e(y)\, d(y^n)$
$= M^n m(B) - \int_0^M ne(y) y^{n-1} \, dy.$
26. Since $\sum_{n=0}^{\infty} |a^n t^n| < \infty$ for $|t| \leqslant 1/2$, we have $\sum_{n=0}^{\infty} \int_0^1 |a_n (f(x))^n| \, dx < \infty$.
From the last exercise we have, for each n,

$$\int_0^1 f^n \, dx = 1/2^n - \int_1^{1/2} n y^{n-1} e(y)\, dy.$$

So multiplying by a_n and adding for $n = 0, 1, \ldots$, we get the result by Theorem 11, p. 64.

HINTS AND ANSWERS TO EXERCISES: CHAPTER 10

1. The non-uniqueness can be seen from $\emptyset = X \times \emptyset = \emptyset \times Y$. However if the rectangle $P \times Q = R \times S$ is non-empty, then P, Q, R, S are non-empty. If $y \in Q$, then for $x \in P$, $(x, y) \in P \times Q = R \times S$, so $x \in R$ and $y \in S$, so $P \subseteq$

$R, Q \subseteq S$, and as the opposite inclusion follows similarly, the representation is unique.

2. Yes, as $A \times \emptyset = \emptyset \times \emptyset \; (= \emptyset)$ satisfies Definition 3, p. 176.
3. If either A or B is empty, $A \times B = \emptyset \in \mathcal{S} \times \mathcal{T}$. If neither A nor B is empty, $A \times B \in \mathcal{S} \times \mathcal{T}$ iff both A and B are both measurable, by Theorem 4, p. 178.
4. $(\chi_V)_x(y) = \chi_V(x, y) = 1$ iff $(x, y) \in V$, iff $y \in V_x$, iff $\chi_{V_x}(y) = 1$. Similarly for $(\chi_V)^y$.
5. The function $f^*(x, y) = f(x)$ is $\mathcal{S} \times \mathcal{T}$-measurable, as is $g^*(x, y) = g(y)$. So $f^*g^* = fg$ is measurable.
6. Take $X = Y = \mathsf{R}$, $\mathcal{S} = \mathcal{T} = \mathcal{M}$, $\mu = \nu = m$. Let A be a non-measurable set in X, B a non-empty set of measure zero in Y. Then $A \times B \subseteq \mathsf{R} \times B$, a set of measure zero. But $A \times B \notin \mathcal{M} \times \mathcal{M}$, by Theorem 4, p. 178.
7. (i) Let $\{\phi_n\}$ be a sequence of measurable simple functions $\phi_n \leqslant f$, $\phi_n \uparrow f$. Then 0_{ϕ_n} is measurable for each n and $0_f = \bigcup_{n=1}^{\infty} 0_{\phi_n}$. (ii) $\int f \,\mathrm{d}x = \lim \int \phi_n \,\mathrm{d}x = \lim\, (m \times m)(0_{\phi_n}) = (m \times m)(0_f)$. (iii) Write $G_n = [(x, y) \colon y = f(x), f(x) \leqslant n]$. Since $G = \bigcup_{n=1}^{\infty} G_n$, it is sufficient to show that each G_n is measurable. Let $X_n = [x \colon (x, y) \in G_n]$. So $X_n = [x \colon f(x) \leqslant n]$ is measurable. Choose sequences of measurable simple functions on X_n, $\{\phi_k\}$ and $\{\psi_k\}$, $\phi_k \leqslant f$, $\psi_k \geqslant f$, $\phi_k \uparrow f$, $\psi_k \downarrow f$. Then 0_{ϕ_k} and 0_{ψ_k} are measurable. But $G_n = \bigcap_{k=1}^{\infty} 0_{\psi_k} - \bigcup_{k=1}^{\infty} 0_{\phi_k}$. (iv): By (iii) we can apply Theorem 6, p. 179, to G. Let ϕ be as in Theorem 6, then clearly $\phi(x) = 0$ for each x, so $(m \times m)(G) = \int \phi \,\mathrm{d}x = 0$ by Definition 9, p. 181.
8. Immediate consequence of Theorem 9, p. 182, as f, being bounded, is integrable.
9. By Exercise 5, fg is measurable, so the result follows from Theorem 8, p. 182.
10. We wish to show that $\int_0^a \mathrm{d}x \int_x^a |f(t)|/t \,\mathrm{d}t < \infty$. Interchanging the order of integration, this integral becomes
$$\int_0^a \mathrm{d}t \int_0^t |f(t)|/t \,\mathrm{d}x = \int_0^a |f(t)| \,\mathrm{d}t < \infty.$$
By the Corollary to Theorem 9, $g \in L(0, a)$ and the iterated integrals are equal. Integrating as above (without modulus signs) gives $\int_0^a f \,\mathrm{d}t = \int_0^a g \,\mathrm{d}t$.
11. For $y \neq 0$, $\int_0^1 e^{-y} \sin 2xy \,\mathrm{d}x = e^{-y}(\sin^2 y)/y \in L(0, \infty)$. Now $|e^{-y} \sin 2xy|$

$\leqslant e^{-y}$, integrable over $[x: 0 \leqslant x \leqslant 1] \times [y: 0 \leqslant y < \infty]$. So

$$\int_0^\infty dy \int_0^1 e^{-y} \sin 2xy\, dx = \int_0^1 dx \int_0^\infty e^{-y} \sin 2xy\, dy.$$

But $\int_0^\infty e^{-y} \sin 2xy\, dy = 2x/(1+4x^2)$. So

$$\int_0^\infty (e^{-y} \sin^2 y)/y\, dy = \int_0^1 2x/(1+4x^2)\, dx = \tfrac{1}{4} \log 5.$$

12. $\int_0^1 |e^{-xy} \sin 2y|\, dx = |\sin 2y|\,(1-e^{-y})/y \in L(0,N)$ for each $N > 0$. So

$$\int_0^N dy \int_0^1 e^{-xy} \sin 2y\, dx = \int_0^1 dx \int_0^N e^{-xy} \sin 2y\, dy.$$

Now $\int_0^N e^{-xy} \sin 2y\, dy = 2/(x^2+4) - x/(x^2+4)\, e^{-xN} \sin 2N - 2e^{-xN}/$ $(x^2+4) \cos 2N$. So the second repeated integral $=$ arc tan $\tfrac{1}{2} + o(1)$ as $N \to \infty$, by Theorem 10, p. 63. Hence

$$\int_0^N (\sin 2y/y - \sin 2ye^{-y}/y)\, dy = \text{arc tan}\, \tfrac{1}{2} + o(1) \text{ as } N \to \infty.$$

But $\lim \int_0^N \sin 2y/y\, dy = \pi/2$ (by a standard contour integral).

So $\int_0^\infty e^{-y} \sin 2y/y\, dy$ equals $\pi/2 -$ arc tan $\tfrac{1}{2} =$ arc tan 2.

13. To obtain $\int_0^y dx \int_0^x (x-t)^{\alpha-1} f(t)\, dt$, consider the iterated integral of the modulus of this function, in the opposite order, viz.

$$\int_0^y |f(t)| \left(\int_t^y (x-t)^{\alpha-1}\, dx \right) dt = \int_0^y |f(t)| \frac{(y-t)^\alpha}{\alpha}\, dt < \infty.$$

So the order of integration may be changed and the desired integral equals

$$\int_0^y f(t) \left(\int_t^y (x-t)^{\alpha-1}\, dx \right) dt = \alpha^{-1} g_{\alpha+1}(y).$$

14. We have $\int_0^a x^{-1} F(x)\, dx = \int_0^a dx \int_0^x x^{-1} f(t)\, dt = \int_0^a dt \int_t^a x^{-1} f(t)\, dx$

$= \int_0^a (\log a - \log t) f(t)\, dt = \int_0^a f(x) \log 1/x\, dx - F(a) \log 1/a$, where the charge of order of integration is justified if either of the iterated integrals is finite, the integrands being non-negative.

15. For $y \neq 0, \int_{-1}^{1} f(x, y)\, dx = 0$, and for $x \neq 0, \int_{-1}^{1} f(x, y)\, dy = 0$, so the iterated integrals are both zero. But f integrable on $[-1,1] \times [-1,1]$ would imply f integrable on $[0,1] \times [0,1]$ which would give, by Theorem 9, p. 182, $\int_0^1 dx \int_0^1 f(x, y)\, dy < \infty$. However this integral is

$$\int_0^1 \left(\frac{1}{2x} - \frac{x}{2(x^2+1)}\right) dx = \infty.$$

16. Substitute $y = \tan x$ to get (for $x > 0$)

$$\int_0^1 \frac{x^2 - y^2}{(x^2+y^2)^2}\, dy = \frac{1}{1+x^2}$$

and the first integral. By symmetry

$$\int_0^1 dx \int_0^1 \frac{x^2 - y^2}{(x^2+y^2)^2}\, dy = \int_0^1 dy \int_0^1 \frac{y^2 - x^2}{(y^2+x^2)^2}\, dx = \frac{\pi}{4},$$

so we obtain the second integral. To show that f is not integrable directly, integrate f^+ over $[0,1] \times [0,1]$ to get $\int_0^1 dx \int_0^x \frac{x^2 - y^2}{(x^2+y^2)^2}\, dy$, by Theorem 7, p. 181. The same substitution as above reduces this to $\int_0^1 1/2x\, dx = \infty$.

17. From a diagram it is clear that $\int f(x, y)\, dx = 0$, $(0 < y < 1)$ but $\int f(x, y)\, dy = 2$ if $1/2 \leqslant x < 1$, $\int f(x, y)\, dy = 0$, $0 < x < 1/2$. Theorem 9 is not contradicted as $\int |f(x, y)|\, dx\, dy = \infty$, by inspection.

18. For each y we have $\int f\, d\mu = -2^{-y-1}$, so $\int d\nu \int f\, d\mu = -1/2$. But

$$\int d\mu \int f(x, y)\, d\nu = 3/2.$$

However it is easily seen that $\int f^+\, d(\mu \times \nu) = \int f^-\, d(\mu \times \nu) = \infty$.

19. Clearly $\mathcal{S}(\mathcal{E} \times \mathcal{F}) \subseteq \mathcal{S} \times \mathcal{T}$. Also, given a fixed $F \in \mathcal{F}, E \times F \in \mathcal{S}(\mathcal{E} \times \mathcal{F})$ for $E \in \mathcal{E}$. As in Theorem 10, p. 186, we take countable union of these sets and take complements in the set $X \times F$ to get $D \times F \in \mathcal{S}(\mathcal{E} \times \mathcal{F})$ for any $D \in \mathcal{S}$. Keeping D fixed and proceeding similarly we get $D \times K \in \mathcal{S}(\mathcal{E} \times \mathcal{F})$, for any $K \in \mathcal{T}$. But this gives $\mathcal{S} \times \mathcal{T} \subseteq \mathcal{S}(\mathcal{E} \times \mathcal{F})$ and the result.

20. This follows from the limiting arguments of Theorem 11, p. 187, starting from the fact that $g(u) = \chi_E(ru)$ is Borel measurable for the sets E considered there.

21. Let B be an open set in $\mathsf{R}^2 - [0]$ and let $a \in B$. Then $\exists\, \delta, 0 < \delta < |a|$, such that $[x\colon |a - x| < \delta] \subseteq B$. We may take for E the set $[ru\colon |a| - \delta/2 < r <$

$|a| + \delta/2, u \in A]$ where A is the open arc in S_1 centred at $a/|a|$ and subtending an angle 2 arc tan $(\delta\sqrt{3}/(\delta + 2|a|))$.

22. In Theorem 7, p. 181, take $f(x,y) = e^{-(x^2+y^2)}$ to get

$$\int_0^\infty e^{-x^2}\,dx \int_0^\infty e^{-y^2}\,dy = \int_0^\infty \int_0^\infty e^{-(x^2+y^2)}\,dx\,dy.$$

By Theorem 11, p. 187, this equals

$$\int_0^\infty r\,dr \int_0^{\pi/2} e^{-r^2}\,d\theta = \pi/2 \int_0^\infty e^{-r^2}\, r\,dr = \pi/4,$$

where we may make the change of variable using Theorem 12, p. 167, with $k(x) = 2x\ (x > 0)$, $K(x) = x^2$, $g(x) = e^{-x}$, $a = 0$, $b = N$, and letting $N \to \infty$.

So $\int_0^\infty e^{-x^2}\,dx = \sqrt{\pi}/2$.

23. By definition $\Gamma(z) = \int_0^\infty e^{-x}\, x^{z-1}\,dx$; so for each real y, by Example 25, p. 75, $\int_0^\infty e^{-y-x}\, x^{z-1}\,dx = \int_y^\infty e^{-x}(x-y)^{z-1}\,dx = \int_0^\infty f(x,y)\,dx$, say, where $f(x,y) = e^{-x}(x-y)^{z-1}$ if $0 < y < x$, $f = 0$ otherwise. Then as in the proof of Theorem 13, p. 190, $y^{w-1} f(x,y)$ is measurable in the quadrant $x > 0, y > 0$. Also

$$\int_0^\infty dy \int_0^\infty |y^{w-1} f(x,y)|\,dx = \int_0^\infty |e^{-y}\, y^{w-1}|\,dy \int_0^\infty |e^{-x}\, x^{z-1}|\,dx < \infty.$$

So $y^{w-1} f \in L^1(m \times m)$ and hence

$$\int_0^\infty dx \int_0^\infty f(x,y)\, y^{w-1}\,dy = \int_0^\infty y^{w-1}\, e^{-y}\,\Gamma(z)\,dy = \Gamma(z)\Gamma(w).$$

But $$\int_0^\infty f(x,y)\, y^{w-1}\,dy = e^{-x} \int_0^x y^{w-1}(x-y)^{z-1}\,dy =$$

$$e^{-x}\, x^{z+w-1} \int_0^1 u^{w-1}(1-u)^{z-1}\,du,$$

on substituting $y = ux$ and using Theorem 14, p. 169. So integrating with respect to x gives the result.

24. We wish to find $\int_0^1 e^{ixt}\,dL(t)$. Consider the sequence of step functions $\{M_n\}$ on $[0,1]$ such that each function M_n is left-continuous and monotone increasing and M_n increases by 2^{-n} at each point $2\sum_{j=1}^{n} 3^{-j}\,\epsilon_j$ ($\epsilon_j = 0$ or 1), $M_n(0) = 2^{-n}$, $M_n(1) = 1$. Then $M_n = L$ on the set $\bigcup_k I_{n,k}$ of Chapter 1,

p. 24, and $|M_n - L| \leqslant 2^{-n}$. So $M_n \to L$ uniformly and so, as in the proof of Theorem 11, p. 000, we have $\lim \int e^{ixt}\, dM_n(t) = \int e^{ixt}\, dL(t)$. But

$$\int e^{ixt}\, dM_n(t) = 2^{-n} \Sigma \exp\left(ix\left(\frac{2\epsilon_1}{3} + \ldots + \frac{2\epsilon_n}{3^n}\right)\right)$$

where the sum is taken over all 2^n possible n-tuples $(\epsilon_1, \ldots, \epsilon_n)$. So the right-hand side equals

$$\prod_{k=1}^{n} \frac{1}{2}\left(1 + \exp \frac{2ix}{3^k}\right) = \prod_{k=1}^{n} \cos \frac{x}{3^k} \exp \frac{ix}{3^k} =$$

$$\exp\left(\frac{ix}{2}(1 - 3^{-n})\right) \prod_{k=1}^{n} \cos \frac{x}{3^k},$$

and the result follows.

25. Consider the real and imaginary parts of f and apply the Riemann-Lebesgue Lemma (Exercise 44, p. 75), taking $\beta = \pi$, $\phi(t) = \cos t$ or $\sin t$ as appropriate.

26. Take $f = \chi_{[0,]}$. Then $\hat{f}(0) = 1 = \int_0^1 |f|\, dt$.

27. We need to show that $\mathcal{K} f \in L^2(-\infty, \infty)$; the linearity is obvious. By Hölder's inequality, p. 115, we have

$$\left(\int |K(x, t) f(t)|\, dt\right)^2 \leqslant \int |K(x, t)|^2\, dt \int |f|^2\, dt.$$

Integrate both sides with respect to x, to obtain $\mathcal{K} f \in L^2(-\infty, \infty)$ as the integral of the left-hand side is seen to be finite.

References

[1] Austin, D. 'A Geometric Proof of the Lebesgue Differentiation Theorem', *Proc. Amer. Math. Soc.*, **16**, 220–221 (1965).

[2] Brooks, J.K. 'The Lebesgue Decomposition Theorem for Measures', *Amer. Math. Monthly*, **78**, 660–661 (1971).

[3] Goldberg, R.R. *Fourier Transforms* (Cambridge University Press, 1961).

[4] Goodey, P.R. 'Hausdorff Measure Functions', *Proc. Camb. Phil. Soc.*, **74**, 87–96 (1973).

[5] Halmos, P.R. *Measure Theory* (Van Nostrand, Princeton, New Jersey, 1950).

[6] Hardy, G.H., Littlewood, J.E. and Polya, G. *Inequalities* (2nd Edition, Cambridge University Press, 1952).

[7] Kahane, J.-P. and Salem, R. *Ensembles Parfaits et Séries Trigonométriques* (Hermann, Paris, 1963).

[8] Kamke, E. *Theory of Sets* (Dover, New York, 1950).

[9] Kestelman, H. 'On the Functional Equation $f(x + y) = f(x) + f(y)$', *Fund. Math.*, **34**, 144–147 (1946).

[10] Rogers, C.A. *Hausdorff Measures* (Cambridge University Press, 1970).

[11] Rotman, B. and Kneebone, G.T. *The Theory of Sets and Transfinite Numbers* (Oldbourne Press, London, 1968).

[12] Royden, H.L. *Real Analysis* (2nd Edition, Macmillan, New York, 1968).

[13] Rudin, W. *Principles of Mathematical Analysis* (McGraw-Hill, New York, 1964).

[14] Rudin, W. *Real and Complex Analysis* (McGraw-Hill, New York, 1966).

[15] Taylor, A.E. and Lay, D.C. *Introduction to Functional Analysis* (2nd Edition, John Wiley, New York, 1980).

[16] Simmons, G.F. *Introduction to Topology and Modern Analysis* (McGraw-Hill, New York, 1963).

[17] Yosida, K. *Functional Analysis* (3rd Edition, Springer-Verlag, Heidelberg, 1971).

Index